MODEL AND DESIGN OF BIPOLAR AND MOS CURRENT-MODE LOGIC

Model and Design of Bipolar and MOS Current-Mode Logic

CML, ECL and SCL Digital Circuits

by

Massimo Alioto
University of Siena,
Italy

and

Gaetano Palumbo
University of Catania,
Italy

Springer

A C.I.P. Catalogue record for this book is available from the Library of Congress.

ISBN 1-4020-2878-4 (HB)
ISBN 1-4020-2888-1 (e-book)

Published by Springer,
P.O. Box 17, 3300 AA Dordrecht, The Netherlands.

Sold and distributed in North, Central and South America
by Springer,
101 Philip Drive, Norwell, MA 02061, U.S.A.

In all other countries, sold and distributed
by Springer,
P.O. Box 322, 3300 AH Dordrecht, The Netherlands.

Printed on acid-free paper

springeronline.com

Printed in the Netherlands.

To our:

Maria Daniela and (mom) Giusi

Michela, Francesca & Chiara

CONTENTS

ACKNOWLEDGMENT

The authors wish to thank Prof. Salvatore Pennisi for its help during the correction of the draft.

We would like to thank our families and parents for their endless support and interest in our careers.

Massimo Alioto
Gaetano Palumbo

PREFACE

Current-Mode digital circuits have been extensively analyzed and used since the early days of digital ICs. In particular, bipolar Current-Mode digital circuits emerged as an approach to realize digital circuits with the highest speed. Together with its speed performance, CMOS Current-Mode logic has been rediscovered to allow logic gates implementations which, in contrast to classical VLSI CMOS digital circuits, have the feature of low noise level generation. Thus, CMOS Current-Mode gates can be efficiently used inside analog and mixed-signal ICs, which require a low noise silicon environment. For these reasons, until today, many works and results have been published which reinforce the importance of Current-Mode digital circuits.

In the topic of Current-Mode digital circuits, the authors spent a lot of effort in the last six years, and their original results highly enhanced both the modeling and the related design methodologies. Since the fundamental Current-Mode logic building block is the classical differential amplifier, the winning idea, that represents the starting point of the authors' research, was to change the classical point of view typically followed in the investigation and design of Current-Mode digital circuits. In particular, they properly exploited classical paradigms developed and used in the analog circuit domain (a topic in which one of the authors maturated a great experience). The change of perspective allowed to collect many results in the domain of Current-Mode digital circuits. Such results represent a complete set of tools to be used during the modeling and design process of these high-performance digital circuits, that are accurate, but so simple to be even used in a pencil and paper approach.

The main focus of this book will be to provide the reader with a deep understanding of modeling and design strategies of Current-Mode digital circuits, as well as to organize in a coherent manner all the authors' results in the domain of Current-Mode digital circuits. Hence, the book allows the reader not only to understand the operating principle and the features of bipolar and MOS Current-Mode digital circuits, but also to design optimized digital gates.

The book can be used as a reference to practicing engineers working in this area and as text book to senior undergraduate, graduate and postgraduate students (already familiar with electronic circuits and logic gates) who want to extend their knowledge and cover all aspects of the analysis and design of Current-Mode digital circuits. Thus the prerequisites to a well understanding of the book are basic electronics and familiarity with digital circuits design.

Although the material is presented in a formal and theoretical manner, much emphasis is devoted to a design perspective. Indeed, the book can be a valid reference for high-performance digital circuits IC designers.

To further link the book's theoretical aspects with practical issues, and to provide the reader with an idea of the real order of magnitude involved assuming actual technologies, numerical examples together with SPICE simulations are included in the book.

The outline of the text is as follows:

An introduction to the operating principles of Bipolar and MOS transistors is presented in the first invited chapter, where circuit models are also introduced and developed with emphasis on those more suited for Current-Mode digital circuits. Then the Current-Mode logic along with their typical applications are discussed in the second chapter.

In Chapter 3 the various techniques to define the topology of Current-mode gates (series gates) are reviewed, by highlighting their strength and weakness.

All the remaining chapters organically report the results developed by the authors. In particular, the propagation delay models for bipolar Current-Mode gates are discussed in Chapter 4. Both a simple and an accurate model is derived for CML and ECL inverter, MUX/XOR and D latch. The resulting models, which are simple, help the reader to gain an in-depth view of the fundamental parameters affecting the speed performance of a single gate. By using these models, gate-level design strategies to achieve the best performance in terms of speed is achieved in Chapter 5. After a brief introduction, the approach is customized for CML and ECL inverter, MUX/XOR and D latch. Moreover, a comparison between the CML and the ECL logic is also carried out.

Chapter 6 changes the context moving into the MOS technology domain. Timing models for the SCL inverter without and with buffer, MUX/XOR and D latch are developed and discussed in detail. The models consider in depth the peculiarity of modern CMOS processes, and hence include submicron effects. Indeed, they are analytically derived by using standard BSIM3v3 model parameters. Starting from these delay models and considering the static behavior of the SCL gates, design strategies are derived in detail for the inverter without and with buffer, MUX/XOR and D latch.

Chapter 8 completes modeling and design aspects for special classes of circuits blocks, such as ring oscillators, frequency dividers and low-voltage gate topologies. Moreover, this chapter extends the optimized design procedure to chains of gates.

Chapter 1

DEVICE MODELING FOR DIGITAL CIRCUITS

Gianluca Giustolisi and Rosario Mita

This chapter will deal with the operation and modeling of semiconductor devices in order to give the reader a basis for understanding, in a simple and efficient manner, the operation of the main building blocks of digital circuits.

1.1 PN JUNCTION

A semiconductor is a crystal lattice structure with free electrons and/or free holes or, which is the same, with negative and/or positive carriers. The most common semiconductor is silicon which, having a valence of four, allows its atoms to share four free electrons with neighboring atoms thus forming the covalent bonds of the crystal lattice.

In intrinsic silicon, thermal agitation can endow a few electrons with enough energy to escape their bonds. In the same way, they leave an equal number of holes in the crystal lattice that can be viewed as free charges with an opposite sign. At room temperature, we have $1.5\cdot 10^{10}$ carriers of each type per cm^3. This quantity is referred to as n_i and is a function of temperature as it doubles for every 11 °C increase in temperature [S81], [MK86].

This intrinsic quantity of free charges is not sufficient for the building of microelectronic devices and must be increased by doping the intrinsic silicon. This means adding negative or positive free charges to the pure material. Several doping materials can be used to increase free charges. Specifically, when doping pure silicon with a pentavalent material (that is, doping with atoms of an element having a valence of five) we have almost one extra free electron that can be used to conduct current for every one atom of impurity. Likewise, doping the pure silicon with atoms having a valence of three, gives us almost one free hole for every impurity atom. A pentavalent atom donates electrons to the intrinsic silicon and is known as a donor. In contrast, a trivalent atom accepts electrons and is known as an acceptor. Typical pentavalent impurities, also called n-type dopants, are arsenic, As, and phosphorus, P, while the most used trivalent impurity, also called p-type dopant, is boron, B. Silicon doped with a pentavalent impurity is said to be n-type silicon, while silicon doped with a trivalent impurity is called p-type silicon.

If we suppose that a concentration N_D (N_A) of donor (acceptor) atoms is used to dope the silicon, with $N_D >> n_i$ ($N_A >> n_i$), the concentration of free electrons (holes) in the n-type (p-type) material, n_n (p_p), can be assumed as

$$n_n \approx N_D \qquad (p_p \approx N_A) \tag{1.1}$$

Since some free electrons (holes) recombine with holes (electrons), the concentration of holes (electrons) in the n-type (p-type) material, p_n (n_p), is also reduced to

$$p_n \approx \frac{n_i^2}{N_D} \qquad (n_p \approx \frac{n_i^2}{N_A}) \tag{1.2}$$

Joining a p-type to an n-type semiconductor as in Fig. 1.1 makes a pn junction, or most commonly a diode. The p-side terminal is called anode (A) while the n-side terminal is called cathode (K).

Note that the p-type section is denoted with p+, meaning that this side is doped more heavily (in the order of 10^{20} carriers/cm^3) than its n-type counterpart (in the order of 10^{15} carriers/cm^3), that is $N_A >> N_D$. This is not a limitation since most pn junctions are built with one side more heavily doped than the other.

Close to the junction, free electrons on the n side are attracted by free positive charges on the p side so they diffuse across the junction and recombine with holes. Similarly, holes on the p side are attracted by electrons on the n side, diffuse across the junction and recombine with free electrons on the n side.

This phenomenon leaves behind positive ions (or immobile positive charges) on the n side, and negative ions (or immobile negative charges) on the p side, thus creating a depletion region across the junction where no free carriers exist. Moreover, since charge neutrality obliges the total amount of charge on one side to be equal to the total amount of charge on the other, the width of the depletion region is greater on the more lightly doped side, that is, in our case where $N_A >> N_D$, we have $x_n >> x_p$.

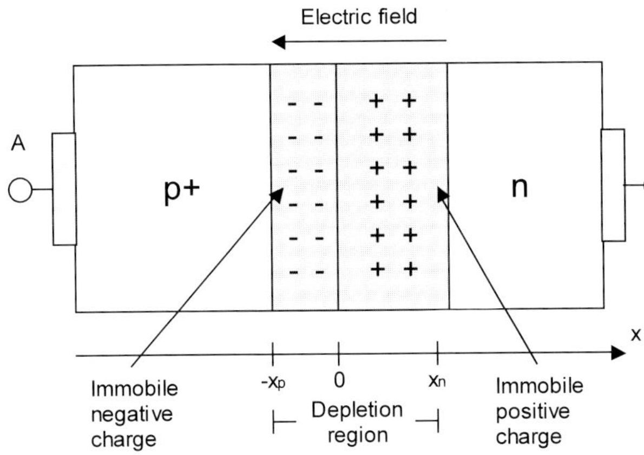

Fig. 1.1. pn junction.

Due to immobile charges, an electric field appears from the n side to the p side and generates the so-called built-in potential of the junction. This potential prevents further net movement of free charges across the junction under open circuit and steady-state conditions. It is given by [S81], [MK86]

$$\Phi_0 = V_T \ln\left(\frac{N_A N_D}{n_i^2}\right) \tag{1.3}$$

V_T being the thermal voltage defined as

$$V_T = \frac{kT}{q} \tag{1.4}$$

where T is the temperature in degrees Kelvin (≈ 300 K at room temperature), k is the Boltzmann's constant ($1.38 \cdot 10^{-23}$ JK^{-1}) and q is the charge of an electron ($1.602 \cdot 10^{-19}$ C). At room temperature, V_T is approximately equal to 26 mV. Typical values of the built-in potential are around 0.9 V.

Under open circuit and steady-state conditions, it can be shown that the widths of depletion regions are given by the following equations

$$x_n = \left[\frac{2\varepsilon_{si}\varepsilon_0\Phi_0}{q}\frac{N_A}{N_D(N_A+N_D)}\right]^{1/2} \approx \left[\frac{2\varepsilon_{si}\varepsilon_0\Phi_0}{qN_D}\right]^{1/2} \tag{1.5a}$$

$$x_p = \left[\frac{2\varepsilon_{si}\varepsilon_0\Phi_0}{q}\frac{N_D}{N_A(N_A+N_D)}\right]^{1/2} \approx \left[\frac{2\varepsilon_{si}\varepsilon_0\Phi_0 N_D}{qN_A^2}\right]^{1/2} \tag{1.5b}$$

where ε_0 is the permittivity of free space ($8.854 \cdot 10^{-12}$ F/m), ε_{si} is the relative permittivity of silicon (equal to 11.8) and where the approximations hold if $N_A >> N_D$.

Dividing (1.5a) by (1.5b) yields

$$\frac{x_n}{x_p} = \frac{N_A}{N_D} \tag{1.6}$$

which justifies the fact that x_n is greater than x_p if p-type semiconductor is more heavily doped than n-type.

The charge stored in the depletion region, per unit device area, is found by multiplying the width of the depleted area by the concentration of the immobile charge, which can be considered equal to q times the doping concentration. So for both the sides of the device we have

$$Q^+ = qN_D x_n = \left(2q\varepsilon_{si}\varepsilon_0\Phi_0\frac{N_A N_D}{N_A+N_D}\right)^{1/2} \tag{1.7a}$$

$$Q^- = qN_A x_p = \left(2q\varepsilon_{si}\varepsilon_0\Phi_0\frac{N_A N_D}{N_A+N_D}\right)^{1/2} \tag{1.7b}$$

Note that the charge stored on the n side equals the charge stored on the p side, as is expected due to the charge neutrality.

All the above equations are valid in the case of abrupt junctions. For graded junctions, that is where the doping concentration changes smoothly from p to n, a better model for the charge can be described by changing the exponent in (1.5) and (1.7) by (1-m) [AM88], where m is a technology dependent parameter (typical m values are around 1/3).

1.1.1 Reverse Bias Condition

By grounding the anode and applying a voltage V_R to the cathode, we reverse-bias the device. Under such a condition the current flowing through the diode is mainly determined by the junction area and is independent of V_R. In many cases this current is considered negligible and the device is modeled as an open circuit. However, the device also has a charge stored in the junction that changes with the voltage applied and causes a capacitive effect, which cannot be ignored at high frequencies. The capacitive effect is due to the so-called junction capacitance.

Specifically, when the diode is reverse biased as in Fig. 1.2, free electrons on the n side are attracted by the positive potential V_R and leave behind positive immobile charges. Similarly, free holes in the p region move towards the anode leaving behind negative immobile charges. This means that the depletion region increases and that the built-in potential increases exactly by the amount of applied voltage, V_R.

Given that the built-in potential is increased by V_R, both the width and the charge of the depletion region can be found by substituting the term Φ_0+V_R to Φ_0 in (1.5) and (1.7), respectively. In particular the charge stored results as

$$Q^+ = Q^- = \left[2q\varepsilon_{si}\varepsilon_0(\Phi_0 + V_R)\frac{N_A N_D}{N_A + N_D}\right]^{1-m} \tag{1.8}$$

This charge denotes a non-linear charge-voltage characteristic of the device, modeled by a non-linear capacitor called junction capacitance.

For small changes in the applied voltage around a bias value, V_R, the capacitor can be viewed as a small-signal capacitance, C_j, whose expression is found by differentiating[1] (1.8) with respect to V_R

$$C_j = \frac{dQ^+}{dV_R} = \frac{C_{j0}}{\left(1+\frac{V_R}{\Phi_0}\right)^m} \tag{1.9}$$

where

$$C_{j0} = \left(\frac{q\varepsilon_{si}\varepsilon_0}{2\Phi_0}\frac{N_A N_D}{N_A + N_D}\right)^{1-m} \tag{1.10}$$

is a capacitance per unit of area and depends only on the doping concentration.

For large changes of the reverse voltage across the junction capacitance, as it happens mostly in digital circuits, small-signal capacitance must be

---

[1] All the derivatives are evaluated at the quiescent operating point.

replaced by large-signal equivalent linear capacitance which displaces equal charge over voltage swing of interest [R96]

$$\overline{C}_j = \frac{\Delta Q^+}{\Delta V_R} = \frac{Q^+(V_2) - Q^+(V_1)}{V_2 - V_1} \tag{1.11}$$

where V_2 and V_1 are the highest and the lowest reverse voltages applied on the pn junction, respectively.

Evaluating (1.8) for $V_R = V_2$ and $V_R = V_1$, substituting in (1.11) follows that

$$\overline{C}_j = K_{eq} C_{j0} \tag{1.12}$$

where C_{j0} is again given by (1.10) and

$$K_{eq} = \Phi_0^m \frac{(\Phi_0 - V_1)^{1-m} - (\Phi_0 - V_2)^{1-m}}{(1-m)(V_2 - V_1)} \tag{1.13}$$

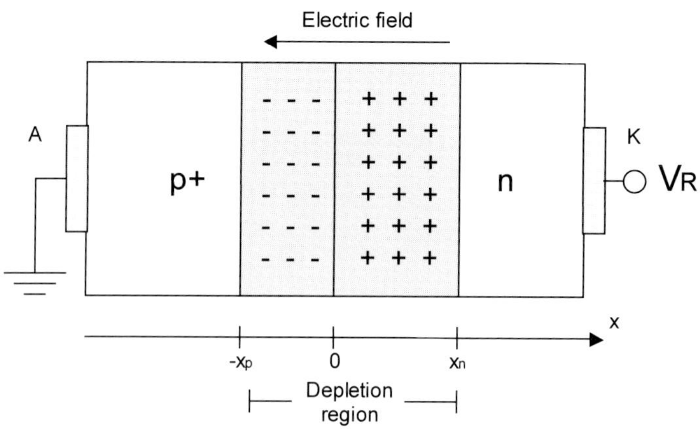

Fig. 1.2. Reverse-biased pn junction.

1.1.2 Forward Bias Condition

With reference to Fig. 1.3, by grounding the cathode and applying a voltage V_D to the anode, we forward-bias the device. Under this condition the built-in potential is reduced by the amount of voltage applied.

Consequently, the width of the depletion regions and the charge stored in the junction are reduced, too.

If V_D is large enough, the reduction in the potential barrier ensures the electrons in the n side and the holes in the p side are attracted by the anode and the cathode, respectively, thus crossing the junction. Once free charges cross the depletion region, they become minority carriers on the other side and a recombination process with majority carriers begins. This recombination reduces the minority carrier concentrations that assume a decreasing exponential profile. The concentration profile is responsible for the current flow near the junction, which is due to a diffusive phenomenon that is called diffusion current. On moving away from the junction, some current flow is given by the diffusion current and some is due to majority carriers that, coming from the terminals, replace those carriers recombined with minority carriers or diffused across the junction. This latter current is termed drift current.

This process causes a current to flow through the diode that is exponentially related to voltage V_D as follows

$$I_D = A_D J_S \exp\left(\frac{V_D}{V_T}\right) \tag{1.14}$$

where A_D is the junction area and J_S the scale current density which is inversely proportional to the doping concentrations. The product $A_D J_S$ is often expressed in terms of a scale current and denoted as I_S.

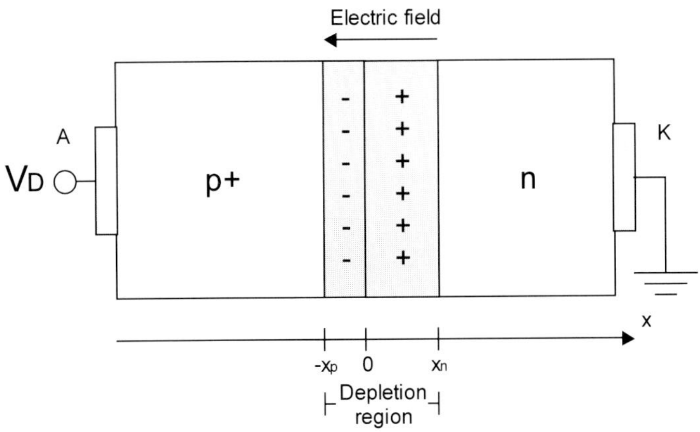

Fig. 1.3. Forward-biased pn junction.

As far as the charge stored in the device is concerned, we have two contributions under the forward bias condition. The first is given by the charge stored in the depletion region that yields a junction capacitance that can be expressed by (1.9) or (1.12) for small signal variation or large signal swing of the forward voltage applied, respectively. In any case, this contribution is very small and the junction capacitance negligible.

The second contribution takes into account the charge due to minority carrier concentrations close to the junction that are responsible for the diffusion current. This component yields a diffusion capacitance, C_d, which is proportional to the current I_D as follows [S81], [MK86]

$$C_d = \tau_T \frac{I_D}{V_T} \tag{1.15}$$

where τ_T is a technology parameter known as the transit time of the diode.

Observe that C_d is a small-signal capacitance which is valid around the given bias voltage. Therefore, diffusion capacitance must be corrected when a large forward voltage swing is applied across the pn junction because I_D is strongly depending by forward voltage applied (1.14).

Similar to the junction capacitance, a large-signal equivalent diffusion capacitance can be defined as follows [R96]

$$\overline{C}_d = \frac{\tau_T}{V_2 - V_1} A_D J_S e^{\frac{V_H}{V_L}} = \frac{V_T}{V_2 - V_1}\left[C_d(V_2) - C_d(V_1)\right] \tag{1.16}$$

1.2 BIPOLAR-JUNCTION TRANSISTORS

Bipolar transistors or BJT were widespread until the end of seventies when MOS technology started to become popular thanks to the fact that a larger number of transistors could be put together in a single integrated circuit. Although bipolar digital circuits designs occupy a small slice of the digital market, they still are the technology choice in very high performance applications. With respect to MOS, bipolar transistors have the advantage of a larger transconductance factor, g_m, and a larger output resistance, r_c, so they exhibit better performance in terms of current driving capability and achievable voltage gain. Moreover, the very high unity-gain frequency of bipolar transistors (in the order of 20-50 GHz), make them suitable for high-speed digital applications.

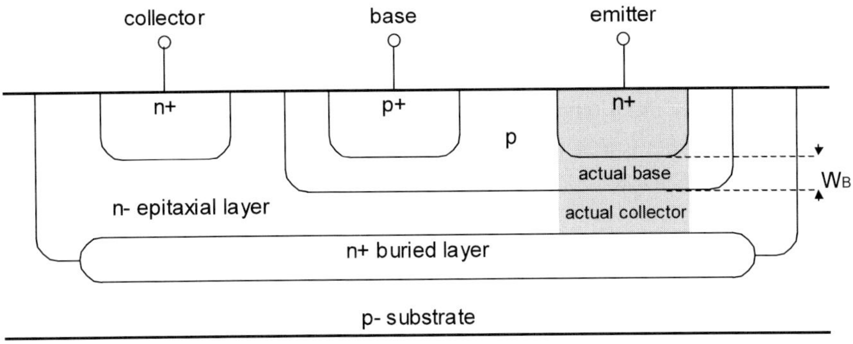

Fig. 1.4. Bipolar-Junction Transistor cross section.

A typical simplified BJT cross-section is shown in Fig. 1.4, where the so-called npn vertical transistor is depicted. It can be seen as two back-to-back diodes because it is made up of two n-regions separated by a p-region called base. The actual base region is the gray p-region in the figure whose width, W_B, is small with respect to the other proportions and in modern bipolar processes is between 0.5-0.8 μm. This region has a medium doping concentration, in the order of 10^{17} carriers/cm^3. The emitter is the heavily doped n+ region in the figure. It has a width of a few μm and its doping concentration is in the order of 10^{21} carriers/cm^3. Finally, the actual collector region is the gray n- epitaxial layer in the figure. The collector doping concentration is in the order of 10^{15} carriers/cm^3. To reduce the resistive path that connects the actual collector region to the collector contact, a heavily doped buried layer is grown below the device. The gray area represents the region where the so-called transistor effect takes place and is the actual npn transistor. Since this area extends vertically, the transistor is said to be vertical. Finally, note that, unlike for MOSFETs, the transistor is not symmetric.

1.2.1 Basic Operation

The operation mode of the transistor changes depending upon the voltages applied over the device terminals. In digital circuits, the transistor operates by preference cut-off mode (i.e., both junctions are reverse-biased either) or in forward-active mode (i.e., the emitter-base junction is forward-biased and the collector-base junction is reverse-biased). Operation in saturation region (i.e., both junctions in forward-bias condition) is avoided, as the circuit performance tends to deteriorate.

In cut-off mode, both diodes are reverse-biased, therefore currents into the terminals are saturation current which is extremely small and the device is considered to be in the off-state.

To understand the forward-active operation of a bipolar transistor let us consider the simplified scheme in Fig. 1.5 where the emitter terminal is connected to ground. The base-emitter junction acts as a diode and a current flows if the junction is forward biased. In such a situation, that is $V_{BE} > 0$, a current of majority carriers (holes in this case) flows from the base region across the base-emitter junction. Meantime, a current of electrons flows from the emitter across the base-emitter junction and enters the base thus diffusing towards the base-collector junction. Due to the different doping levels electrons that diffuse into the base are much more than just holes that diffuse into the emitter [S81], [MK86].

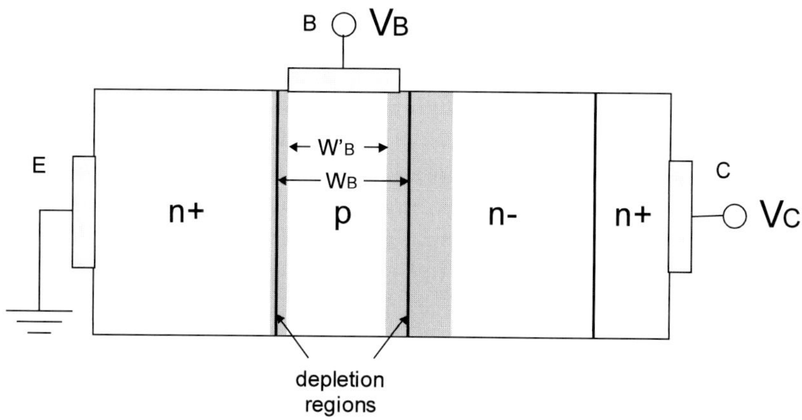

Fig. 1.5. Simplified scheme of a BJT.

If V_C is larger than 0.2-0.3 V, the excess of electrons in the base is subject to a negative electric field imposed by the collector voltage. When those electrons appear at the base-collector junction, they are pushed into the collector region. Since the base width, W_B, is small, electrons coming from the emitter do not have the possibility to recombine with holes in the base and almost all are pushed into the collector.

In such a situation, the small base current is mainly determined by holes while electrons coming from the emitter mainly determine the large collector current. Consequently, the emitter current is the sum of those two contributions.

The collector current, I_C, is caused by the base-emitter voltage and, as for a diode, it has an exponential relationship that is

$$I_C = A_E J_{CS} \exp\left(\frac{V_{BE}}{V_T}\right) \tag{1.17}$$

where A_E is the emitter area and J_{CS} is a constant term that represents a current density and is inversely proportional to the base width, W_B, and its doping concentration. The product $A_E J_{CS}$ is often expressed in terms of the current scale factor I_{CS}.

In addition, the base current is exponentially related to the base-emitter voltage and has an expression similar to (1.17). Consequently, at a first approximation, the ratio between the collector and the base current is constant and independent of both voltages and currents. This ratio is commonly referred to as β_F, that is

$$\beta_F = \frac{I_C}{I_B} \tag{1.18}$$

Due to the small amount of base current with respect to the large collector current, the value of β_F is typically between 50 and 200.

The ratio between the collector current and the emitter current, I_E, is denoted with α_F and results as

$$\alpha_F = \frac{I_C}{I_E} \tag{1.19}$$

Since $I_E = I_C + I_B$, the constant α_F can be expressed in terms of β_F, that is

$$\alpha_F = \frac{\beta_F}{\beta_F + 1} \approx 1 \tag{1.20}$$

which is close to unity for high values of β_F.

In saturation mode, the collector-emitter voltage, V_{CE}, approaches the value of about 0.2-0.3 V, commonly referred to as V_{CEsat}, and the base-collector junction becomes forward biased at a voltage $V_{CB,on}$≈0.5 V. In such a situation, holes from the base start to diffuse into the collector, and the collector current is no longer related to the base current by (1.18). Specifically, the base-emitter junction behaves like a diode whose current I_B

exponentially depends on V_{BE} while the base-collector junction behaves like a voltage source whose value is set to V_{CEsat}.

1.2.2 Early Effect or Base Width Modulation

In (1.17) the collector current is independent of the collector voltage. However, this is true only at a first order approximation since the dependence in fact exists. Referring to Fig. 1.5, we note that the effective base width, W_B', that should be used for evaluating J_{CS}, is different from the designed base width, W_B, due to the presence of two depletion regions. The base-emitter depletion region is caused by a forward biasing. Therefore, it is small and almost independent of the voltage applied. In contrast, reverse biasing creates the base-collector depletion region, which, consequently, is larger and strongly depends on the voltage applied. Specifically, the collector voltage modulates the base-collector depletion region, thus decreasing and influencing the effective base width, W_B'.

To take this effect into account, a corrective term is introduced in (1.17) that becomes [GT86], [GM93]

$$I_C = A_E J_{CS} \exp\left(\frac{V_{BE}}{V_T}\right)\left(1+\frac{V_{CE}}{V_A}\right) \tag{1.21}$$

where the constant V_A is commonly referred to as the *Early Voltage* and has a typical value between 50 and 100 V. In most of the applications, especially in digital circuits, this effect is negligible.

1.2.3 Charge Effects in the Bipolar Transistor

As described previously, depending upon the operation region, the base-emitter and base-collector junctions can be biased in forward or reverse mode. This fact impacts on the charge stored in depletion layer and, in turns, on the equivalent capacitance. In particular, when these junctions are in reverse-biased condition and assuming large changes of the voltage, their dynamic behaviors can be modeled by means of a linearized junction capacitance given by

$$\overline{C}_j = K_{eq} C_{j0} \tag{1.22}$$

where C_{j0} is the zero-bias value of the junction capacitance (C_{j0be} and C_{j0bc} for base-emitter and base-collector junction, respectively) and K_{eq} is the correctness term, defined in (1.13), which depends on the junction grading coefficient (m_{be} and m_{bc}) and the built-in potentials (Φ_{0be} and Φ_{0bc}).

Observing the BJTs cross-section depicted in Fig. 1.4, it can be seen that the collector is isolated from the substrate by diode a reversed-biased which can be modeled, once again, by a parasitic nonlinear junction capacitance. It is worth noting that this capacitive effect might dominate the performance of the transistor due to the large junction area that leads to a very high capacitance, C_{j0cs}.

In the active region, the base-emitter junction is forward-biased so, like in a forward-biased diode, the stored charge is given by two contributions. The first one takes into account the charge in the small depletion region that leads to a negligible junction capacitance. The second contribution is given by the minority carrier concentrations accumulated both the base and the emitter. As in a forward-biased diode, this contribution leads to a small-signal diffusion capacitance, C_{dbe}, expressed by

$$C_{dbe} = \tau_{Tb} \frac{I_C}{V_T} \tag{1.23}$$

where τ_{Tb} is a technology parameter commonly referred to as the base-transit-time constant.

In a digital context, the small-signal diffusion capacitance should be replaced by a large-signal equivalent diffusion capacitance whose expression is given by

$$\overline{C}_{dbe} = \frac{\tau_T}{V_2 - V_1} A_D J_S e^{\frac{V_H}{V_L}} = \frac{V_T}{V_2 - V_1}\left[C_{dbe}(V_2) - C_{dbe}'(V_1)\right] \tag{1.24}$$

However, since in the active-region the voltage swing across the forward-biased junction is very small, for manual analysis, it is convenient to use the small-signal model expressed by (1.23).

Finally, in saturation region, since both the junctions are forward-biased, both junction capacitances can be neglected and a new diffusion capacitance, C_{dbc}, arises due to the carriers injected from the collector. This capacitive value is higher than that given by (1.23) and greatly affects the switching-time when a transistor must be driven out of the saturation region. This large capacitive contribution slows the BJT's transient behavior and justifies the fact that, in digital circuits, BJTs are to operate either in cut-off or in forward-active mode.

1.2.4 Small Signal Model

As already mentioned, bipolar transistors exhibit high performance if and only if they are switched from the cut-off to forward-active region. For this reason it is useful to define a small-signal model valid in both cases. In particular, the cut-off model will be an extension of the forward-active model assuming zero currents into the terminal.

This model is similar to the hybrid-π model and is shown in Fig. 1.6. The most important parameter is the voltage-controlled current source, $g_m v_{be}$, whose transconductance, g_m, is defined as

$$g_m = \frac{\partial I_C}{\partial V_{BE}} = \frac{\partial}{\partial V_{BE}} I_{CS} \exp\left(\frac{V_{BE}}{V_T}\right) = \frac{I_C}{V_T} \tag{1.25}$$

The small signal resistance, r_{be}, is defined as

$$r_{be} = \left(\frac{\partial I_B}{\partial V_{BE}}\right)^{-1} = \left(\frac{1}{\beta_F}\frac{\partial I_C}{\partial V_{BE}}\right)^{-1} = \frac{\beta_F}{g_m} \tag{1.26}$$

The resistor r_o models the Early effect and is defined as

$$r_o = \left(\frac{\partial I_C}{\partial V_{CE}}\right)^{-1} = \left[\frac{\partial}{\partial V_{CE}} I_{CS} \exp\left(\frac{V_{BE}}{V_T}\right)\left(1+\frac{V_{CE}}{V_A}\right)\right]^{-1} \approx \frac{V_A}{I_C} \tag{1.27}$$

All the above dc small-signal components are intrinsic of any BJT components since they depend on the npn structure itself.

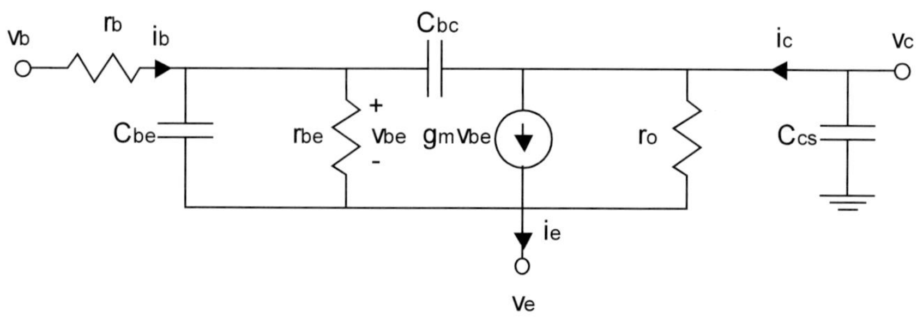

Fig. 1.6. Small signal model for a BJT in active region.

The model in Fig. 1.6 also includes a base resistance, r_b, which comes out in real implementations. Specifically, r_b, models the resistive path that exists between the effective transistor base region (i.e. the gray area in Fig. 1.4) and the base contact (i.e. the p+ doped region). This path presents a small ohmic resistance of a few tens or hundreds of ohm. With respect to r_{be}, r_b has a small value and, in low frequency operations, it can be neglected since the base-emitter voltage is practically equal to v_{be}. In high-frequency circuits (i.e. in RF applications), part of the base current flows across C_{be} thus reducing the effective impedance in the base-emitter branch. Because of the presence of r_b, v_{be} can be significantly different from the base-emitter voltage applied thus affecting transistor properties considerably. In practice, r_b cannot be neglected if a high-speed circuit is being analyzed or designed.

Note that there is also an ohmic resistance in series with the actual collector (whose value is lowered by the n+ buried layer) but its presence is not as crucial as the base resistance is.

As far as the capacitive contribution is concerned, we have two main intrinsic capacitors, C_{be} and C_{bc}, as well as capacitor, C_{cs}, which exists in integrated implementations only. Specifically, capacitor C_{be}, is the base-emitter capacitor and is expressed by

$$\overline{C}_{be} \approx \tau_{Tb} \frac{I_C}{V_T} + K_{eq} C_{j0be} \tag{1.28}$$

which reduces to $K_{eq}C_{j0be}$ in cut-off region, and C_{bc}, which represents the base-collector capacitive contribution, is expressed by

$$\overline{C}_{bc} = K_{eq} C_{j0bc} \tag{1.29}$$

Capacitor C_{cs}, comes out from the reverse-biased pn region realized by the collector-substrate junction and is modeled by the following expression

$$\overline{C}_{cs} = K_{eq} C_{j0cs} \tag{1.30}$$

1.3 MOS TRANSISTORS

Currently, Metal-Oxide-Semiconductor Field-Effect Transistors (MOSFETs or simply MOS transistors) are the most commonly used components in digital integrated circuit implementations since their characteristics make them more attractive than other devices such as, for example, BJTs. Specifically, their simple realization and low cost, the possibility of having complementary devices with similar characteristics, their small geometry and, consequently, the feasibility of integrating a large number of devices in a small area, their infinite input resistance at the gate terminal and the faculty of building digital cells with no static dissipation, all motivate the great success of MOS transistors in modern technologies.

A simplified cross section of an n-channel MOS (n-MOS) transistor is shown in Fig. 1.7. It is built on a lightly doped p-type substrate (p-) that separates two heavily doped n-type regions (n+) called source and drain. A dielectric of silicon oxide and a polysilicon gate are grown over the separation region. The region below the oxide is the transistor channel and its length, that is the length that separates the source and the drain, is the channel length, denoted by *L*. In present digital MOS technologies the channel length is typically lower than 0.18 μm. In a p-channel MOS (p-MOS) all the regions are complementary doped.

There is no physical difference between the source and the drain as the device is symmetric, the notations source and drain only depend on the voltage applied. In an n-MOS the source is the terminal at the lower potential while, in a p-MOS, the source is the terminal at the higher potential.

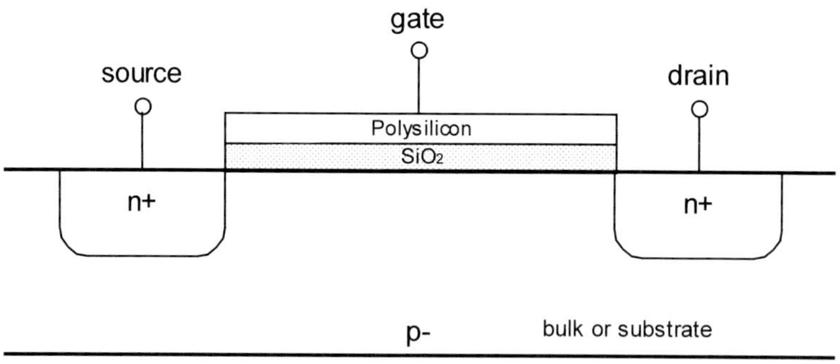

Fig. 1.7. Simplified cross section of an n- MOS transistor.

1.3.1 Basic Operation

To understand the basic operation of MOS transistors we shall analyze the behavior of an n-MOS depending on the voltages applied at its terminals.

If source, drain and substrate are grounded, the device works as a capacitor. Specifically, the gate and the substrate above the SiO_2 interface are two plates electrically insulated by the silicon oxide.

If we apply a negative voltage to the gate, negative charges will be stored in the polysilicon while positive charges will be attracted to the channel region thus increasing the channel doping to p+. This situation leads to an accumulated channel. Source and drain are electrically separated because they form two back-to-back diodes with the substrate. Even if we positively bias either the source or the drain, only a negligible current (the leakage current) will flow from the biased n+ regions to the substrate.

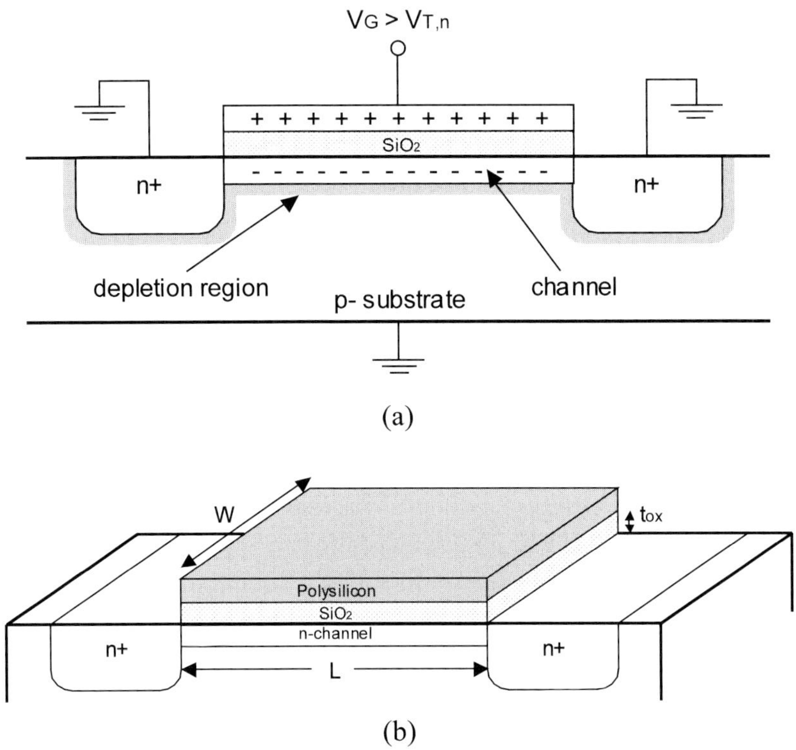

Fig. 1.8. Cross section of an n-MOS transistor when the channel is present: a) bidimensional, b) tridimensional.

By applying a positive voltage to the gate, positive charges will be stored in the gate. Below the silicon oxide, if the gate voltage is small, positive free charges of the p- substrate will be repelled from the surface thus depleting the channel area. A further increase in the gate voltage leads to negative free charges being attracted to the channel that thereby becomes an n region. In this condition the channel is said to be inverted.

The gate-source voltage for which the concentration of electrons under the gate equals the concentration of holes in the p- substrate far from the gate is said to be the transistor threshold voltage, $V_{T,n}$.

At a first approximation, if the gate-source voltage, V_{GS}, is below the threshold voltage, no current can exist between the source and the drain and the transistor is said to be in the cutoff region. In contrast, if the gate-source voltage is greater than the threshold voltage, an n channel joins the drain and the source and a current can flow between these two electrically connected regions.

Actually, for gate voltages around $V_{T,n}$, the charge does not change abruptly and a small amount of current can flow even for small negative values of $V_{GS} - V_{T,n}$. This condition is termed weak inversion and the transistor is said to work in subthreshold region.

When the channel is present, as in Fig. 1.8, the accumulated negative charge is proportional to the gate source voltage and depends on the oxide thickness, t_{ox}, since the transistor works as a capacitor. Specifically, the charge density of electrons in the channel is given by [S81]-[MK86]

$$Q_n = C_{ox}\left(V_{GS} - V_{T,n}\right) \tag{1.31}$$

where C_{ox} is the gate capacitance per unit area defined as

$$C_{ox} = \frac{\varepsilon_{ox}\varepsilon_0}{t_{ox}} \tag{1.32}$$

and ε_{ox} is the relative permittivity of the SiO_2 (ε_{ox} is approximately 3.9).

The total capacitance and the total charge are obtained by multiplying both the equations (1.31) and (1.32) by the device area, as follows

$$C_{gs} = WLC_{ox} \tag{1.33a}$$

$$Q_{T-n} = WLC_{ox}\left(V_{GS} - V_{T,n}\right) \tag{1.33b}$$

where W is the channel width of the MOS transistor as depicted in Fig. 1.8.

1.3.2 Triode or Linear Region

Increasing the drain voltage, V_D, causes a current to flow from the drain to the sourcc through the channel. A drain voltage different from zero will modify the charge density but for small V_D the channel charge will not change appreciably and can be expressed by (1.31) again. Under this condition, the device operates as a resistor of length L, width W with a permittivity proportional to Q_n. Therefore, the relationship between voltage V_{DS} and the drain-source current, I_D, can be written as [LS94]

$$I_D = \mu_n Q_n \frac{W}{L} V_{DS} \tag{1.34}$$

where μ_n is the mobility of electrons near the silicon surface.

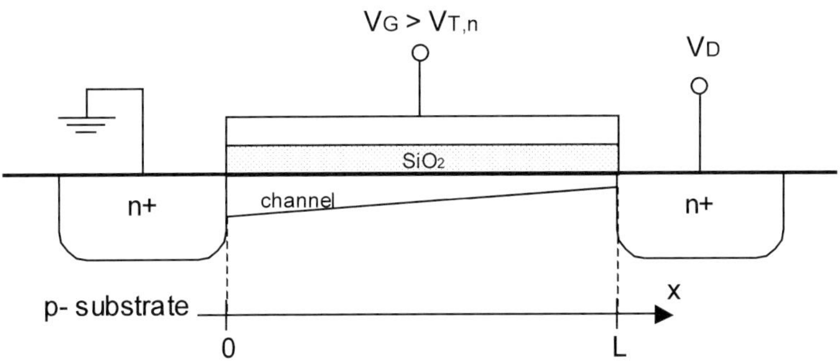

Fig. 1.9. MOS transistor channel for large V_D.

Substituting (1.31) in (1.34) yields

$$I_D = \mu_n C_{ox} \frac{W}{L} \left(V_{GS} - V_{T,n}\right) V_{DS} \tag{1.35}$$

Larger drain voltages modify the charge density profile in the channel. Specifically, referring to Fig. 1.9, we can express the channel charge density as a function of channel length. For $x = 0$, that is, close to the source, (1.31) holds, while for $x = L$, that is, close to the drain, we have

$$Q_n(L) = C_{ox}\left(V_{GD} - V_{T,n}\right) \tag{1.36}$$

Assuming a linear profile, the charge density has the following expression

$$Q_n(x) = \frac{Q_n(L) - Q_n(0)}{L} x + Q_n(0) \tag{1.37}$$

The current can be expressed in a form similar to (1.34) but with a different charge expression. If the charge density profile is linear, the average charge density can be used instead. The average charge density results in

$$\overline{Q}_n = \frac{Q_n(L) + Q_n(0)}{2} = C_{ox}\left(V_{GS} - V_{T,n} - \frac{V_{DS}}{2}\right) \tag{1.38}$$

and substituting this value in (1.34) leads to

$$I_D = \mu_n \overline{Q}_n \frac{W}{L} V_{DS} = \mu_n C_{ox} \frac{W}{L}\left(V_{GS} - V_{T,n} - \frac{V_{DS}}{2}\right) V_{DS} \tag{1.39}$$

The current I_D is linearly related to V_{GS} and has a quadratic dependence on V_{DS}. Under this condition the device is said to operate in triode or linear region. Note also that (1.39) is reduced to (1.35) for small values of V_{DS}.

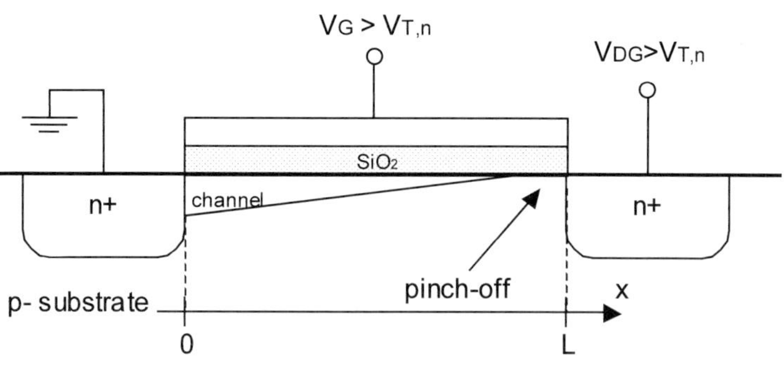

Fig. 1.10. MOS transistor channel for $V_{DG} > V_{T,n}$.

1.3.3 Saturation or Active Region

A further increase of V_D, can lead to the condition of a gate-drain voltage equal to $V_{T,n}$. In this case the charge density close to the drain, $Q_n(L)$, becomes zero and current I_D reaches its maximum value. This condition is shown in Fig. 1.10.

At a first approximation, the current does not change over this point with V_{DS} since the charge concentration in the channel remains constant and the electron carriers are velocity saturated. Under this condition the transistor is said to work in saturation or active region.

Denoting V_{DSsat} as the drain source voltage when the charge density $Q_n(L)$ becomes zero, we can find an equivalent relationship that expresses the pinch-off condition by substituting $V_{DG} = V_{DS} - V_{GS}$ into $V_{DG} > V_{Tn}$. Specifically, we get

$$V_{DS} > V_{DSsat} \tag{1.40}$$

where

$$V_{DSsat} = V_{GS} - V_{T,n} \tag{1.41}$$

Substituting the value $V_{DS} = V_{DSsat}$ defined in (1.41) into (1.39) gives the current expression in the pinch-off case and results as

$$I_D = \frac{\mu_n C_{ox}}{2} \frac{W}{L} \left(V_{GS} - V_{T,n}\right)^2 \tag{1.42}$$

As mentioned above, (1.42) is valid at a first approximation. In fact, increasing V_D yields an increase in the pinch-off region as well as a decrease in channel length. This effect is commonly known as channel length modulation. To take this effect into account, a corrective term is used to complete (1.42) which becomes

$$I_D = \frac{\mu_n C_{ox}}{2} \frac{W}{L} \left(V_{GS} - V_{T,n}\right)^2 \left[1 + \lambda\left(V_{DS} - V_{DSsat}\right)\right] \tag{1.43}$$

The parameter λ is referred to as the channel length modulation factor and, at a first approximation, it is inversely proportional to the channel length, L.

1.3.4 Body Effect

All the equations derived above were based on the assumption that the source and the substrate (or the bulk) were connected together. Although this is a rather common condition, in general the voltage of these two terminals can be different. In this event a second order effect occurs commonly referred to as the *body effect* [GM93]. A different voltage between the source and the bulk is modeled as an increase in the threshold voltage, which assumes the following expression [GM93], [LS94], [JM97]

$$V_{T,n} = V_{T,n0} + \gamma\left(\sqrt{V_{SB} + 2|\phi_F|} - \sqrt{2|\phi_F|}\right) \tag{1.44}$$

with V_{SB} being the source-bulk voltage, $V_{T,n0}$ the threshold voltage with zero V_{SB}, ϕ_F the Fermi potential of the substrate and γ a constant referred to as the body-effect constant. The Fermi potential is defined as [S81]

$$|\phi_F| = \frac{kT}{q}\ln\left(\frac{N_A}{n_i}\right) \tag{1.45}$$

while the value of γ depends on the substrate doping concentration as follows [S81]

$$\gamma = \frac{\sqrt{2qN_A\varepsilon_{si}}}{C_{ox}} \tag{1.46}$$

1.3.5 p-channel Transistors

For a p-channel transistor we can use the same equations derived in the previous sections, provided that a negative sign is placed in front of every voltage variable.

Therefore, V_{GS} becomes V_{SG}, V_{DS} becomes V_{SD}, V_{Tn} becomes $-V_{T,p}$, and so on. Note that in a p-MOS transistor the threshold voltage is negative. The condition for a p-MOS to be in saturation region is now $V_{SD} > V_{SG} + V_{T,p}$. Current equations (1.39) and (1.43) still hold but the current now flows from the source to the drain.

1.3.6 Charge Effects in Saturation Region

Charge effects of a MOS transistor in saturation region include several capacitive effects, each of them with its own physical meaning that can be understood by analyzing the detailed n-MOS cross section in Fig. 1.11.

The most important capacitor is the gate-source capacitor whose value is given by two different terms. The first term takes into account the capacitive effect between the gate and the channel, which is electrically connected to the source. At a first approximation, the corresponding capacitor, C_{g_ch}, is a linear capacitor that depends on the oxide thickness as well as on the device area. It can be demonstrated that its value is approximately given by [R96], [LS94]

$$C_{g\_ch} \approx \frac{2}{3} WLC_{ox} \tag{1.47}$$

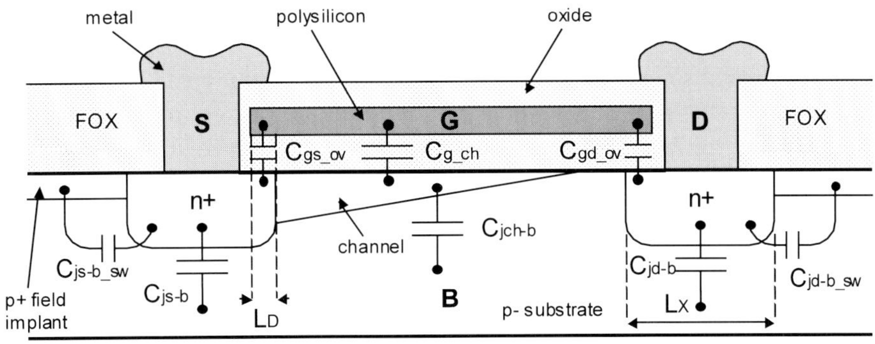

Fig. 1.11. Detailed n-MOS cross-section.

The second term that contributes to the gate-source capacitance is given by the overlap that exists between the gate and the source n+ region. This overlap is unavoidable and results from the fact that during the fabrication process the doping element also spreads horizontally. Naming L_D the overlap diffusion length, the resulting parasitic capacitor, C_{gs_ov}, is given by

$$C_{gs\_ov} = WL_D C_{ox} \tag{1.48}$$

Hence, the capacitor C_{gs} is expressed by the sum of (1.47) and (1.48), that is

$$C_{gs} \approx W\left(\frac{2}{3}L + L_D\right)C_{ox} \tag{1.49}$$

The same boundary effect that determines the gate-source overlap capacitance yields the gate-drain capacitance that is given by

$$C_{gd} = C_{gd\_ov} = WL_D C_{ox} \tag{1.50}$$

The second largest capacitor is the source-bulk capacitor, which can be split into three contributions all of them given by the depletion capacitances of reverse biased pn junctions. The first, C_{sb}, takes into account the junction capacitance between the n+ source area and the bulk. Assuming that voltages move rapidly over a wide range, its expression is similar to (1.12) that is

$$\overline{C}_{sb} = K^{I}_{eq} W L_X C_{j0,sb} \tag{1.51}$$

where K^{I}_{eq} is the correctness term, defined in (1.13), and $C_{j0,sb}$ is defined as the zero-voltage source-bulk junction capacitance per unit area.
The second contribution is responsible for $C_{jch\text{-}b}$ and takes into account the depletion region between the channel and the bulk. Even in this case we have an expression similar to (1.51) that is

$$\overline{C}_{jch-b} = K^{I}_{eq} W L C_{j0,sb} \tag{1.52}$$

where K^{I}_{eq} is again the correctness which at a first approximation, is equal to that given in (1.51).

The third term is referred to as the source-bulk sidewall capacitance and is denoted as $C_{sb,sw}$. This capacitance is due to the presence of a highly p+ doped region (field implant) that exists under the thick field oxide (FOX) and prevents the leakage current from flowing between two adjacent transistors. The value of $C_{sb,sw}$ can be particularly large if the field implant is heavily doped as in modern technologies. The expression of $C_{sb,sw}$ is then

$$\overline{C}_{sb,sw} = K^{I}_{eq} (W + 2L_X) C_{j0,sb\_sw} \tag{1.53}$$

where the term $(W + 2L_X)$ is the perimeter of the source junction, excluding the side adjacent to the channel, and C_{j0,sb_sw} is the zero-voltage junction capacitance per unit length.

Consequently, the source-bulk capacitance is given by the sum of (1.51), (1.52) and (1.53), that is

$$\overline{C}_{jsb} = K^{I}_{eq} \left[W (L + L_X) C_{j0,sb} + (W + 2L_X) C_{j0,sb\_sw} \right] \tag{1.54}$$

The fourth capacitor in the model in Fig. 1.11 is the drain-bulk capacitor, C_{db}. This is similar to the source-bulk capacitance except for the fact that the channel does not make any contribution. Therefore equations similar to (1.51) and (1.53) can be written as follows

$$\overline{C}_{db} = K_{eq}^{II} W L_X C_{j0,db} \tag{1.55a}$$

$$\overline{C}_{db,sw} = K_{eq}^{II} (W + 2L_X) C_{j0,db\_sw} \tag{1.55b}$$

and the drain-bulk capacitance results as

$$\overline{C}_{db} = K_{eq}^{II} \left[W L_X C_{j0,db} + (W + 2L_X) C_{j0,db\_sw} \right] \tag{1.56}$$

1.3.7 Charge Effects in Triode Region

Charge effects in triode region are not easy to determine because the channel is directly connected to both the source and drain resulting in a distributed RC network over the whole length of the device. Moreover, because of the capacitive nature of junction capacitances, the capacitive elements are highly non-linear. This is another factor making the model quite complicated for management by hand analysis.

A simplified model, which is quite accurate for small V_{DS} and for long-channel devices, can be obtained by evaluating the total channel charge contribution and by assuming half of this contribution to be referred to the source and half to the drain [LS94].

Specifically, since the total gate-channel capacitance, is given by

$$C_{g\_ch} = W L C_{ox} \tag{1.57}$$

gate-source and gate-drain capacitances, including the overlap contribution, can be modeled as

$$C_{gs} = C_{gd} = W \left(\frac{L}{2} + L_D \right) C_{ox} \tag{1.58}$$

Unfortunately, the evaluation of C_{gd} in the linear region cannot be carried out in the same manner as for long-channel devices, since the decomposition of channel capacitance into gate–source and gate–drain capacitances no longer applies in submicron technologies [CH99]. As a result, a more accurate capacitance model, is given by

$$C_{gd} = W \left(\frac{3}{4} A_{\mathrm{bulk,max}} L + L_D \right) C_{ox} \tag{1.59}$$

where $A_{\text{bulk,max}}$ is a BSIM3v3 parameter slightly greater than unity, as will be shown in Section 6.2.1. Obviously, the gate-source capacitance contribution is evaluated as

$$C_{gs} = W\left(\frac{4-3A_{\text{bulk,max}}}{4}L + L_D\right)C_{ox} \tag{1.60}$$

so that the sum of C_{gd} and C_{gs} equals the total gate capacitance, $W(L+2L_D)C_{ox}$.

Also, the channel-bulk contribution is shared between the source and the drain and capacitances C_{sb} and C_{db} become

$$\overline{C}_{sb} = K_{eq}^{I}\left[W\left(L_X + \frac{L}{2}\right)C_{j0,sb} + (W+2L_X)C_{j0,sb\_sw}\right] \tag{1.61a}$$

$$\overline{C}_{db} = K_{eq}^{II}\left[W\left(L_X + \frac{L}{2}\right)C_{j0,sb} + (W+2L_X)C_{j0,sb\_sw}\right] \tag{1.61b}$$

1.3.8 Charge Effects in Cutoff Region

In the cutoff region, since the channel is not present, both C_{gd} and C_{gs} are due only to the overlap contribution, that is

$$C_{gs} = C_{gd} = WL_D C_{ox} \tag{1.62}$$

Source-bulk and drain-bulk capacitances are similar to those given in (1.61) with the difference that the channel does not make any contribution, that is

$$\overline{C}_{sb} = K_{eq}^{I}\left[WL_X C_{j0,sb} + (W+2L_X)C_{j0,sb\_sw}\right] \tag{1.63a}$$

$$\overline{C}_{db} = K_{eq}^{II}\left[WL_X C_{j0,sb} + (W+2L_X)C_{j0,sb\_sw}\right] \tag{1.63b}$$

The fact that no channel exists, generates a new capacitor, C_{gb}, which connects the gate and the bulk. Its value is given by the oxide capacitance multiplied by the device area, that is

$$C_{gb} = WLC_{ox} \tag{1.64}$$

1.3.9 Small Signal Model

Although in almost all digital circuits signals move in wide voltage ranges, there are several digital configurations (e.g. the Source Coupled Logic family) where the MOS device is biased in saturation region and a small signal is applied and processed as a digital value. In these cases it may be useful to consider a low frequency small signal model.

1.3.9.1 Saturation region

The low-frequency small signal model for a MOS transistor operating in the active region is shown in Fig. 1.12.

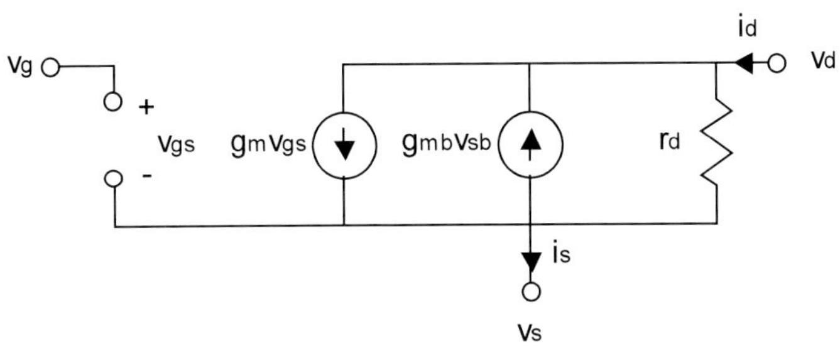

Fig. 1.12. Low-Frequency small signal model for a MOS transistor in active region.

The most important small signal component is the dependent current generator, $g_m v_{gs}$, whose transconductance, g_m, is defined as

$$g_m = \frac{\partial I_D}{\partial V_{GS}} = \mu_n C_{ox} \frac{W}{L}\left(V_{GS} - V_{T,n}\right) \tag{1.65}$$

Solving (1.42) with respect to V_{GS}-$V_{T,n}$ and substituting the result to (1.65), leads to the well-known expression for the transconductance

$$g_m = \sqrt{2\mu_n C_{ox} \frac{W}{L} I_D} \tag{1.66}$$

The second dependent current source, $g_{mb}v_{sb}$, accounts for the body effect and its transconductance is defined as

$$g_{mb} = -\frac{\partial I_D}{\partial V_{SB}} = -\frac{\partial I_D}{\partial V_{T,n}}\frac{\partial V_{T,n}}{\partial V_{SB}} \tag{1.67}$$

The first derivative in (1.67) results as

$$\frac{\partial I_D}{\partial V_{T,n}} = -\mu_n C_{ox}\frac{W}{L}\left(V_{GS} - V_{T,n}\right) = -g_m \tag{1.68}$$

while the second one comes out by deriving (1.44) with respect to V_{SB}, thus yielding

$$\frac{\partial V_{T,n}}{\partial V_{SB}} = \frac{\partial}{\partial V_{SB}}\left[V_{T,n0} + \gamma\left(\sqrt{V_{SB} + 2|\phi_F|} - \sqrt{2|\phi_F|}\right)\right] = \frac{\gamma}{2\sqrt{V_{SB} + 2|\phi_F|}} \tag{1.69}$$

Therefore, substituting (1.68) and (1.69) in (1.67) we get

$$g_{mb} = \frac{\gamma g_m}{2\sqrt{V_{SB} + 2|\phi_F|}} \tag{1.70}$$

Note that this value is nonzero even if the quiescent value of V_{SB} equals zero. Specifically, the body effect arises only if a small signal, v_{sb}, is present between the source and the bulk terminals. In general g_{mb} is 0.1–0.2 times g_m and can be neglected in a non-detailed analysis.

The last model parameter is the resistor r_d, which takes into account the channel length modulation or, which is the same, the dependence of the drain current on V_{DS}. It is related to the large signal equations by

$$\frac{1}{r_d} = \frac{\partial I_D}{\partial V_{DS}} \tag{1.71}$$

Substituting in (1.71) the current expression in (1.43) results as

$$\begin{aligned}\frac{1}{r_d} = \frac{\partial I_D}{\partial V_{DS}} &= \frac{\partial}{\partial V_{DS}}\frac{\mu_n C_{ox}}{2}\frac{W}{L}\left(V_{GS} - V_{T,n}\right)^2\left[1 + \lambda\left(V_{DS} - V_{DSsat}\right)\right] \\ &= \lambda\frac{\mu_n C_{ox}}{2}\frac{W}{L}\left(V_{GS} - V_{T,n}\right)^2 \approx \lambda I_D\end{aligned} \tag{1.72}$$

and finally

$$r_d \approx \frac{1}{\lambda I_D} \tag{1.73}$$

1.3.9.2 Triode region

The low frequency small signal model for a MOS in triode region is a resistor whose value can be determined by deriving (1.39) with respect to V_{DS}, that is

$$\frac{1}{r_d} = \frac{\partial I_D}{\partial V_{DS}} = \mu_n C_{ox} \frac{W}{L}\left(V_{GS} - V_{T,n} - V_{DS}\right) \tag{1.74}$$

If V_{DS} is small (1.74) is often approximated by

$$r_d \approx \frac{1}{\mu_n C_{ox} \frac{W}{L}\left(V_{GS} - V_{T,n}\right)} \tag{1.75}$$

1.3.9.3 Cut-off region

In the cutoff region the resistance r_d is assumed to be infinite so the equivalent model is purely capacitive.

1.3.10 Second Order Effects in MOSFET Modeling

The main second order effects that should be taken into account when determining a MOS large signal model are reported in this section. Their effects are always present and are especially prominent in short-channel devices.

In the following we shall neglect the subscript n, which referred to n-MOS transistors.

1.3.10.1 Channel length reduction due to overlap

Referring to Fig. 1.11, we see that designed channel, L, is reduced due to the overlap. Assuming a symmetric device with equal overlap, L_D, at both the source and the drain, the amount of reduction is equal to $2L_D$, that is, the effective channel length, L_{eff}, is equal to

$$L_{eff} = L - 2L_D \tag{1.76}$$

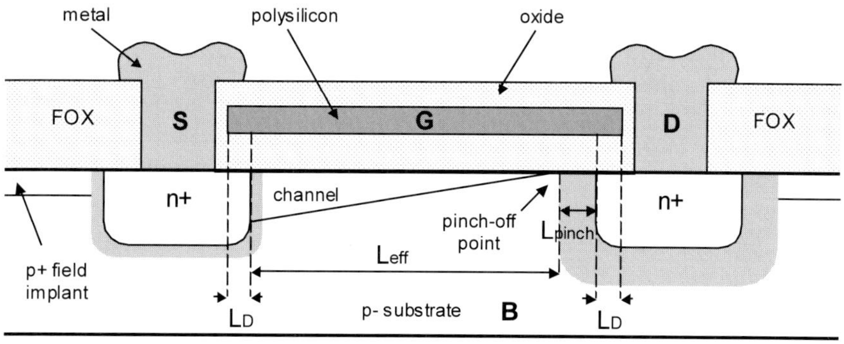

Fig. 1.13. Channel length modulation.

Obviously, the influence of the overlap is greater in short channel devices as it strongly affects the real channel. As a consequence, in all the previous equations, (1.76) should be used for the channel length.

A similar equation holds for the width, W, as well ($W_{eff} = W - 2W_D$).

1.3.10.2 Channel length modulation

The channel length modulation was discussed in previous sections and was modeled by the channel length modulation factor, λ, in (1.43). In this model, the pinch-off point was assumed to be close to the drain end.

A more effective modeling would take into consideration the fact that, in practice, the pinch-off point moves towards the source as V_{DS} increases due to the variation in the drain depletion region. As a consequence, the effective channel length, L_{eff}, is further reduced as shown in Fig. 1.13. Defining L_{pinch} as the distance between the drain end and the pinch-off point we get

$$L_{eff} = L - 2L_D - L_{pinch} \tag{1.77}$$

The value of L_{pinch} is a function of V_{DS}, V_{DSsat} and the doping concentration of the channel. Substituting, for example, (1.77) in (1.42) we observe that, due to a shorter channel, the drain current increases with V_{DS}.

Obviously, this effect is particularly evident in short-channel devices [JM97].

1.3.10.3 Mobility reduction due to vertical electric field

As known, the mobility, μ, relates the electrical field, E, to the drift velocity of carriers, v_d, as [S81], [MK86]

$$v_d = \mu E \tag{1.78}$$

In our previous model we assumed the mobility to be a constant. Actually the value of this parameter depends on several physical factors, the most important of which is related to the carrier-scattering mechanisms.

The carrier scattering in the channel is greatly influenced by the vertical electric field induced by the gate voltage. Consequently mobility changes with V_{GS}. A semi-empirical equation used to model the mobility reduction due to vertical fields in NMOS transistors is [GM93]

$$\mu_s = \frac{\mu_0}{1+\vartheta\left(V_{GS}-V_{T,n}\right)} \tag{1.79}$$

where μ_s is now the new mobility (or better the new surface mobility), μ_0 is the mobility in the case of low fields and ϑ is the mobility degradation factor whose value can be related to oxide thickness as $2.3/t_{ox}$ where t_{ox} is expressed in nm [GM93]. An analogous expression holds for PMOS transistors.

It can be shown that this effect can be modeled as a series resistance, R_S, in the source of the MOS where

$$R_S \approx \frac{\vartheta}{\mu_0 C_{ox} W/L_{eff}} \tag{1.80}$$

1.3.10.4 Mobility reduction due to lateral electric field

Mobility is further reduced due to the high lateral electric field. Since, at a first approximation, the electric field is proportional to V_{DS}/L_{eff}, this effect is more pronounced in short-channel devices.

The linear relationship (1.78), which relates the drift velocity to the electric field, no longer holds for high fields because the mobility strongly depends on the field itself and decreases as the field increases. Specifically, at high electric fields, the drift velocity of carriers deviates from the linear dependency in (1.78) and even saturates. To account for this physical phenomenon, the mobility, μ_s, in (1.79) is corrected as follows [GM93]

$$\mu_{eff} = \frac{\mu_s}{1 + \frac{\mu_s}{v_{\max}} \frac{V_{DS}}{L_{eff}}} \tag{1.81}$$

where μ_{eff} is the effective mobility, V_{DS}/L_{eff} represents the lateral field and the term v_{max} is the maximum drift velocity of the carriers. A typical value of v_{max} is in the order of a 10^5 m/s.

This velocity limitation can be responsible for the saturation in MOS transistors since a MOS can enter the active region before V_{DS} reaches the value of $V_{GS} - V_{T,n}$. Consequently, (1.41) must be adjusted to account for the carrier saturation velocity.

1.3.10.5 Drain Induced Barrier Lowering (DIBL)

This effect is due to the strong lateral electric field and affects the threshold voltage. The principal model assumes the channel is created by the gate voltage only. Actually, a strong lateral field from the drain can also help to attract electrons towards the surface. Strictly speaking, the drain voltage influences the surface charge and helps the gate voltage to form the channel. This effect is modeled with a reduction in the threshold voltage (that is, a barrier lowering) and is also modeled by modifying (1.44) as [GM93], [JM97]

$$V_{T,n} = V_{T,n0} + \gamma\left(\sqrt{V_{SB} + 2|\phi_F|} - \sqrt{2|\phi_F|}\right) - \sigma_D V_{DS} \tag{1.82}$$

where σ_D is a corrective factor responsible for the dependence of the threshold voltage on V_{DS}. Also the DIBL is more pronounced in short-channel devices and its main effect is a further reduction in the output resistance.

1.3.10.6 Threshold voltage dependency on transistor dimensions

As transistor dimensions are reduced, the fringing field at border edges can also affect the threshold voltage [JM97]. Referring to Fig. 1.14 and without entering into a detailed physical explanation, applying a voltage V_G to the gate creates a channel. However, due to border effects, only charges in the darker trapezoidal area are linked to the gate voltage. The threshold voltage definition in (1.44) refers all the charges in the rectangular area below the silicon to the gate voltage, as in Fig. 1.8. Since the threshold voltage depends on the channel charge linked to the gate voltage, it is apparent that the previous model overestimates the value of $V_{T,n}$. This border effect is not critical in long-channel devices, but in a short-channel transistor it can be significant.

To model this phenomenon, the threshold voltage in (1.44) is modified in

$$V_{T,n} = V_{T,n0} - \gamma\sqrt{2|\phi_F|} + F_s\gamma\sqrt{V_{SB} + 2|\phi_F|} \tag{1.83}$$

where F_s is a corrective factor that represents the ratio between the trapezoidal and the rectangular areas used to model the channel. As a consequence, V_T is less than its original value in (1.44).

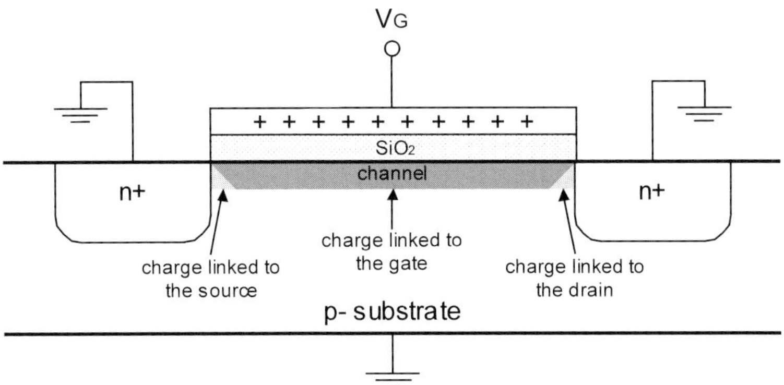

Fig. 1.14. Border effects in MOS transistors.

In a similar way, the threshold voltage depends on transistor width if this dimension becomes comparable to the edge effect regions, that is, in narrow-channel (i.e., with short width) devices.

In this case, after applying a voltage to the gate, border effects deplete a wider region thus increasing the threshold voltage. This effect is modeled by adding the term $F_n(V_{SB} + 2|\phi_F|)$ to the original $V_{T,n}$ in (1.44). F_n is a corrective factor that approaches zero in the case of wide channels.

Taking into account (1.82) and (1.83) the final form for the threshold voltage becomes

$$V_{T,n} = V_{T,n0} - \gamma\sqrt{2|\phi_F|} + F_s\gamma\sqrt{V_{SB} + 2|\phi_F|} - \sigma_D V_{DS} + F_n\left(V_{SB} + 2|\phi_F|\right) \tag{1.84}$$

$$\frac{i_{C1}}{i_{C2}} \cong e^{\frac{v_d}{V_T}} \qquad (2.1)$$

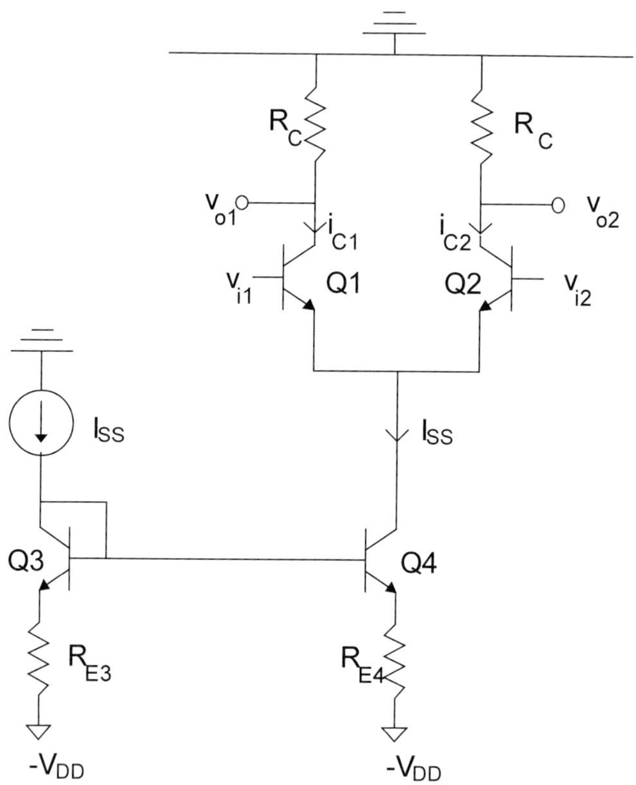

Fig. 2.1. CML inverter.

Some numerical values of (2.1) are reported in Table 2.1. It is worth noting that assuming transistors Q1-Q2 to work in the linear region is always correct in practical cases, as their switching speed is unacceptably reduced when operating in the saturation region [R96], because the transistor base charge that must be provided or extracted during the switching strongly increases in the saturation region.

By inspecting Fig. 2.1 and neglecting the base current, we can write

$$i_{C1} + i_{C2} \cong I_{SS} \qquad (2.2)$$

Hence, the bias current is almost completely steered to one of the two transistors for differential voltages v_d in the order of some V_T, e.g. about 100 mV at room temperature. Therefore, emitter coupled pair Q1-Q2 efficiently implements a voltage-controlled current switch that steers bias current I_{SS} according to the input logic value. Load resistors R_C perform a current-to-voltage conversion and generate the complementary output voltages v_{o1} and v_{o2}.

TABLE 2.1

v_d/V_T	i_{C1}/i_{C2}	v_d/V_T	i_{C1}/i_{C2}
1	2.7	-1	$(2.7)^{-1}$
2	7.4	-2	$(7.4)^{-1}$
3	20.1	-3	$(20.1)^{-1}$
4	54.6	-4	$(54.6)^{-1}$
5	148.4	5	$(148.4)^{-1}$

An in depth analysis of bipolar Current-Mode gates, which includes the static behavior and the noise immunity, is developed in the next section.

2.2 THE BIPOLAR CURRENT-MODE INVERTER: INPUT-OUTPUT CHARACTERISTICS AND NOISE MARGIN

In CML gates, input and output voltages can be differential or single-ended. Their static behavior is discussed in the following subsections in terms of logic swing V_{SWING}, small-signal voltage gain A_V and noise margin *NM*. Analysis is carried out by assuming the current provided to the driven gates to be negligible, as will be justified in Subsection 2.2.3.

2.2.1 Differential input/output

When differential signaling is assumed, the input and output voltage of the CML inverter gate are defined as the difference

$$v_i \cong v_{i1} - v_{i2} \tag{2.3a}$$

$$v_o = v_{o1} - v_{o2} = -R_C\left(i_{C1} - i_{C2}\right) \tag{2.3b}$$

From relationships (2.1) and (2.2), currents i_{C1} and i_{C2} are easily found to be equal to [GM94]

To evaluate noise margin *NM*, it is necessary to introduce some notations by referring to the output transfer characteristics of a generic logic gate in Fig. 2.3, whose points having slope equal to –1 are named (V_{ILmax}, V_{OHmin}) and (V_{IHmin}, V_{OLmax}). According to Fig. 2.4, noise margin at the low level, NM_L, and at the high level, NM_H, are respectively defined as

$$NM_L = V_{OH\min} - V_{IH\min} \tag{2.11a}$$

$$NM_H = V_{IL\max} - V_{OL\max} \tag{2.11b}$$

that are equal when the transfer characteristics is symmetric, and whose minimum value defines noise margin

$$NM = \min(NM_L, NM_H) \tag{2.12}$$

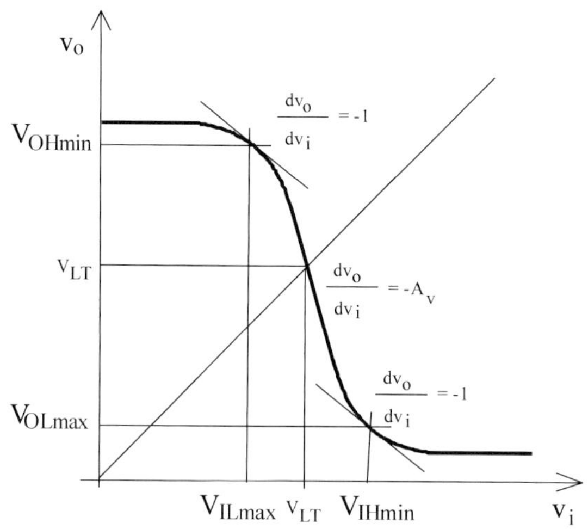

Fig. 2.3. DC output voltage v_o versus input voltage v_i in a generic logic gate.

Parameter V_{IHmin} of a CML inverter can be found by differentiating relationship (2.5) for v_i and setting the result to –1. Solving this equality for v_i leads to

$$V_{IH\min} \cong V_T \ln(\frac{V_{SWING}}{V_T} - 2) \tag{2.13}$$

where V_{ILmax} was assumed to be lower than $-2V_T$ and relationship (2.9) was used.

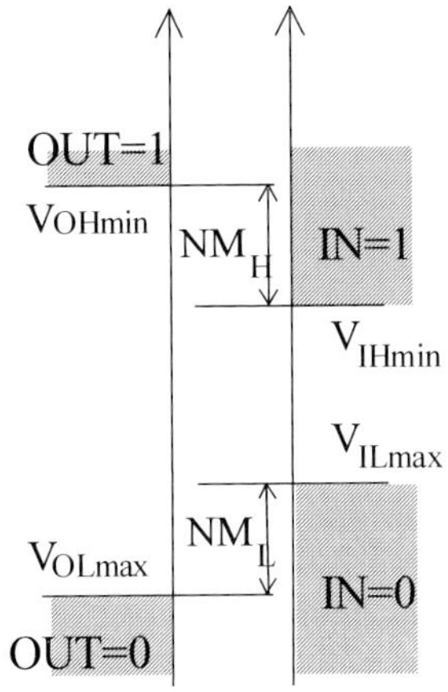

Fig. 2.4. Parameters of the transfer characteristic.

Since relationship (2.5) is symmetric with respect to the logic threshold, the low and high noise margin, given in (2.11a) and (2.11b), result to be equal. Then, introducing approximation $V_{OHmin} \approx R_C I_{SS} = V_{SWING}/2$, the noise margin results equal to

$$NM = NM_L = NM_H \cong \frac{V_{SWING}}{2} - V_T \ln\left(\frac{V_{SWING}}{V_T} - 2\right) \tag{2.14}$$

and, as expected, strongly depends on logic swing.

Some numerical values of the noise margin in (2.14) at room temperature versus the logic swing are reported in Table 2.2. Since minimum acceptable values of noise margin are typically in the order of 100 mV or greater, logic swing must be set at least to 200 mV from Table 2.2 (typical values in high-speed circuits are in the order of 400-500 mV). In practical cases, relationship (2.14) can be used to size logic swing for a noise margin requirement assigned from considerations at the system level.

To improve the noise margin, high values of the logic swing should be used, even though V_{SWING} has an upper bound. This can be seen by observing that, when the input voltage v_i is high, transistor Q1 has its base and collector voltage equal to 0 and $-R_C I_{SS}$, respectively. Therefore, as discussed in Section 1.2.1, the operation in the saturation region is avoided if the base-collector junction is reverse-biased, which is ensured by keeping its voltage lower than $V_{CB,on}$≈0.5 V [GM93]-[MG87], or equivalently by satisfying the following inequality

$$R_C I_{SS} \le V_{CB,on} \approx 0.5\,V \tag{2.15}$$

Therefore from (2.9) the maximum logic swing allowed is about 1 V. This result is based on the assumption that the series collector parasitic resistor r_c (see the SPICE model in Fig. 4.1) is low enough to warrant its neglect. However, relationship (2.15) is easily generalized to arbitrary values of r_c by considering an equivalent resistor at the collector node equal to R_C+r_c, as resistor r_c is in series with R_C.

TABLE 2.2

$R_C I_{SS}$ (mV)	V_{SWING} (mV)	NM (mV)
200	400	134
250	500	178
300	600	223
350	700	269
400	800	315
450	900	362
500	1000	410

2.2.2 Single-ended input/output

In a single-ended CML inverter gate, the base voltage of transistor Q2 in Fig. 2.1 is kept at a constant voltage V_{REF}, implemented by a temperature-compensated voltage reference. The input voltage is v_{i1} and output voltage is v_{o1}, which is equal to

$$v_{o1} = -R_C i_{C1} \tag{2.16}$$

whose high and low voltage levels are obtained when bias current I_{SS} is almost completely steered to transistor Q2 and Q1, respectively

$$V_{OH} = 0 \tag{2.17}$$

$$V_{OL} = -R_C I_{SS} \tag{2.18}$$

and the resulting logic swing is half as that in the differential case

$$V_{SWING} = V_{OH} - V_{OL} = R_C I_{SS} \tag{2.19}$$

that, to avoid operation in the saturation region, cannot greater than about 500 mV, from relationship (2.15).

The base terminal of transistor Q2 is biased to a constant voltage V_{REF}, which must be set to a suitable value that makes transfer characteristics in (2.16) symmetric to maximize noise margin *NM*. This is achieved for a logic threshold V_{LT} at half the entire swing

$$V_{LT} = -\frac{R_C I_{SS}}{2} \tag{2.20}$$

According to (2.20), when v_{i1} is set to V_{LT}, output must also be equal to it, which is achieved when base voltages of Q1-Q2 are equal (since in this case I_{SS} equally divides between Q1-Q2 and hence v_o is equal to (2.20)), thus also voltage V_{REF} must be set equal to V_{LT}. Under this value of V_{REF}, from (2.4a) and (2.16) the transfer characteristics results in

$$v_O = -R_C I_{SS} \frac{e^{\frac{v_i + \frac{R_C I_{SS}}{2}}{V_T}}}{1 + e^{\frac{v_i + \frac{R_C I_{SS}}{2}}{V_T}}} \tag{2.21}$$

which is symmetric with respect to the logic threshold. As an example, plot of relationship (2.21) versus v_i is reported in Fig. 2.5 at room temperature for $R_C I_{SS}$ equal to 0.25 V.

As is well known, the small-signal voltage gain around the logic threshold in (2.20) in the single-ended case is half that in the differential case [GM93]

$$A_V = \frac{1}{2} g_m R_C = \frac{V_{SWING}}{4V_T} \tag{2.22}$$

which is a function of only the logic swing. As an example, when V_{SWING} ranges from 200 mV to 500 mV, which are typical values, voltage gain (2.22) ranges from 2 to 5.

To analytically express noise margin at the high level NM_H in (2.11b), let us evaluate parameter V_{IHmin} by differentiating relationship (2.21) for v_i and setting the result to –1. Solving for v_i leads to

$$V_{IH\min} \approx V_{REF} + V_T \ln(\frac{V_{SWING}}{V_T} - 2)$$

$$= -\frac{R_C I_{SS}}{2} + V_T \ln(\frac{V_{SWING}}{V_T} - 2) \quad (2.23)$$

where V_{ILmax} was assumed to be lower than $-2V_T$ and relationship (2.20) was used.

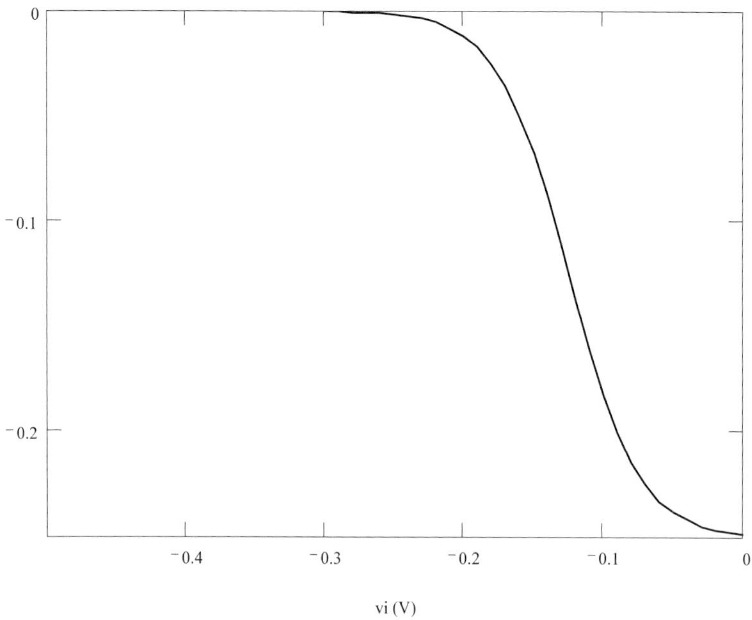

Fig. 2.5. Transfer characteristics of a CML inverter with single-ended input/output for $R_C I_{SS}$=0.25 V.

By approximating V_{OHmin}≈0, from (2.23) noise margin results in

$$NM = NM_L = NM_H \cong \frac{V_{SWING}}{2} - V_T \ln(\frac{V_{SWING}}{V_T} - 2) \quad (2.24)$$

since voltage V_{REF} was chosen to let DC characteristics in (2.21) symmetric with respect to the logic threshold. It is useful to observe that the noise margin of the single-ended CML gate in (2.24) has the same expression as that of the differential one, but logic swing and its maximum value (2.15) in the former (see eq. (2.19)) is half of the latter in (2.9) for an assigned value of $R_C I_{SS}$. For this reason, for an assigned value of $R_C I_{SS}$, noise margin in the single-ended case is about half as that of the differential gate (more precisely, the former is slightly lower than half of the latter). Some numerical values of the noise margin (2.24) at room temperature versus the logic swing are reported in Table 2.3 for the single-ended gate.

TABLE 2.3

$R_C I_{SS}$ (mV)	V_{SWING} (mV)	NM (mV)
200	200	55
250	250	73
300	300	92
350	350	113
400	400	134
450	450	156
500	500	178

To achieve acceptable values of noise margin, that are typically in the order of 100 mV or greater, logic swing must be set at least to 350-400 mV from Table 2.3. In practical cases, relationship (2.24) is used to size logic swing for a given noise margin requirement. Finally, it is interesting to note that the noise margin of both differential and single-ended CML gates linearly decreases as increasing temperature through factor V_T, which is proportional to it.

2.2.3 Considerations on the non-zero input current

In the previous subsections, differential and single-ended inverter gates have been analyzed in terms of their static behavior. Such analysis has been carried out by assuming that the static input current required by the driven gates is negligible. Actually, the input current of a driven gate (equal to the base current of its input transistor) is non-zero only when the previous gate's output node voltage is high (i.e., equal to zero), since in the opposite case the input transistor is in the cut-off region and draws a null current. In the former case, the input current of a driven gate is equal to the emitter current I_{SS} of input transistor divided by (β_F+1) from (1.18). Thus, when a single-ended CML gate drives N equal gates, the high output voltage becomes

$$V_{OH} = -R_C N \frac{I_{SS}}{\beta_F + 1} \approx -R_C N \frac{I_{SS}}{\beta_F} \tag{2.25}$$

while the low output level is still given by relationship (2.18), thus logic swing results in

$$V_{SWING} = R_C I_{SS} \left[1 - \frac{N}{\beta_F} \right] \tag{2.26}$$

which can be substituted into relationship (2.24) to evaluate the effective noise margin achieved. From (2.26), its reduction with respect to the ideal case with a zero gate input current in Section 2.2.2 is negligible for a fan-out N sufficiently low with respect to β_F. For practical values of β_F and N, effect of the gate input current on NM is usually negligible.

Regarding differential gates, both V_{OH} and $|V_{OL}|$ are reduced by the same amount given in (2.25), thus logic swing turns out to be

$$V_{SWING} = 2R_C I_{SS} \left(1 - \frac{N}{\beta_F} \right) \tag{2.27}$$

for which the same considerations reported above still hold.

2.2.4 Remarks and comparison of differential/single-ended gates

In the following, comparison of the two topologies is carried out to justify why differential gates are generally preferred to single-ended ones.

Analysis in Sections 2.2.1 and 2.2.2 has shown that differential CML gates have a better noise immunity by a factor slightly greater than 2, for a given value of $R_C I_{SS}$. As will be shown in Chapter 5, product $R_C I_{SS}$ significantly affects speed and its trade-off with power consumption. In particular, it will be shown that a low value of factor $R_C I_{SS}$ is advantageous in terms of speed when a low bias current is used and in terms of the speed-power trade-off when a high bias current is adopted. Therefore, for a given noise margin, differential gates allow for a reduction of $R_C I_{SS}$ by a factor greater than 2 compared to single-ended gates, thereby improving by the same factor the speed in a low power design or the power efficiency in a high performance design.

There are further arguments concerning immunity to supply noise that make differential gates superior than single-ended counterparts. To understand this point, let us consider a supply noise voltage v_{noise} superimposed to the ground voltage of a CML gate. By assuming v_{noise} to be a small signal, its effect on output voltages v_{o1} and v_{o2} can be evaluated by means of small-signal analysis. To be more specific, the resulting variation of v_{o1} (v_{o2}) is given by the voltage divider between resistor R_C and the equivalent small-signal resistor seen from the collector of transistor Q1 (Q2) to ground. Since in practical cases the latter is much greater than the former, v_{noise} is almost entirely transferred to both output nodes, i.e. the small-signal component of v_{o1} and v_{o2} is equal to v_{noise}. Therefore, in a single-ended CML gate the output noise voltage is essentially equal to the supply noise, while in a differential gate its effect on output is insignificant, since v_{noise} affects both voltages v_{o1} and v_{o2} approximately in the same manner. In any case, due to the very low logic swing, the effect of supply noise on the output voltage is reduced by using the ground voltage as the high supply voltage, since it is usually less noisy than other available voltage sources, as depicted in Fig. 2.1.

Regarding the supply noise that we may superimpose to the negative supply voltage, it does not have any effect on output voltage of a CML gate from inspection of Fig. 2.1, since I_{SS} is an ideal current source (even if it were not ideal, the effect of the noise signal would be a common mode signal with no effect on the differential output). Moreover, differential gates do not need a reference voltage, which in single-ended gates reduces the noise margin (due to unavoidable fluctuations and tolerances on its voltage) and requires a further power consumption contribution.

In both differential and single-ended topologies, the output logic level can be inverted with no extra circuitry, since in the former the inversion is simply obtained by exchanging output nodes, while in the latter it is achieved by taking output signal from the opposite output node. This increased degree of freedom often allows for reducing the gate count when implementing a given logic function, as will be discussed in Chapter 3.

From the previous observations, that can be easily extended to more complex gates, it is apparent that differential gates are largely preferred in practical applications, since they have a better noise immunity, speed or a more favorable trade-off with power consumption. Therefore, in the following sections only differential gates will be considered.

2.3 THE BUFFERED BIPOLAR CURRENT-MODE (ECL) INVERTER

The Emitter Coupled Logic (ECL) inverter is obtained by adding an output buffer (i.e., a common collector stage) to each output node of the CML inverter, as shown in Fig. 2.6. In particular, output buffers are biased by a constant current I_{CC}, while the internal CML gate is biased by the current I_{SS}. Traditionally, output buffers are introduced to enhance the driving capability of the gate, and their effects on the gate delay will be dealt with in Chapters 4 and 5. Alternatively, output buffers may be added to down-shift the common-mode output voltage by a base-emitter voltage $V_{BE3,4}$, as will be shown in Section 2.5.3.

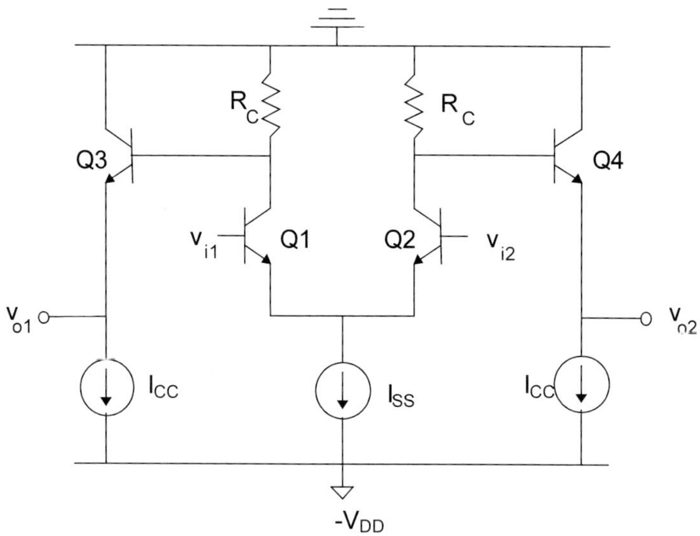

Fig. 2.6. ECL inverter topology.

In the differential case, the input voltage is defined by relationship (2.3a) as for the CML inverter, while the output voltage is given by

$$v_o = v_{o1} - v_{o2} = \lfloor -R_C i_{C1} - V_{BE,3} \rfloor - \lfloor -R_C i_{C2} - V_{BE,4} \rfloor = -R_C (i_{C1} - i_{C2}) \quad (2.28)$$

which is equal to the transfer characteristics of the differential CML inverter in (2.3b). As a consequence, the differential ECL inverter has the same static behavior as the CML gate, i.e. the same logic swing, small-signal gain and noise margin already evaluated in Section 2.2.1. The only difference is that both output nodes voltages are down-shifted by the constant voltage $V_{BE3,4}$.

In the single-ended case, input and output voltages are v_{i1} and v_{o1}, thus one of the two buffers can be avoided (if the opposite value of output is not required), and voltage v_{i2} is set to a constant value V_{REF} that lies in the middle of the logic swing, as already discussed in Section 2.2.2. Being both voltage levels V_{OH} and V_{OL} down-shifted by $V_{BE3,4}$ with respect to those of the CML inverter in Section 2.2.2, logic swing and noise margin of the ECL inverter are still given by (2.9) and (2.14). It is worth noting that, due to the shift of V_{OH} and V_{OL}, from (2.20) the logic threshold and the reference voltage V_{REF} must be set to

$$V_{REF} = V_{LT} = -\frac{R_C I_{SS}}{2} - V_{BE3,4} \tag{2.29}$$

In practical cases, differential ECL gates are preferred to single-ended ones for the same reasons clarified in Section 2.2.4.

2.4 THE MOS CURRENT-MODE INVERTER

The MOS Current-Mode inverter gate is based on the source-coupled pair of NMOS transistors biased by a constant current I_{SS} implemented by an NMOS current mirror, as shown in Fig. 2.7. In this figure, a positive supply voltage V_{DD} is used, as required by the analog circuitry that is usually integrated within the same chip [KA92].

The operation of the differential MOS Current-Mode inverter is similar to that of the bipolar CML counterpart. In particular, assuming transistor operation in the saturation region, currents i_{D1} and i_{D2} of transistors M1-M2 can be expressed as a function of the differential input voltage $v_i = v_{i1} - v_{i2}$ as [SS91]

$$i_{D1}(v_i) = \begin{cases} 0 & \text{if } v_i < -\sqrt{\dfrac{2I_{SS}}{\mu_n C_{OX} \dfrac{W_n}{L_n}}} \\ \dfrac{I_{SS}}{2} + \dfrac{v_i}{2}\sqrt{\mu_n C_{OX}\dfrac{W_n}{L_n} I_{SS} - \left(\mu_n C_{OX}\dfrac{W_n}{L_n}\dfrac{v_i}{2}\right)^2} & \text{if } |v_i| \le \sqrt{\dfrac{2I_{SS}}{\mu_n C_{OX} \dfrac{W_n}{L_n}}} \\ I_{SS} & \text{if } v_i > \sqrt{\dfrac{2I_{SS}}{\mu_n C_{OX} \dfrac{W_n}{L_n}}} \end{cases} \tag{2.30a}$$

$$i_{D2}(v_i) = I_{SS} - i_{D1}(v_i) \tag{2.30b}$$

being parameters W_n and L_n the effective NMOS transistor channel width and length, C_{OX} the oxide capacitance per area, μ_n the NMOS carrier mobility. From (2.30), the bias current I_{SS} is completely steered to one of the two output branches for a magnitude of the differential input voltage greater than $\sqrt{2I_{SS}/\mu_n C_{OX}(W_n/L_n)}$. The current steered in the MOS Current-Mode inverter is converted into the differential output voltage v_o=v_{o1}-v_{o2} through the PMOS active load M3-M4.

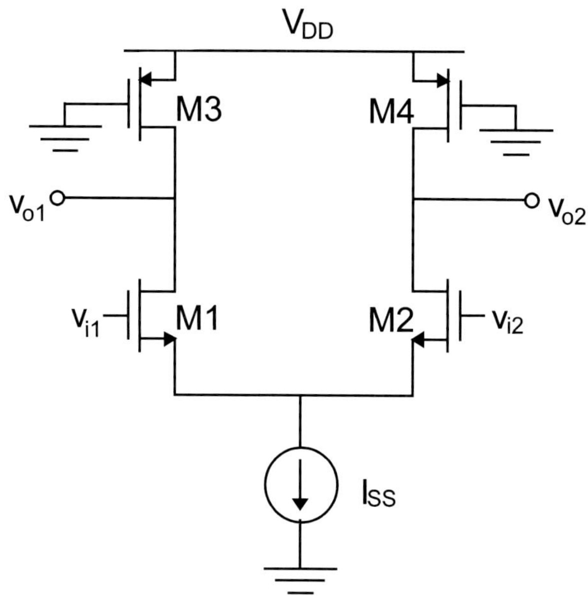

Fig. 2.7. Source-Coupled inverter gate.

Since modeling of MOS transistors is not straightforward as for bipolar devices, the output transfer characteristics of the circuit in Fig. 2.7 will be evaluated through a simplified analysis and a linearization of the PMOS load, as will be shown in the following.

2.4.1 Static modeling of the PMOS active load

The current-to-voltage conversion in the circuit in Fig. 2.7 is performed by the two PMOS transistors M3-M4, both of which have a source-gate voltage equal to V_{DD} and a much smaller source-drain voltage (in the order of hundreds of millivolts). Therefore, transistors M3-M4 work in the triode region, and can thus be modeled as an equivalent linear resistor R_D.

To evaluate R_D, let us consider the expression of the drain current i_D of a PMOS transistor working in the triode region used in the BSIM3v3 MOSFET model, which represents the standard model for deep submicron CMOS technologies [CH99]

$$i_D = \frac{I_{DSAT0}}{1 + R_{DS}\frac{I_{DSAT0}}{V_{SD}}} \tag{2.31}$$

where the parameter R_{DS}=(R_{DSW}*1E-6)/W depends on the empiric model parameter R_{DSW}, which accounts for source/drain parasitic resistor, and heavily affects the I-V relationship in today's CMOS processes with lightly-doped drain (LDD). It is worth noting that R_{DS} does not represent a physical resistor, but only a corrective factor.

Ratio I_{DSAT0}/V_{SD} in relationship (2.31) can be evaluated by considering the expression of current I_{DSAT0} valid for both NMOS and PMOS transistors working in the linear region

$$I_{DSAT0} = \mu_{eff} C_{OX} \frac{W}{L}\left(\left|V_{GS}\right| - \left|V_T\right| - A_{bulk}\frac{\left|V_{DS}\right|}{2}\right)\left|V_{DS}\right| \tag{2.32}$$

whose parameters are reported in the following with the subscript p if referred to a PMOS transistor, and subscript n in the case of an NMOS transistor. In (2.32), parameter V_T is the threshold voltage (not the thermal voltage), V_{GS} and V_{DS} are the source-gate and source-drain voltages, and μ_{eff} is the effective carrier mobility defined as [CH99]

$$\mu_{eff} = \frac{\mu_0}{1 + (U_A + U_C\left|V_{SB}\right|)\left(\frac{\left|V_{GS}\right| + \left|V_T\right|}{T_{OX}}\right) + U_B\left(\frac{\left|V_{GS}\right| + \left|V_T\right|}{T_{OX}}\right)^2} \cdot \frac{1}{1 + \frac{\left|V_{DS}\right|}{E_{SAT} L}} \tag{2.33}$$

where E_{SAT} is the critical electric field at which carrier velocity becomes saturated, U_A, U_B and U_C are model parameters, V_{SB} is the source-bulk voltage, and T_{OX} is oxide thickness. It is worth noting that, in the denominator of (2.33), the terms including V_{GS} model the mobility degradation due to the vertical electric field in the MOS transistor, while those including V_{DS} model the carrier velocity saturation due to the lateral electric field. In particular, in the case of the active load PMOS transistors, we have to set $V_{SG}=V_{DD}$ and $V_{SB}=0$ in the expression (2.33) of the effective mobility $\mu_{eff,p}$.

Parameter A_{bulk} in (2.32) is slightly greater than the unity and is given by

$$A_{bulk} = \frac{1}{1+K_{ETA}|V_{SB}|}\left\{1+\frac{K_{10X}}{2\sqrt{\phi_S-|V_{SB}|}}\left[\frac{A_0 L}{L+2\sqrt{X_J X_{dep}}}\right.\right.$$
$$\left.\left.\cdot\left(1-A_{GS}\left(|V_{GS}|-|V_T|\right)\left(\frac{L}{L+2\sqrt{X_J X_{dep}}}\right)^2\right)+\frac{B_0}{W+B_1}\right]\right\} \tag{2.34}$$

which depends on W, L and various other BSIM3v3 model parameters, and can be simplified by considering its maximum value, $A_{bulk,max}$. This can be obtained by setting W to its minimum value in (2.34) and maximizing the resulting function with respect to L, with straightforward calculations. As an example, for the 0.35-μm CMOS process with main parameters reported in Table 2.4 and V_{DD}=3.3 V, the PMOS transistor parameter $A_{bulk,max,p}$ results in 1.34.

TABLE 2.4

C_{OX}	4.6 fF/μm^2
$\mu_{no}C_{OX}$	175 μA/V^2
$\mu_{po}C_{OX}$	60 μA/V^2
$(W/L)_{min}$	0.6 μm/0.3 μm
V_{Tn0} (short channel)	0.54 V
V_{Tp0} (short channel)	-0.72 V
maximum V_{DD}	3.3 V

From these considerations, by assuming V_{SD} to be small, terms $A_{bulk,p}V_{SD}/2$ and $V_{SD}/E_{SAT,p}L_p$ can be neglected in (2.33). Therefore (2.32) becomes

$$I_{DSAT0} = \frac{V_{SD}}{R_{int}} \tag{2.35}$$

where we have defined

$$R_{\text{int}} = \frac{1}{\mu_{eff,p} C_{OX} \frac{W_p}{L_p} \left(V_{DD} - \left| V_{T,p} \right| \right)} \tag{2.36}$$

which represents the "intrinsic" resistance of the PMOS transistor in the triode region (equal to that derived in (1.75) for a long-channel MOS transistor), since it expresses the behavior of the MOS transistor in the triode region without accounting for the parasitic drain/source resistance.

Now, the expression of i_D in (2.31) can be simplified by expanding it in Taylor series truncated at the first-order term

$$i_D = I_{DSAT0} \left(1 - \frac{R_{DS}}{R_{\text{int}}} \right) \tag{2.37}$$

From (2.37), the equivalent resistance of the PMOS transistors $R_D = V_{SD}/i_D$ results in

$$R_D = \frac{R_{\text{int}}}{1 - \frac{R_{DS}}{R_{\text{int}}}} \tag{2.38}$$

2.4.2 Input-output characteristics

The output voltage transfer characteristics $v_o(v_i)$ of the MOS Current-Mode inverter can be evaluated by substituting the equivalent resistance R_D expressed in (2.38) into the circuit in Fig. 2.7. Thus, the differential output voltage v_o is equal to

$$v_o = v_{o1} - v_{o2} = -R_D \left(i_{D1} - i_{D2} \right) \tag{2.39}$$

which is formally equal to the transfer characteristics (2.3b) of the bipolar Current-Mode inverter. By evaluating transistor currents through relationship (2.30) and substituting them into (2.38), the output transfer characteristics results in

2.4.3 Evaluation of the noise margin

Due to the symmetry of output characteristics (2.40), the noise margin NM is equal to NM_L or equivalently to NM_H, which is evaluated by calculating input value V_{IHmin} that makes the derivative of (2.40) equal to -1, and the associated value of output voltage V_{OHmin}. After simple calculations, V_{IHmin} results to [APP02]

$$V_{IH\min} = \sqrt{\frac{2I_{SS}}{\mu_{eff,n}C_{OX}\frac{W_n}{L_n}} - \frac{I_{SS}}{2\mu_{eff,n}C_{OX}\frac{W_n}{L_n}}\frac{1}{A_V^2}\left(\sqrt{1+8A_V^2}+1\right)}$$

$$\cong \sqrt{\frac{2I_{SS}}{\mu_{eff,n}C_{OX}\frac{W_n}{L_n}}\left(1-\frac{1}{\sqrt{2}A_V}\right)} \tag{2.46}$$

where (2.44) was used and $A_V >> 1/\sqrt{8}$ was assumed. Therefore, by approximating V_{OHmin} to V_{OH} in (2.42b), the noise margin results to

$$NM = V_{OH\min} - V_{IH\min} = R_D I_{SS}\left(1-\frac{\sqrt{2}}{A_V}\sqrt{1-\frac{1}{\sqrt{2}A_V}}\right)$$

$$\cong \frac{V_{SWING}}{2}\left(1-\frac{\sqrt{2}}{A_V}\right) \tag{2.47}$$

where $A_V >> 1/\sqrt{2}$ was assumed. By inspection of (2.47), the noise margin of an SCL gate is proportional to half the logic swing, and roughly equal to it if A_V is in the order of 4÷5.

2.4.4 Validation of the static model

The approximate model of static parameters V_{SWING}, A_V and NM discussed in the previous subsections was compared to simulation results obtained for an SCL inverter by using a 0.35-µm CMOS process, whose main parameters are summarized in Table 2.4 [APP02]. To this end, several DC simulations were performed with V_{DD}=3.3 V, I_{SS} ranging from 5 μA to 100 μA, by choosing the transistors aspect ratio to get A_V ranging from 2 to 7, and $R_D I_{SS}$ from 200 mV to 700 mV (i.e., a maximum value slightly lower than $V_{T,n}$).

The simulated and predicted results are plotted in Figs. 2.9a, 2.9b and 2.9c, in which the scattering plots of the logic swing, the magnitude of the voltage gain and the resulting noise margin are respectively reported.

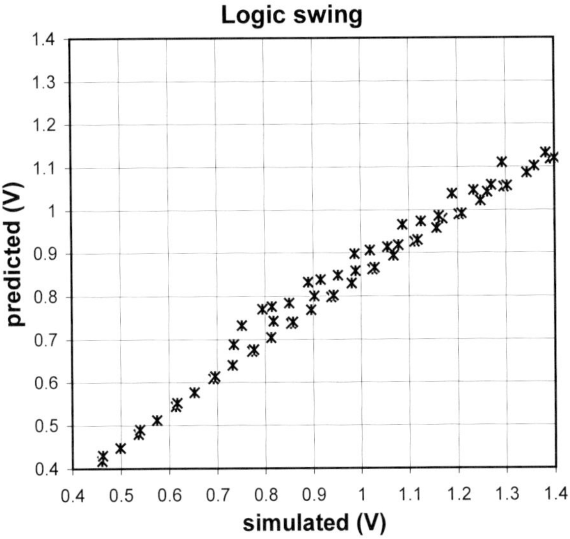

Fig. 2.9a. Scattering plot of predicted vs. simulated logic swing.

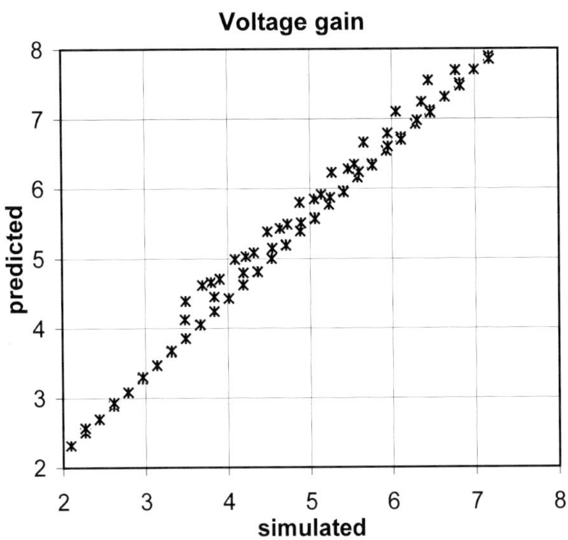

Fig. 2.9b. Scattering plot of predicted vs. simulated voltage gain magnitude.

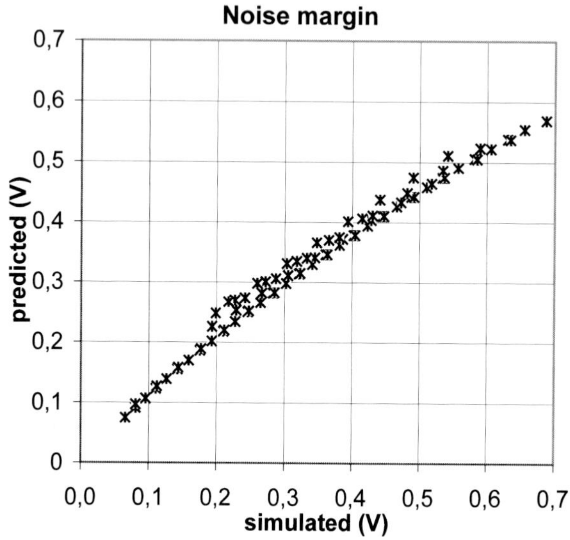

Fig. 2.9c. Scattering plot of predicted vs. simulated noise margin.

From Figs. 2.9a-2.9c, it is evident that the predicted values are close to the simulated ones. More specifically, the maximum error between the model results and the simulated ones for V_{SWING}, A_V and NM is equal to 24.7 %, 25.9 % and 24.7 %, respectively, and in typical cases the error is significantly lower, as can be deduced from the average and standard deviation values of the error reported in Table 2.5. It is worth noting that the model always underestimates V_{SWING} and overestimates A_V.

TABLE 2.5

model error	maximum (%)	Average (%)	standard deviation (%)
V_{SWING}	24.7	14.7	5
A_V	25.9	12.9	4
NM	24.7	7.6	5.7

2.4.5 The buffered MOS Current-Mode inverter and remarks

As already discussed for bipolar ECL gates, to improve the SCL gate driving capability or to shift the common-mode value of the output nodes voltage, an output buffer can be added to each output node, as shown in Fig. 2.10, where $v_o=v_{o1}-v_{o2}$ is the differential output voltage of the gate, and $v_{i,buf1}$ and $v_{i,buf2}$ are the input voltages of the two buffers, respectively.

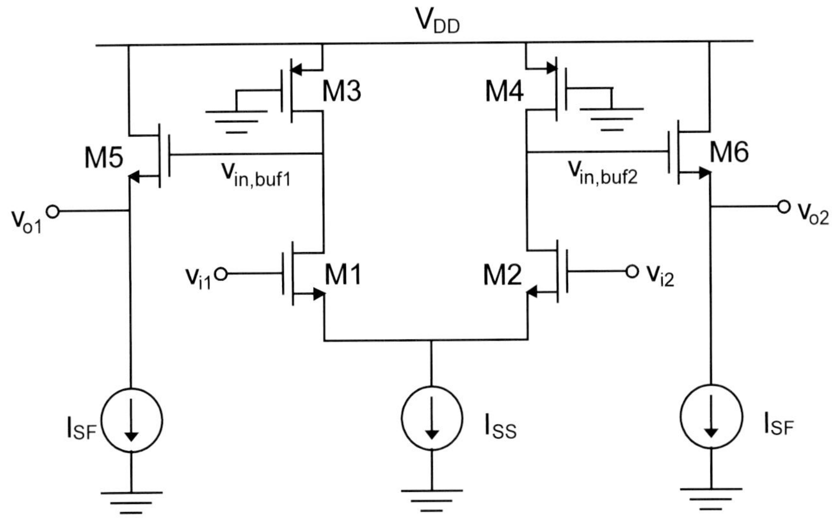

Fig. 2.10. SCL inverter gate with output buffers.

The output buffer is a source-follower stage biased by the current source I_{SF}, which by inverting (1.42) down-shifts output voltages of the internal SCL gate by a gate-source voltage V_{GS} of transistors M5-M6 set by current I_{SF}, according to

$$V_{GS,56} = V_{T,n} + \sqrt{\frac{I_{SF}}{\frac{\mu_{eff,buf} C_{OX}}{2} \frac{W_{buf}}{L_{buf}}}} \qquad (2.48)$$

being W_{buf}/L_{buf} the aspect ratio of buffer transistors M5-M6. The effective mobility $\mu_{eff,n}$ in (2.33) can be simplified into

$$\mu_{eff,buf} = \frac{\mu_0}{1+\left[U_A + U_C\left(V_{DD} - V_{T,n}\right)\right]\left(\frac{2V_{T,n}}{T_{OX}}\right) + U_B\left(\frac{2V_{T,n}}{T_{OX}}\right)^2} \cdot \frac{1}{1+\frac{\left|V_{T,n} + \frac{1}{2}V_{SWING}\right|}{E_{SAT}L}} \tag{2.49}$$

where V_{DS} and V_{SB} are approximated by their maximum values (i.e., $V_{GS,56}+V_{SWING}/2$ and V_{DD}-V_{GS}, respectively), and $V_{GS,56}$ is underestimated by $V_{T,n}$.

The static parameters V_{SWING}, A_V and NM of the SCL inverter in Fig. 2.10 can be derived by properly modifying the results obtained in the previous section. Indeed, the small-signal gain of the common-drain stage (i.e., the ratio $v_o/(v_{i,buf1}$-$v_{i,buf2}))$ is equal to [LS94]

$$\frac{v_o}{v_{i,buf1} - v_{i,buf2}} = \frac{1}{1+\frac{g_{mb,buf}}{g_{m,buf}}} \tag{2.50}$$

where $g_{mb,buf}$ and $g_{m,buf}$ are respectively the body effect transconductance and the transistor transconductance, and their ratio is almost constant and close to unity[1]. Thus, the voltage gain A_V and the logic swing V_{SWING} are obtained by multiplying those of an SCL gate in (2.43)-(2.44) by (2.50). Obviously, these parameters must be substituted in (2.47) to achieve the noise margin of the circuit in Fig. 2.10.

Predicted values of V_{SWING}, A_V and NM were compared to simulation results, in the same conditions as in Section 2.4.4. Simulations reveal that the logic swing, voltage gain and noise margin agree well with the model and the error is due only to the internal SCL gates, due to the good accuracy of relationship (2.50).

For the sake of completeness, transistor transconductance $g_{m,buf}$ in (2.50) can be evaluated by using its long-channel expression (1.66) and substituting into it the effective mobility $\mu_{eff,buf}$ in (2.49).

---

[1] Typical values of $g_{mb,buf}/g_{m,buf}$ range from 0.1 to 0.2; for the 0.35-μm CMOS process used, ratio $g_{mb,buf}/g_{m,buf}$ is equal to 0.13, leading to $v_o/(v_{i,buf1}$-$v_{i,buf2})$=0.88.

Until now, SCL gates have been assumed to include a PMOS active load. Nevertheless, other kinds of load could be considered, such as a physical resistor R_D or a diode-connected NMOS/PMOS. The first solution is not feasible since resistors need a wide silicon area to be integrated, and in addition they are affected by a parasitic capacitance greater than the PMOS transistor for practical values required (as an example, for the 0.35-µm CMOS process considered, this occurs for R_D greater than 1 kΩ, which is lower than typical values used). The diode-connected transistor load has other drawbacks, among which the loss of a threshold voltage in output levels and the floating output node for a high output level [R96]. Furthermore, the MOS diode load is slower than the PMOS active load for practical bias currents (for the process considered, this occurs for I_{SS} greater than 1 µA, which is lower than typical values). For these reasons, only the PMOS active load will be considered in the following.

2.5 FUNDAMENTAL CURRENT-MODE LOGIC GATES

Differential Current-Mode gates are usually implemented according to the series-gating approach [T89] [W90], i.e. by properly stacking source-coupled transistor pairs, as will be clarified in the following subsections. Since MOS and bipolar Current-Mode gates have the same principle of operation, i.e. the switching of the source/emitter-coupled transistor pair, bipolar gates will be mainly focused in the remainder of the section. Results and topologies can be immediately extended to MOS Current-Mode gates by substituting n-channel devices to npn transistors in bipolar gates.

2.5.1 Principle of operation of Current-Mode gates: the series gating concept

As depicted in Fig. 2.11, Current-Mode bipolar (MOS) gates are made up of a bias current source which is steered to one of the two output resistors by a network consisting of a bipolar npn (n-channel) transistors network, according to its input signals value. To allow a correct operation, in which I_{SS} entirely flow through only one of the two output resistors and the two output nodes are thus opposite, there must be only a unique conductive path from I_{SS} to output nodes for all possible input values. It is worth noting that current I_{SS} does not entirely reach one of the two output nodes, since each emitter-coupled pair outputs a (collector) current that is reduced by the common-base current gain α_F with respect to the input (emitter) current. Being $\alpha_F \approx 1$ in practical cases, a switching emitter coupled pair is biased by a current approximately equal to I_{SS}, thus the noise margin in a complex gate for all possible inputs is the same as that of a simple inverter in Section 2.2. This allows for extending results in Sections 2.2-2.3 to arbitrary CML/ECL gates. The same observation holds for MOS Current-Mode logic, since each source-coupled pair generates an output (drain) current equal to the input (source) current.

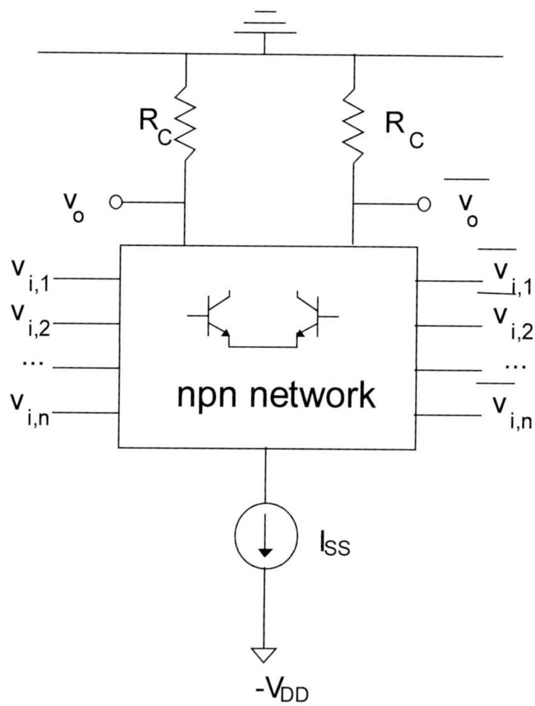

Fig. 2.11. General topology of a CML gate.

To understand how bipolar series gates are implemented, let us analyze the basic AND (·) and OR (+) operation of two input signals, A and B. In particular, observe that the function $A \cdot B$ is simply achieved by stacking two emitter coupled pairs as in Fig. 2.12a, since current $i_{A \cdot B}$ is high[2] only when both transistors Q1-Q3 are ON, which occurs if $A=B=1$. The current obtained $i_{A \cdot B}$ is then converted into an output voltage through a resistor R_C. It is worth noting that the current-to-voltage conversion is an inverting operation, i.e.

[2] i.e. approximately equal to I_{SS}, as is obtained from $i_{C3}=\alpha_{F,1}\alpha_{F,3}I_{SS} \approx I_{SS}$, by assuming the common-base current gain α_F to be about unity.

the output voltage is low when the output current is high. Since a differential output voltage is required, the other unused branches (i.e. collector of Q2-Q4 in Fig. 2.12a) must be connected to generate the opposite to $i_{A\cdot B}$ current, that in Fig. 2.12a is referred to as $\overline{i_{A\cdot B}}$ and is approximately equal to I_{SS}-$i_{A\cdot B}$. This branch is then connected to the other output resistor R_C to perform the current-to-voltage conversion. The complement of the function (e.g. to compensate the inversion introduced in the current-to-voltage conversion) is trivially obtained by exchanging the two output nodes.

The *OR* operation can be still performed by the stacked topology in Fig. 2.12a by rearranging signals through De Morgan laws. In particular, since $A+B=\overline{\overline{A}\cdot\overline{B}}$, the logic OR of *A* and *B* is simply obtained by complementing (i.e. exchanging differential signals) output, as well as inputs *A* and *B* (see Fig. 2.12b). In cases where signals *A*, *B* are applied to emitter coupled pairs Q1-Q2 and Q3-Q4 which are not stacked, their OR can be simply performed by connecting the collector of Q1 and Q3 (and that of Q2 and Q4 to obtain the complement). The resulting topology is reported in Fig. 2.11c.

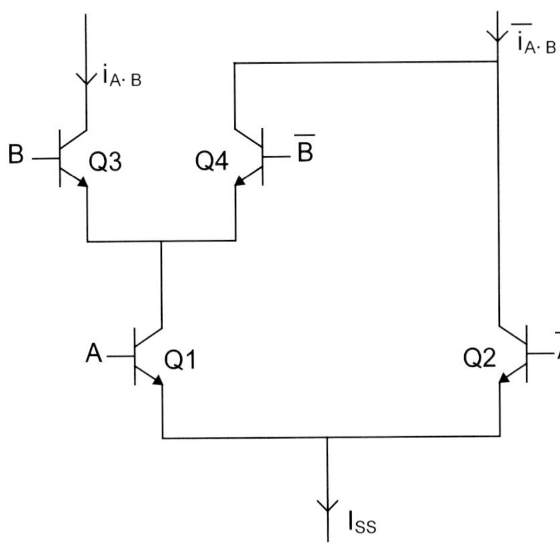

Fig. 2.12a. Logic AND between two input signals, *A* and *B*.

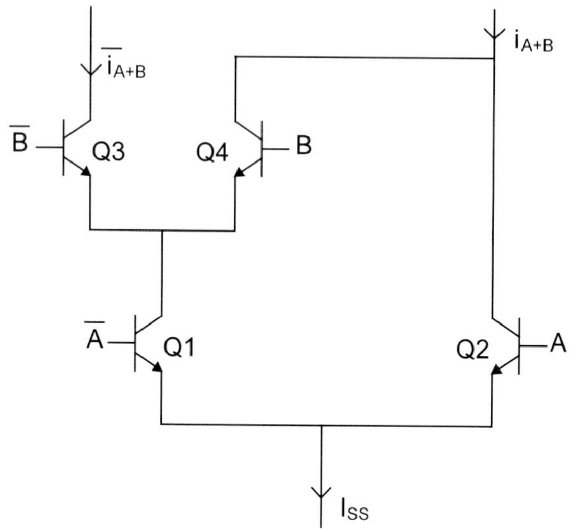

Fig. 2.12b. Logic OR between two input signals, *A* and *B* associated with stacked transistor pairs.

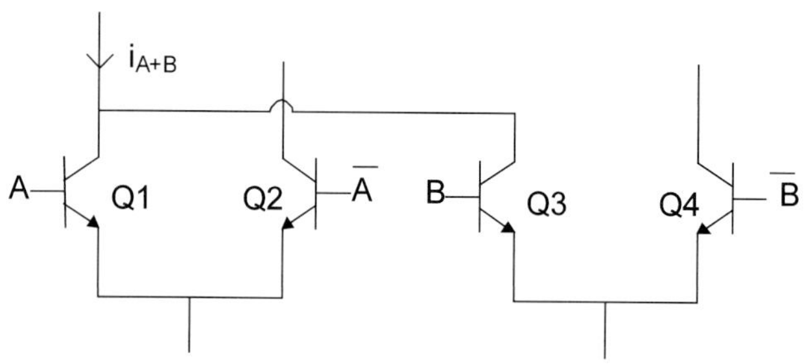

Fig. 2.12c. Logic OR between two input signals, *A* and *B*, associated with non-stacked transistor pairs.

In the following subsection, some of the most used Current-Mode series gates are introduced. After analyzing the operation of such gates, issues related to the minimum supply voltage required for a correct functioning are addressed.

2.5.2 Some examples of Current-Mode series gates

By applying the concepts developed in the previous subsection, the bipolar AND2 gate has the topology reported in Fig. 2.13. From this circuit, the NAND2 gate is obtained by inverting the output nodes, the NOR2 gate is achieved by inverting the input signals, while the OR2 gate is achieved by inverting both the input and output signals, from the De Morgan laws.

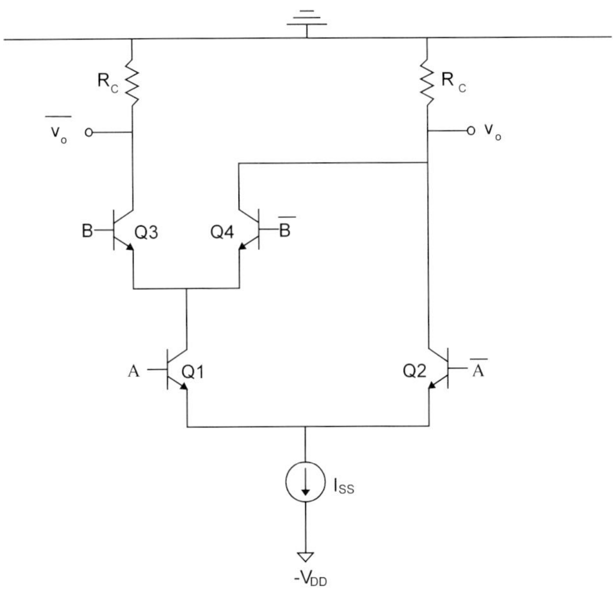

Fig. 2.13. AND2 gate topology.

Operation of the AND gate in Fig. 2.13 can be easily verified by observing that output is high when I_{SS} is steered to the left-hand output node by transistors Q1-Q3, which are active only when A=B=1.

From an applicative point of view, a more important Current-Mode circuit is the XOR gate, as will be discussed in Section 2.6.1. Its topology is easily found by expanding the XOR function (represented by the operator ⊕)

$$A \oplus B = \overline{A} \cdot B + A \cdot \overline{B} \tag{2.51}$$

where each term can be implemented by means of the topology in Fig. 2.12b. The two currents obtained can be OR-ed by summing them according to Fig. 2.12c, i.e. by connecting both the two branches to the output node, while connecting unused branches to the complementary output, as shown in Fig. 2.14. This topology is equivalent to the well-known Gilbert Quad [PM91].

Analogously, the 2:1 multiplexer (MUX) gate with control signal ϕ has the topology in Fig. 2.15, and its operation is easily understood by observing that the output is set by the emitter-coupled pair Q3-Q4 or Q5-Q6, depending on whether current I_{SS} is steered by transistor Q1 or Q2, i.e. when control signal ϕ is high or low.

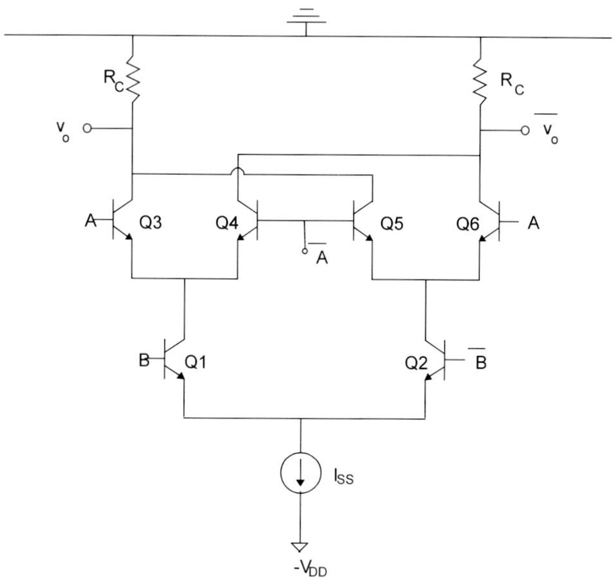

Fig. 2.14. CML XOR gate topology.

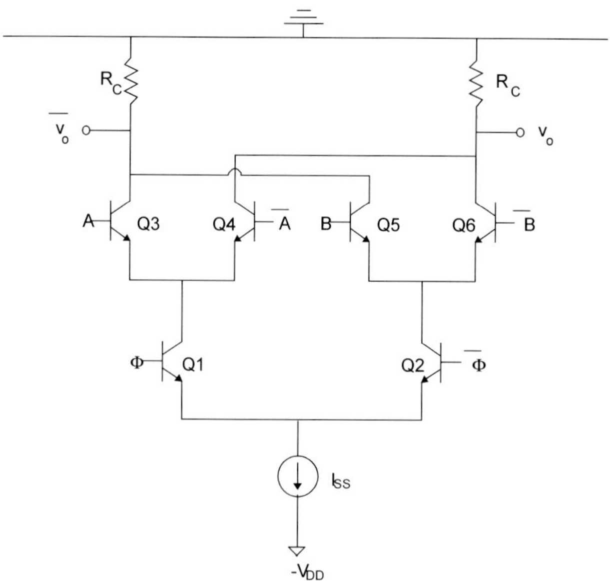

Fig. 2.15. CML MUX gate topology.

The 1:2 demultiplexer (DEMUX) gate with control signal ϕ can be built as in Fig. 2.16 by following the same reasoning. To be more specific, the emitter-coupled pair Q3-Q4 or Q5-Q6 is activated by current I_{SS} steered by transistors Q1-Q2, according to the control signal value. The inactivated transistors pair leaves both of its output nodes at a high voltage.

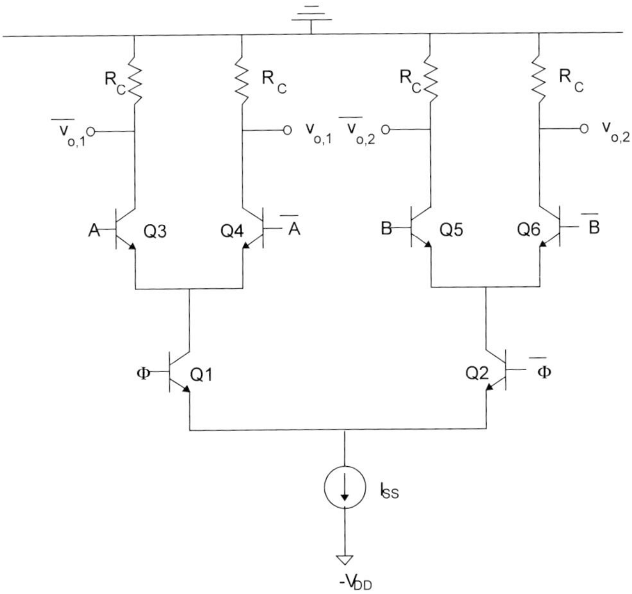

Fig. 2.16. CML DEMUX gate topology.

As an example of sequential blocks, let us consider the D latch gate implementing the logic function

$$OUT = CK \cdot D + \overline{CK} \cdot OUT_{previous} \tag{2.52}$$

where *CK* is usually the clock signal, which enables input *D* when it is high and keeps output at the previous value when it is low. Thus, the bias current must be steered to an emitter-coupled pair driven by input *D* when *CK*=1, and to a bistable sequential circuit when *CK*=0. As is well known [R96], the latter block can be implemented through positive feedback by cascading two inverter gates, as depicted in Fig. 2.17. In practical cases, the two cascaded inverter gates are actually implemented by resorting to the feedback emitter-coupled pair in Fig. 2.17, since each of the two transistors along with a load resistor is an inverting stage. The resulting topology of the Current-Mode D latch is reported in Fig. 2.18.

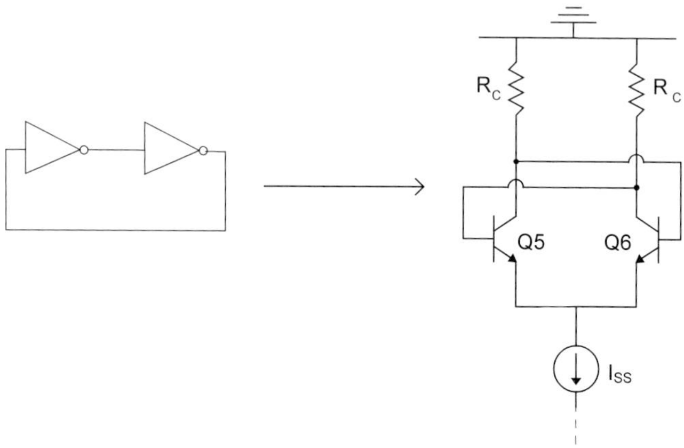

Fig. 2.17. Positive feedback inverter gates implementing the memory circuit.

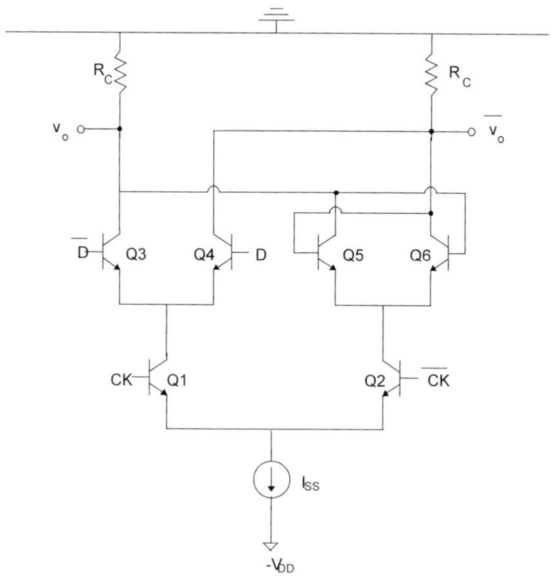

Fig. 2.18. CML D latch gate.

Until now, CML topologies of fundamental gates have been considered, and their ECL counterparts can easily be derived by adding the two output buffers as in Fig. 2.19, as already done for the simple inverter.

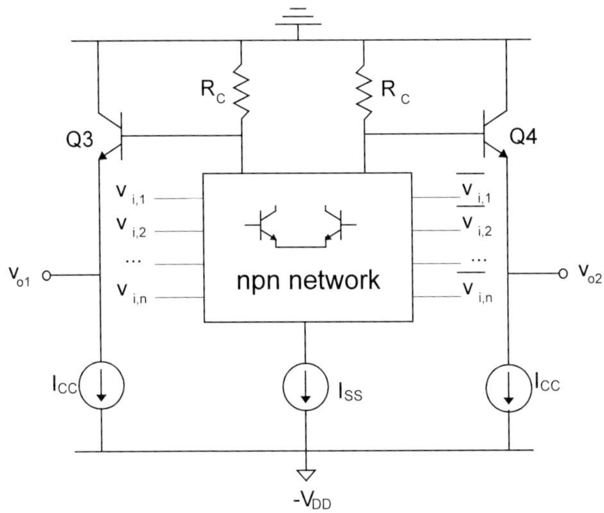

Fig. 2.19. ECL series gates.

2.5.3 Supply voltage limitations in bipolar Current-Mode gates

As discussed in the previous subsections, series gates are implemented by multiple levels of stacked emitter-coupled pairs. The transistor pairs connected to the current source I_{SS} belong to the lower level, while pairs connected to load resistors R_C are at the upper level. Let n the number of levels, i.e. the maximum number of stacked pairs from current source to load resistors.

In CML gates, the input of transistor pairs at the upper level are driven by the output of the previous gate. However, emitter-coupled pairs at lower levels cannot be directly connected to the output node of the driving gate, in order to keep all transistors out of the saturation region. This is a fundamental requirement in CML gates, since saturated transistors have a very low transconductance and a substantial amount of stored base charge, which dramatically slows down the switching of the circuit [R96]. This is not acceptable in bipolar gates, which are to be used in high-speed applications, as will be discussed in Section 2.6.

To avoid the saturation region, the input voltages of lower transistor pairs are progressively reduced through a level shifter circuit with a voltage level shift V_{SHIFT} [T89]. To understand this point, let us consider two contiguous levels i and $(i+1)$ in the generic n-level CML series gate in Fig. 2.20, whose input values v_i are equal to the output voltage v_o of the previous gate reduced by $(i-1)$ times the voltage shift is applied (i.e., $v_i = v_o - (i-1) \cdot V_{SHIFT}$). Let assume transistors Q_i and Q_{i+1} to be ON (due to a high input voltage V_{DD}-$(i-1)\cdot V_{SHIFT}$ and $v_{i+1}=V_{DD}$-$i\cdot V_{SHIFT}$, respectively). The resulting base-collector voltage of the lower transistor Q_{i+1} is[3]

$$V_{BC,i+1}=v_{i+1}-(v_i-V_{BE})=-V_{SHIFT}+V_{BE} \tag{2.53}$$

Thus to maintain it lower than $V_{CB,on}$=0.5 V (which avoids the operation in the saturation region), the voltage level shift must be kept greater than

$$V_{SHIFT}\geq V_{BE}-V_{CB,on}=V_{CE,sat}\approx 0.3\,V \tag{2.54}$$

where V_{BE} was assumed to be about 0.8 V.

In practical cases, the basic level shift is implemented by means of the topology in Fig. 2.21 [R96], [GMC91], where all transistors are biased by the current source I_{CC}. In this figure, which refers to the case with 4 levels, the upper level is directly taken from the output of another CML gate, and the successive levels are downshifted by V_{BE}, $2V_{BE}$ and $3V_{BE}$ (i.e., V_{SHIFT} is set to V_{BE}). The number of npn transistors required in an n-level shifter is equal to (n-1). The main limitation of the circuit in Fig. 2.21 is its slow switching for a high number of levels and for large fan-out values [GMC91]. To overcome this problem, the alternative solution in Fig. 2.22 based on cascaded emitter follower stages is used. The speed increase is achieved at the cost of higher power consumption, since a current source I_{CC} is required for each V_{BE} drop.

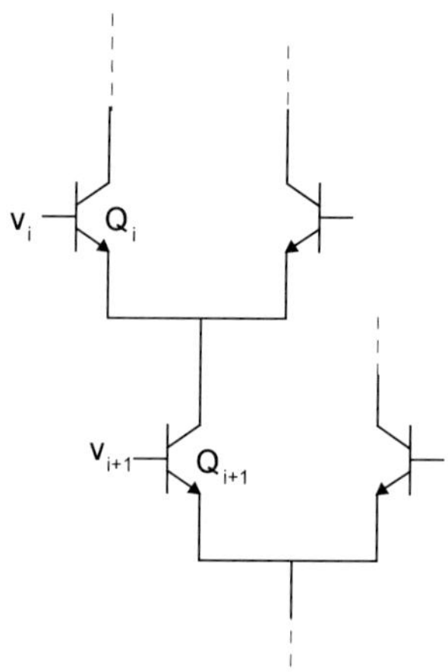

Fig. 2.20. Evaluation of minimum level shift between two adjacent levels.

[3] It is worth noting that the same value is obtained when v_i is low, since in this case the role of transistor Q_i is played by that sharing the emitter node.

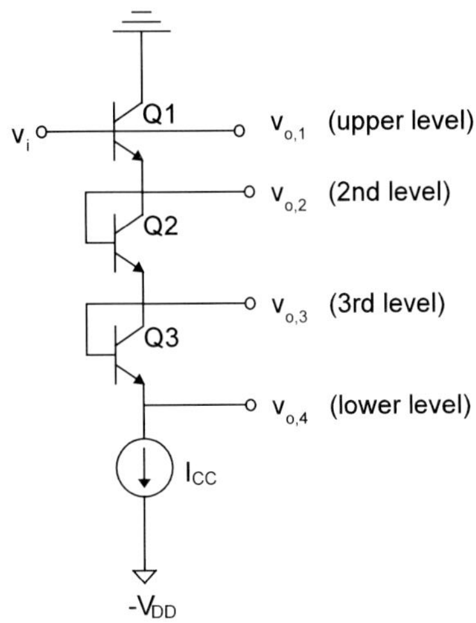

Fig. 2.21. A circuit solution to implement a 4-level shifter.

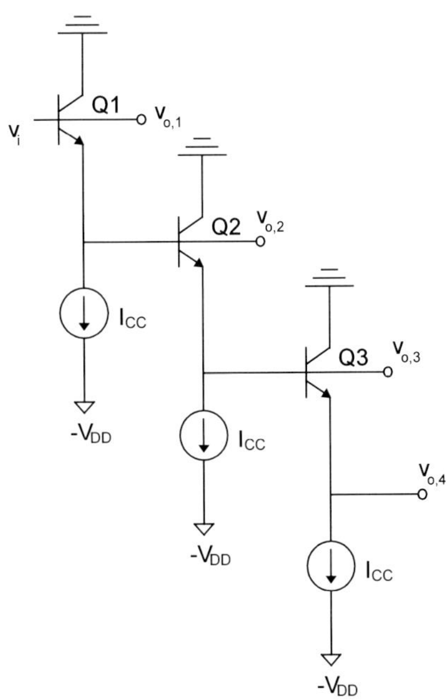

Fig. 2.22. A faster circuit solution to implement a 4-level shifter.

The level shifter circuits presented before can be applied to ECL circuits with only slight modifications. Indeed, the only difference with respect to

CML circuits is that even the upper level is downshifted by a V_{BE} drop. In other terms, an ECL gate is obtained from a CML one by adding one transistor to the level shifter, and taking the input of the former one level below with respect to the latter (e.g., in Figs. 2.21-2.22 the input at the upper level are taken as $v_{o,2}$, that at the second level as $v_{o,3}$ and so on).

In practical cases, the number of logic levels n in series gates is limited by the available supply voltage. This limit can be understood by considering a CML n-level series gate with a level shifter driving its lowest level, whose transistor in the ON state is assumed to be Q2, as depicted in Fig. 2.23. To guarantee the correct behavior of the current mirror, the collector-emitter voltage across transistor Q1 has to be greater than $V_{CE,sat}$. Equivalently, the supply voltage must be greater than the sum of (n-1) base-emitter drops of transistor in the level shifter and that of transistor Q2, other than $V_{CE,sat}$ of transistor Q1 and the small voltage across resistor R_{E1}, as discussed in Section 2.1. Thus, for a given supply voltage V_{DD}, the number of logic levels n has the following upper bound

$$n \leq \frac{V_{DD} - V_{CEsat} - V_{RE,1}}{V_{BE,on}} \approx \frac{V_{DD}}{V_{BE,on}} \tag{2.55}$$

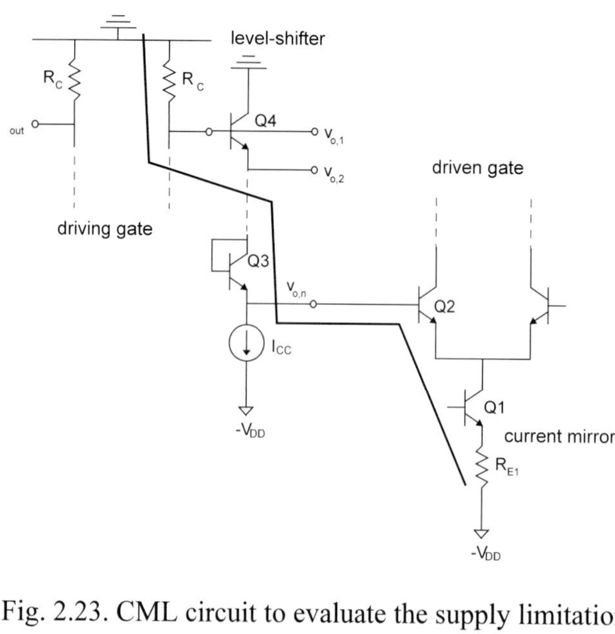

Fig. 2.23. CML circuit to evaluate the supply limitation.

An analogous result is found for ECL gates, which have a greater by unity number of level-shifter V_{BE} drops, thus leading to the following upper bound

$$n \le \frac{V_{DD} - V_{CE,sat}}{V_{BE,on}} - 1 \approx \frac{V_{DD}}{V_{BE,on}} - 1 \tag{2.56}$$

Inspection of relationships (2.55)-(2.56) reveals that the number of logic levels essentially depends on the ratio of supply voltage and V_{BE} drop. While the former tends to be reduced for reasons related to power consumption [CB95a], the latter does not scale even for more advanced technologies [CB95b], [RP96], thus the number of logic levels tends to decrease. For example, the supply voltage in current applications can be as low as 3 V (or slightly lower), thus from (2.55) three logic levels are allowed for CML gates, and two for ECL gates. Therefore, to allow at least two levels of series gating, as required in the implementation of logic functions, it is apparent that supply voltage will no longer scale in the near future. This justifies why series gates are not a suitable circuit solution in applications that require a low supply voltage, in which cases alternative logic styles must be used. Further details will be provided in Chapter 8, where low-voltage logic styles will be analyzed and modeled.

2.5.4 MOS Current-Mode series gates and supply voltage limitations

Since MOS Current-Mode gates have a principle of operation similar to that of bipolar ones, considerations introduced in the previous subsection can easily be generalized to MOS circuits. In particular, implementation of logic functions through MOS Current-Mode gates can be carried out by exploiting the series-gating approach presented in Section 2.5.1. Therefore, topologies of bipolar gates presented in Section 2.5.2 can be extended to MOS technology by simply substituting each npn transistor with an NMOS, and each load resistor by a PMOS active load, as already done for the SCL inverter in Section 2.4.

Like the bipolar counterparts, MOS Current-Mode series gates require level shifters to operate correctly , i.e. to keep NMOS transistors out of the triode region. By following the same procedure as in the previous subsection, the voltage shift between adjacent levels must be greater than

$$V_{SHIFT} \ge V_{GS} - V_{T,n} = V_{DS,sat} \tag{2.57}$$

In practical cases, the two solutions in Figs. 2.21 and 2.22 with npn transistors replaced by NMOS devices are used. Accordingly, the maximum number of series-gating levels allowed for an assigned V_{DD} is given by (2.55)-(2.56) where the NMOS gate-source voltage V_{GS} replaces the base-emitter voltage V_{BE} of npn transistors, respectively for gates without and with output buffers. Since the gate-source voltage reduces for more advanced CMOS processes due to the scaling of the threshold voltage (even slowly for reasons related to the subthreshold current [CB95b], [RP96]), MOS Current-Mode gates are more amenable than bipolar circuits for low-voltage applications.

2.6 TYPICAL APPLICATIONS OF CURRENT-MODE CIRCUITS

Current-Mode logic is currently used in a number of applications, due to its high-speed potential, as well as the reduced switching noise [KKI97], [M97], [K98]. Such properties make Current-Mode logic more suitable than standard CMOS logic in applications ranging from RF circuits to fiber-optic communications and high-resolution mixed-signal CMOS circuits, as discussed in the following.

2.6.1 Radio Frequency applications

A fundamental block in RF applications is the Phase-Locked Loop (PLL), which allows for frequency synthesis, clock generation, data recovery and synchronization [R961], [H96], [LR00], [DS03]. The block diagram of a PLL is shown in Fig. 2.24 (in some applications the input frequency divider is omitted).

The periodic input signal is usually generated by a crystal reference, and its frequency is divided by M by means of the input frequency divider. The phase of the signal obtained is compared to that of the feedback signal through a mixer, which is generally implemented by a Gilbert Cell (i.e. the XOR gate in Fig. 2.14). The charge pump is essentially an amplifier driving an RC loop filter, which is introduced to stabilize the closed-loop circuit.

The filtered signal drives a Voltage Controlled Oscillator (VCO) generating a periodic output signal, whose frequency is divided through the feedback frequency divider by a factor of N. Among these blocks, the phase detector and the frequency dividers are frequently implemented through Current-Mode logic circuits. The VCO can also be implemented by a Current-Mode ring oscillator (whose design is discussed in Chapter 8), when the noise requirement does not necessarily require the use of an LC oscillator. Since the speed of a PLL is mainly limited by the feedback frequency divider other than the VCO, it is essential to properly design Current-Mode gates to maximize their operating frequency.

In the specific case of frequency dividers, they are usually implemented by cascading a high-speed prescaler circuit (typically divide-by-8 circuits) and a low-speed divider, whose modulus can be eventually varied. Techniques to achieve a high speed prescaler and a ring oscillator while consciously managing the trade-off with the power consumption are dealt with in the following chapters.

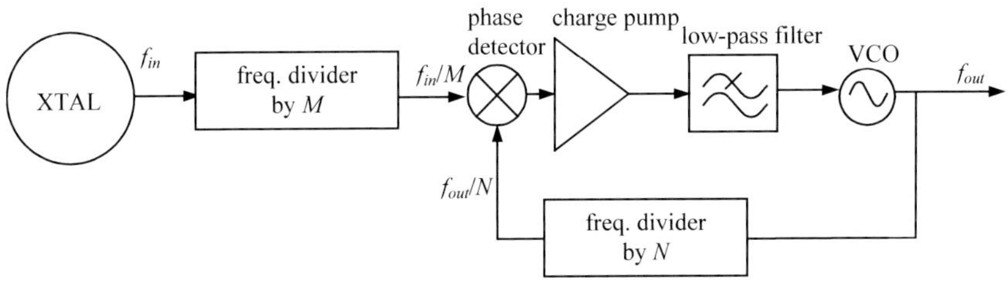

Fig. 2.24. Schematic of a Phase-Locked Loop (PLL).

Frequency dividers are often implemented as a cascade of divide-by-two stages. In general, frequency dividers can be classified into regenerative and static type, depending on the principle of operation of the divide-by-two cell. The regenerative frequency divider (RFD), as shown in Fig. 2.25, is made up of a mixer with a periodic input signal having a frequency f_{in} and an output signal at frequency f_{out} [IIS89], [FBA90], [KUO92], [R98]. The latter one is fed back to the mixer, which generates two sidebands at frequencies f_{in}-f_{out} and $f_{in}+f_{out}$, as well as the harmonics $2(f_{in}\pm f_{out})$, $3(f_{in}\pm f_{out})$, etc. The low-pass filter cuts off all harmonics, excepting that at the lowest frequency f_{in}-f_{out}, that is amplified by the successive amplifier[4]. At the steady state, the input signal frequency f_{in}-f_{out} of the amplifier and its output frequency f_{out} output are equal, thus leading to the divide-by-two behavior

$$f_{out} = \frac{f_{in}}{2} \tag{2.58}$$

[4] Actually, the low-pass filter and the amplifier are not implemented, since the upper cut-off frequency and the voltage gain of the mixer are exploited.

By following the same reasoning, it can easily be verified that the sideband at frequency $f_{in}+f_{out}$ is rejected by the feedback loop.

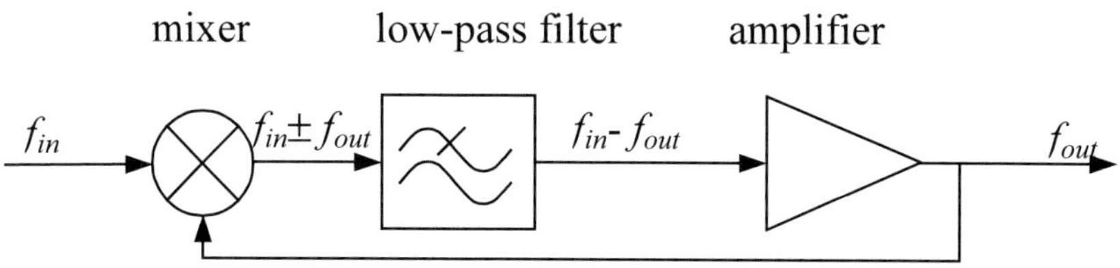

Fig. 2.25. Schematic of a regenerative frequency divider.

The maximum input frequency allowed by the regenerative divider depends on the speed of the mixer, or equivalently by the delay τ_{PD} of a XOR gate. To be more specific, the XOR gate implementing the mixer in Fig. 2.25 must be able to switch every half period of the output signal, i.e. every period of the input signal. Since the input period must be greater than or equal to the XOR delay, the maximum input frequency allowed by the dynamic divider is equal to

$$f_{in,\max}=\frac{1}{\tau_{PD}} \tag{2.59}$$

Strategies to design the Current-Mode XOR gate for a high speed or an efficient trade-off with power consumption will be addressed in Chapters 5 and 7 for the bipolar and CMOS gates, respectively.

Regenerative dividers require a further frequency constraint, in order to correctly operate according to (2.58), which is based on the assumption that harmonics at frequencies $2(f_{in}\pm f_{out})$, $3(f_{in}\pm f_{out})$, etc. are strongly attenuated by the mixer circuit. This can be achieved when all harmonics generated by the mixer are rejected, or equivalently the harmonics at the lowest frequency $2(f_{in}-f_{out})$ leads to an output sideband at a greater than the maximum output frequency $f_{out,max}=f_{in,max}/2$. At the steady state, harmonics $2(f_{in}-f_{out})$ must be equal to the output frequency f_{out} of the amplifier, thus its effect on the output is an undesired sideband at a frequency $3f_{in}/2$. As a consequence, the loop rejects this sideband when the input frequency is kept high enough so that the $3f_{in}/2>f_{in,max}/2$, which imposes a lower limit value $f_{in,min}$ equal to

$$f_{in,\min}=\frac{f_{in,\max}}{3} \tag{2.60}$$

Static frequency dividers are based on cascaded Master-Slave T flip-flops (T-FF), which switch every two edges of the periodic input signal applied to their clock input, thus performing a divide-by-two operation [FBA90], [K91], [KOS91], [IIT95]. In practical cases, T-FFs consist of two cross-coupled feedback D latches driven by opposite clock signals, as depicted in Fig. 2.26. After two clock edges, the input of each latch crosses both gates and turns to the opposite value, due to the inversion associated with the cross coupling.

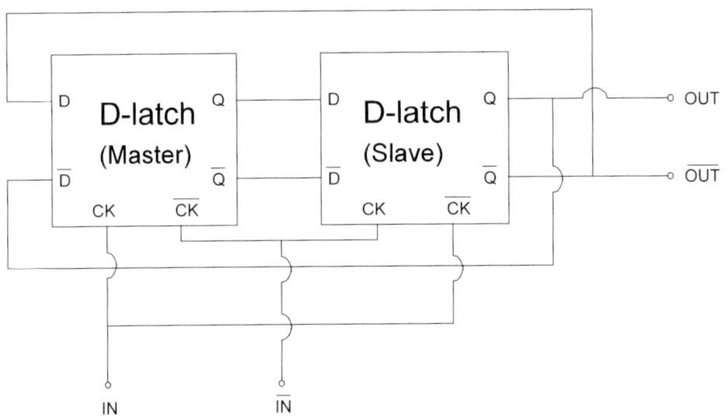

Fig. 2.26. Schematic of a T-FF divide-by-two stage.

In contrast to regenerative dividers, static circuits are able to work at arbitrarily low input frequencies. This greater flexibility is achieved at a cost of a lower speed, as the maximum input frequency allowed is lowered by a factor of 2. To understand this point, let us observe that the time needed by a latch to generate its output after the transition of the input signal is equal to the CK-Q latch delay τ_{PD} (i.e. the delay between the transition of the clock input and output). Thus, the time available to each latch (i.e. half an input period) must be greater than or equal to the CK-Q delay, in order to correctly generate the latch output. As a consequence, the maximum input frequency results to

$$f_{in,\max} = \frac{1}{2\tau_{PD}} \tag{2.61}$$

which is halved compared to (2.59), by assuming delay of XOR gate and D latch to be comparable (actually, as will be shown in Chapters 4-5, the latter is slower, thus the speed advantage of the RFD is greater than 2). Strategies to improve speed in (2.61) and manage the trade-off with the power consumption will be discussed in Chapter 8.

In many RF applications, the frequency divider may also be required to have a programmable modulus *N*, as in the case of dual-modulus prescalers [SPW91], [CHL92], [MSO92], [SMS94], [VK95], [CW98], [HFP01], [KBW01]. In this case, the frequency divider is a Finite State Machine based on D flip-flops (D-FF) and combinational logic that updates the count and sets the modulus. Essentially, its speed depends on the speed of the D-FFs used, which in practical cases are of the Master-Slave type, shown in Fig. 2.27. The signal to be divided drives the clock terminals of the D latches, while the combinational logic properly sets their inputs.

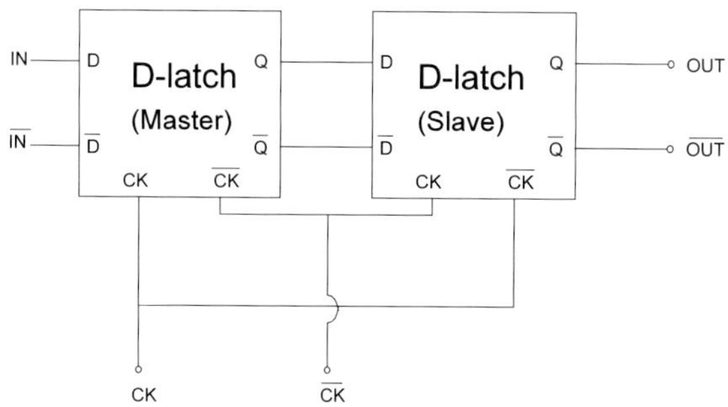

Fig. 2.27. Schematic of a D-FF Master-Slave.

In sequential circuits, flip-flops affect the speed through two timing parameters: the CK-Q delay and the setup time t_{SETUP} [R96]. By definition, the CK-Q delay of the D-FF in Fig. 2.27 is equal to the CK-Q delay τ_{PD} of the single latch, which will be modeled in Chapters 4 and 6 for a bipolar and CMOS latch, respectively. Instead, the flip-flop setup time (i.e. the amount of time before a clock transition in which the inputs must be kept constant) is equal to the D-Q delay of the single latch between the transition of *IN* and the subsequent output (more details are provided in Chapter 8). Indeed, the input of the Slave D-FF at the falling clock edge is the output of the Master D-FF, which is generated after the input transition by the latch D-Q delay. As a consequence, input *IN* must settle to the correct value a D-Q delay before the clock transition, i.e. the flip-flop setup time is equal to the (Master) latch D-Q delay.

2.6.2 Optic-fiber communications

Another important application field of Current-Mode logic is the implementation of integrated circuits for signal multiplexing/demultiplexing in optic-fiber systems, whose typical structure is depicted in Fig. 2.28 [ARL95], [I95], [SR01], [R02].

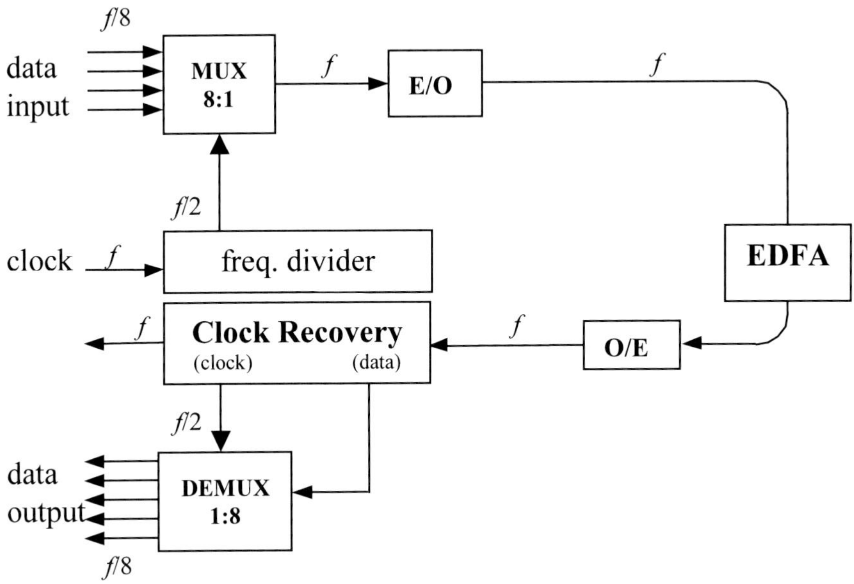

Fig. 2.28. Block diagram of a fiber-optic link.

In Fig. 2.28, the E/O block is a semiconductor laser diode converting electrical signals to optical ones, while photodetector O/E performs the opposite conversion. To exploit the wide bandwidth of the optic-fiber channel, parallel input signals are serially transferred to the optic fiber through a multiplexer (MUX) at a clock rate f, and are amplified by an erbium-doped fiber amplifier (EDFA). Serial data crossing the optic fiber are then transferred in parallel through a demultiplexer (DEMUX). A clock recovery circuit is needed to resynchronize the clock signal of the receiver circuit to that of the transmitter. In practical cases, MUX as well as clock-recovery with DEMUX are implemented in a single chip to achieve a high speed. For current bipolar and CMOS technologies, data rate as high as 40 Gb/s [RM96], [RDR01] and 10 Gb/s [TUF01], [NIE03] have been achieved.

The schematic of an 8:1 MUX is reported in Fig. 2.29a, where each 2:1 MUX is implemented by the cell shown in Fig. 2.29b. Since the circuit in

Fig. 2.29a is driven at half the clock frequency, a retiming D-FF at the output node is added to reduce the jitter contribution due to a different from 50% clock duty cycle [SD93], even at the cost of a lower speed.

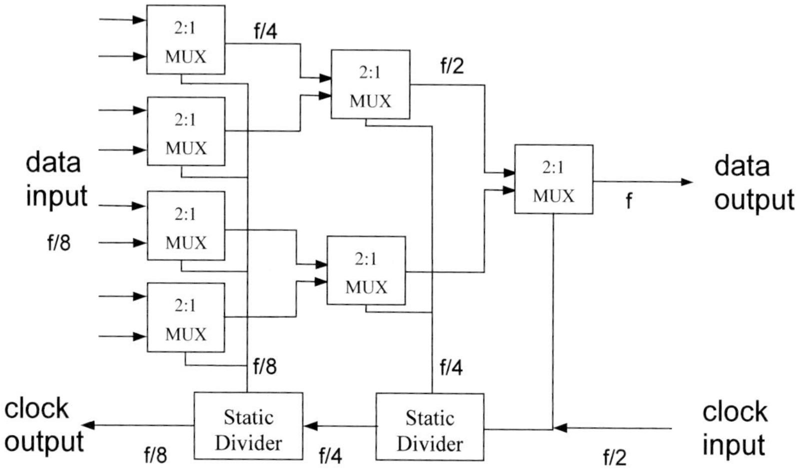

Fig. 2.29a. Block diagram of an 8:1 MUX.

The 8:1 MUX in Fig. 2.29a consists of 2:1 MUX cells connected in a tree-like fashion, according to three logic levels [L95]. The one providing the output signal works at full rate, while the other ones work at a halved speed with respect to the previous one. Such progressively halved operating frequencies are obtained from the (halved) clock signal through divide-by-two dividers, the first of which could be regenerative, while the following ones are static. The 2:1 MUX cell in Fig. 2.29b consists of a Master-Slave D-FF for input IN1 and a Master-Slave-Slave D-FF for input IN2. The latter contains a further latch that delays the arrival of IN2 to the multiplexer by a half period, as required by the alternate selection of IN1 and IN2.

The schematic of a 1:8 DEMUX used in Fig. 2.28 is analogous to that of the MUX in Fig. 2.29a, with 1:2 DEMUX cells being connected in a tree-like manner, as shown in Fig. 2.30a [L96], [LL96], [S96]. The schematic of the fundamental 1:2 DEMUX is depicted in Fig. 2.30b.

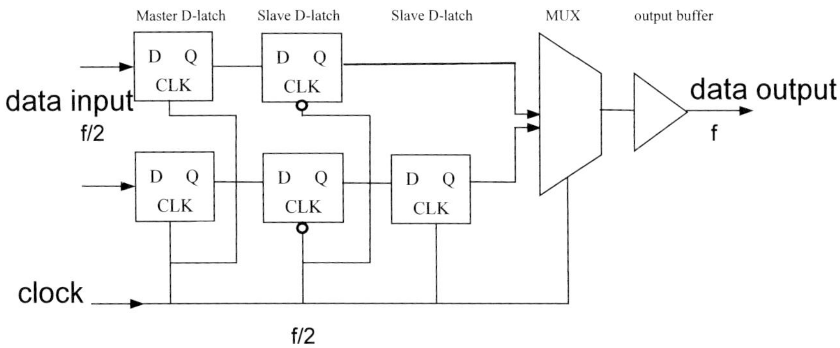

Fig. 2.29b. Schematic of a 2:1 MUX used in the 8:1 MUX.

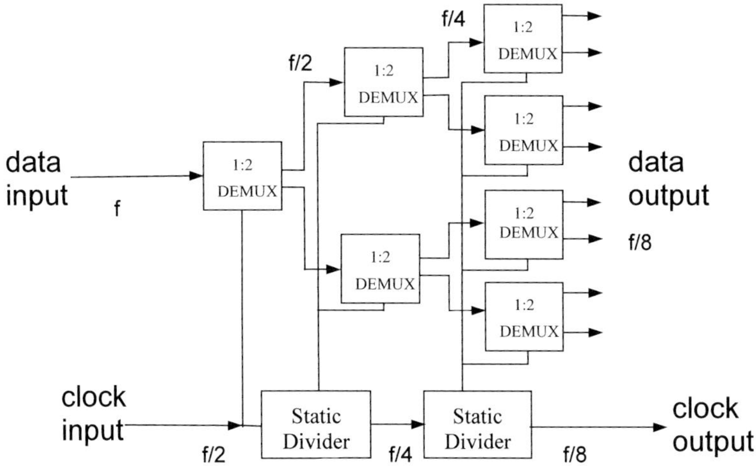

Fig. 2.30a. Block diagram of a 1:8 DEMUX.

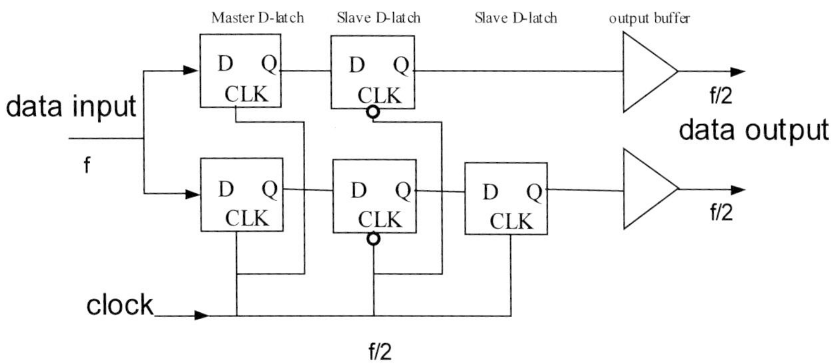

Fig. 2.30b. Schematic of a 1:2 DEMUX used in the 1:8 DEMUX.

In the 1:2 DEMUX cell in Fig. 2.30b, the two output signals are taken from the output buffers, respectively. It is worth noting that a 1:2 DEMUX in Fig. 2.30a belonging to a logic level far from the output can have a much lower speed than the last one. The same consideration holds in MUX cells depicted in Fig. 2.29a.

2.6.3 High-resolution mixed-signal ICs

In general, logic circuits must satisfy assigned constraints in terms of speed, power consumption and silicon area. Moreover, additional requirements on the switching noise must be taken into account in the design process of CMOS mixed-signal ICs, which consist of both a digital and an analog section sharing the same substrate. Mixed-signal circuits are cost-effective in a number of applications, such as video signal processing, magnetic disk recording channel processors, oversampled A/D and D/A converters [KCA90], [KA92]. In such cases, the switching noise generated by the logic gates couples with the analog circuitry and degrades its resolution.

Even though exhibiting an appealing speed performance, noise margin, ease of design, a low static power consumption and a low area, the CMOS static logic style generates a considerable amount of noise due to the supply current spikes needed during the switching of logic gates. Such current spikes determine voltage drops in parasitic resistors and inductors associated with the supply rails, bonding pads, bonding wires, package pins, as well as in the substrate resistance [ACKS93].

Until now, solutions at various levels of abstraction have been proposed to partially attenuate the effect of the switching noise on the operation of the analog circuits in mixed-signal ICs. From the technology point of view, using a silicon-on-insulator (SOI) or a highly-doped epitaxial wafer CMOS technology can reduce the amount of noise coupled with the analog circuitry, though increasing costs.

From a physical point of view, optimal floorplanning and a safe layout style can improve the immunity of analog circuits to switching noise [S02]. Regarding the floorplan, distance of analog blocks from digital one should be maximized. As far as the layout is concerned, some precautions should be taken, such as making substrate almost equipotential through widespread substrate contacts, as well as shielding analog blocks by means of diffused guardbands, as depicted in Fig. 2.31.

From a circuit point of view, analog circuits should be designed by exploiting differential topologies, since they are intrinsically more immune to (common-mode) external noise. Topologies with a high Power Supply

Rejection Ratio (PSRR) should also be used to attenuate the effect of the supply switching noise on analog signals.

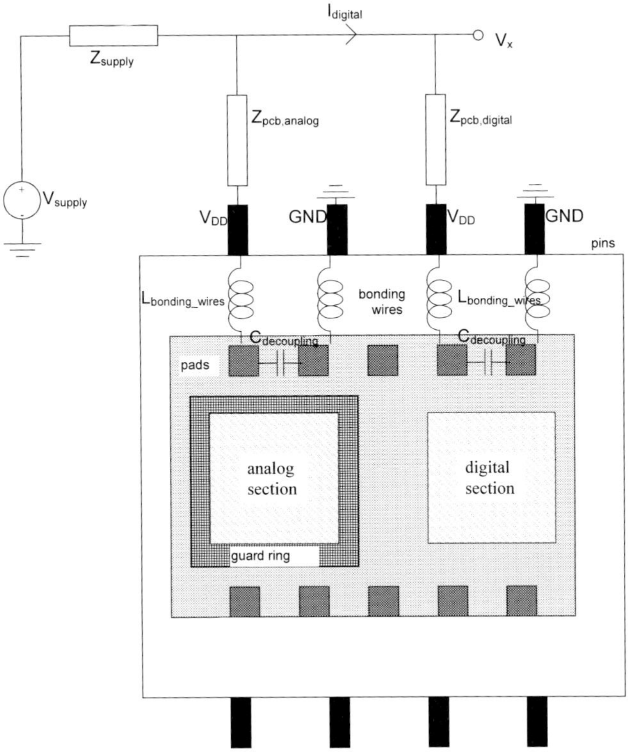

Fig. 2.31. A mixed-signal IC and parasitic effects.

At the system level, the effect of switching noise can be mitigated by reducing the common impedance Z_{supply} of the paths from the supply to the analog and digital sections. This is easily understood by observing that digital circuits basically affect operation of analog section through their supply current $I_{digital}$, which determines a voltage drop on the common impedance Z_{supply} equal to $Z_{supply} \cdot I_{digital}$, which perturbs the analog supply voltage. In practical cases, common impedance Z_{supply} is minimized by using separate analog and digital power distribution networks, as well as separate bonding pads, bonding wires, package pins and printed circuit board runs. Moreover, the effect of the digital section on the analog circuitry is minimized by resorting to multiple pins and bonding wires to reduce their parasitic inductance, as well as to on-chip bypass capacitors, at the expense of silicon area and number of pads and pins.

When a very high resolution must be achieved (e.g. 16-18 bit), remedies that attenuate the transmission of the switching noise to analog circuits discussed above are not sufficient, and integration of both analog and digital blocks on the same chip requires the generated switching noise to be lowered. Thus, CMOS static logic is no longer an amenable solution, and alternative logic styles must be selected [SKD96], [NA97], [KH00]. To be more specific, switching noise is reduced by reducing digital supply current variations, and a logic style drawing a constant supply current would be highly desirable. Such property is achieved in Current-Mode logic, since each gate constantly requires a supply current equal to its bias current, as already discussed in the previous subsections. For this reason, CMOS Source Coupled Logic has been successfully used in high-resolution mixed-signal circuits [DKS90], [LWO91], [KDN91], [F97], [JMS97]. A quantitative evaluation of the switching noise produced by CMOS Current-Mode logic will be discussed in Section 3.5.

As discussed in the previous sections, Current-Mode logic gates can be implemented in a differential or in a single-mode fashion, of which the former allows for a better performance in terms of speed and power efficiency. Moreover, differential implementation allows logic gates for rejecting (common mode) supply disturbances, since they have the same effect on output nodes. This improvement in immunity to noise allows for reducing the noise margin requirement, and can represent an advantage in cases where the logic swing reduction is beneficial, which are addressed in Chapter 7.

It is worth noting that differential bipolar Current-Mode gates have the same features as CMOS gates in terms of switching noise and immunity to supply noise. This is exploited in RF circuits operating at very high frequencies, at which impedance of parasitic capacitances significantly lowers, thus making transmission of spurious signals easier. Thanks to a better noise immunity, noise margin (and hence logic swing) can be reduced, thereby enhancing speed and power-delay trade-off performance, as will be discussed in Chapter 5.

Until now, Current-Mode logic gates have been presented and analyzed in terms of their principle of operation and static behavior, as well as from an application point of view, reporting some specific and widely used logic gates. In the next chapter, techniques to implement arbitrary logic functions with a minimum amount of hardware are reviewed.

Chapter 3

DESIGN METHODOLOGIES FOR COMPLEX CURRENT-MODE LOGIC GATES

In this chapter, general methodologies to map a given logic function into a series gate are discussed and applied to practical examples. After introducing some basic concepts, a graphic method [CJ89], its analytical formulation and a strategy based on VEMs [MKA92] are described by assuming a given input variables ordering. Finally, guidelines to properly choose the input ordering are discussed.

3.1 BASIC CONCEPTS ON THE DESIGN OF A SERIES GATE

Let us consider a given n-variable logic function $F(X_1 \ldots X_n)$ of input signals $X_1, \ldots, X_n$. According to Fig. 2.11, this function is implemented in a series gate by properly choosing the topology of its npn/NMOS network, which consists of stacked emitter/source coupled transistor pairs. In the following, for the sake of brevity we will refer to only bipolar gates, and extension to MOS circuits is immediately obtained by substituting an NMOS transistor to each npn bipolar transistor.

In general, the npn network in Fig. 2.11 must provide a unique conductive path from I_{SS} to the output nodes for all possible input values, as discussed in Section 2.5.1. For this reason, emitter-coupled pairs are not allowed to share their emitter node with other transistor pairs (indeed, if two transistor pairs shared their emitter, the two conducting transistors - one for each pair -would draw the same current, thus determining two different current paths). From this observation, the current source I_{SS} must be connected to the emitter of only one transistor pair Q1-Q2. Analogously, each of the two collector nodes of Q1-Q2 is connected to the emitter of another transistor pair which is stacked to the first one, i.e. it lies in an upper logic level. By iterating this reasoning, a tree of stacked emitter-coupled pairs is obtained, whose transistor pairs at the highest level have their collector node connected to one of the two output nodes. Note that the collector nodes of two different transistors may be connected to the same transistor pair.

To understand how the npn network topology is related to the function $F(X_1 \ldots X_n)$ to be implemented, let us assume in the following that input X_1 is applied to transistors belonging to the highest level (i.e. the first one), input X_2 is applied to the immediately lower level (i.e. the second level) and so on, up to input X_n associated with the lowest level[1] (i.e. the last one), as depicted in Fig. 3.1.

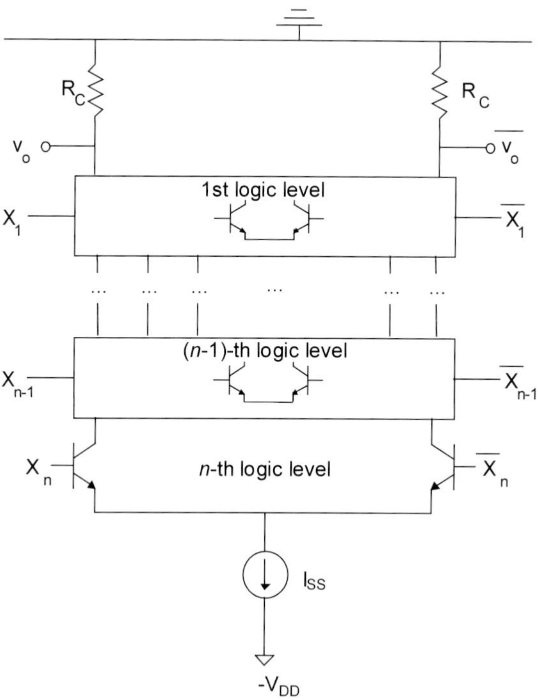

Fig. 3.1. Correspondence among input signals and logic levels.

By assuming $\alpha_F \approx 1$, the collector current of the generic transistor Qa in a series gate can be associated with a logic level I_{Qa} equal to 0 (when it is

[1] In some specific cases, which will be considered in Section 3.5, more than one input might be associated with the same level.

equal to zero) or 1 (when it is equal to I_{SS}). The same consideration holds for the logic level I_O (or $\overline{I_O}$) associated with the current flowing through the load resistance connected to the output node v_o (or $\overline{v_o}$). It is worth noting that, since the output voltage v_o is low when the output current logic level I_O is high, the logic level V_O associated with output voltage is the complement of that associated with the output current

$$V_O = \overline{I_O} \tag{3.1}$$

From this relationship, a simple method can be derived to evaluate the function $F(X_1 \ldots X_n)$ implemented by an assigned npn network topology, as will be discussed in the following subsection.

3.1.1 Evaluation of function $F(X_1,\ldots,X_n)$ implemented by a given topology

In general, in an n-level series gate, the collector current I_{Qa} of transistor Qa at the j-th level is high only if a conductive path between this node and the current source exists. This conductive path, which will be referred to as an *active path* in the following, consists of at most (n-j+1) stacked transistors driven by input signals $x_j \ldots x_n$ which are all at the high level (in the following, lower case variables x_i will be used to refer to a *literal*, X_i, or its complement, $\overline{X_i}$)

$$\left(x_j \cdot x_{j+1} \cdot \ldots \cdot x_n\right)_{active\_path} = 1 \tag{3.2}$$

From (3.2), an active path is unambiguously associated with the product of literals $x_j \ldots x_n$ resulting to 1 (i.e. the set of input signals driving the transistors of the active path), since there is always one active path at a time for each input value. For the sake of clarity, some examples are reported in Fig. 3.2. According to these considerations, I_{Qa} is equal to 1 if the first, or the second, etc. among all possible paths from the collector node of transistor Qa to the current source is active. Therefore, I_{Qa} is analytically expressed by the OR of the products of literals $x_j \ldots x_n$ driving the transistors of all possible paths starting from the collector of transistor Qa to the current source.

By reiterating the same reasoning for the current $\overline{I_O}$ flowing through the load resistance connected to the output node $\overline{v_o}$, from (3.1) it follows that function $F(X_1 \ldots X_n)$ is equal to the OR of the products of literals $x_1 \ldots x_n$ associated with (i.e., driving the transistors of) all possible paths from the output node $\overline{v_o}$ to the current source. Obviously, the same result is obtained by analytically expressing the current I_O flowing through the load resistance connected at the output node v_o, and then complementing the result, from eq. (3.1). In practical cases, of the two currents $\overline{I_O}$ and I_O, it is convenient to evaluate that having a lower number of paths to the current source, when identifying the function implemented by a given topology.

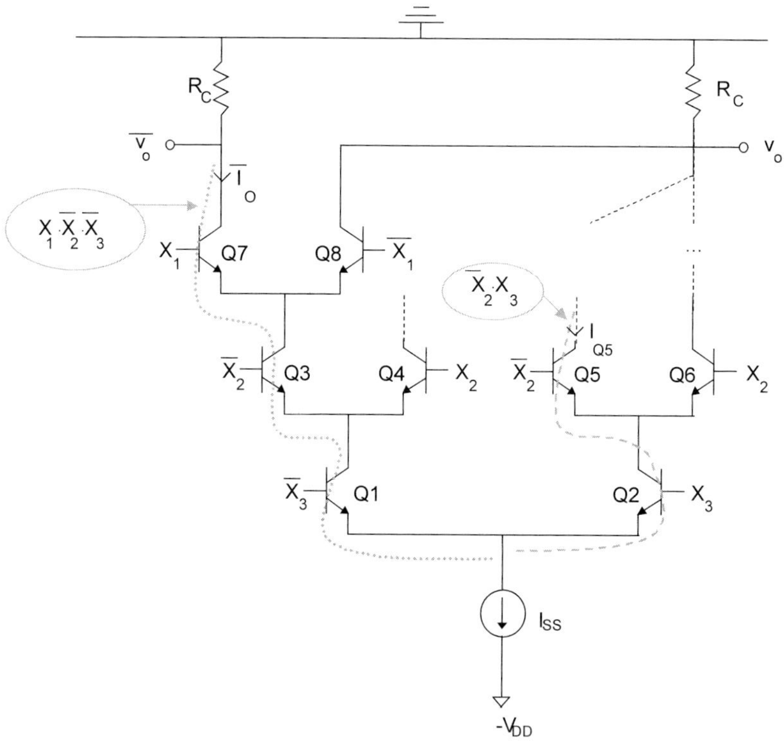

Fig. 3.2. Correspondence of product of literals and active paths.

As an example, let us evaluate the function implemented by the series gate in Fig. 2.13. To simplify calculations, it is convenient to evaluate the boolean expression of current $\overline{I_O}$ rather than I_O, since in the former case there is only one possible active path (i.e. Q3-Q1), while in the latter case there are two possible active paths (i.e. Q4-Q1 and Q2). The only possible active path Q3-Q1 connecting the output node $\overline{v_o}$ and the current source consists of transistors driven by signals B and A, which is associated with the product AB, hence function $F(A, B)$ results in

$$F(A,B)=\overline{I_O}=A\cdot B \tag{3.3a}$$

where relationship (3.1) was used. Of course, the same expression is obtained by expressing current I_O and then complementing it. Indeed, the first path Q4-Q1 is associated with the product $A\cdot\overline{B}$, while the second path Q2 is associated with $\overline{A}$, thus function I_O results in the OR of these terms, yielding

$$F(A,B)=\overline{I_O}=\overline{A\cdot\overline{B}+\overline{A}}=\left(\overline{A\cdot\overline{B}}\right)\cdot A=\left(\overline{A}+B\right)\cdot A=A\cdot B \tag{3.3b}$$

where the De Morgan laws were applied.

Even though the case of a combinational series gate has been discussed until now, extension to sequential blocks is straightforward. Indeed, in practical cases, Current-Mode sequential blocks are frequently D latches (Fig. (2.18)) or Master-Slave D-FF (Fig. (2.27)), which have already been described. In addition, in the infrequent cases where combinational logic is embedded into a sequential series gate, the same topology as the D latch gate is still used, since it is sufficient to replace the only combinational emitter coupled pair Q3-Q4 in Fig. (2.18) by the desired combinational function $F(X_1 \ldots X_n)$, as depicted in Fig. 3.3.

According to Fig. 3.3, the resulting series gate has $(n+1)$ levels. For the above considerations, only combinational functions will be discussed in the following, without loss of generality.

Until now, the identification of the function implemented by an assigned circuit topology has been addressed. However, in practical design cases, one has to identify the topology which implements a given boolean function $F(X_1 \ldots X_n)$. Basic concepts on this design aspect will be introduced in the following subsection.

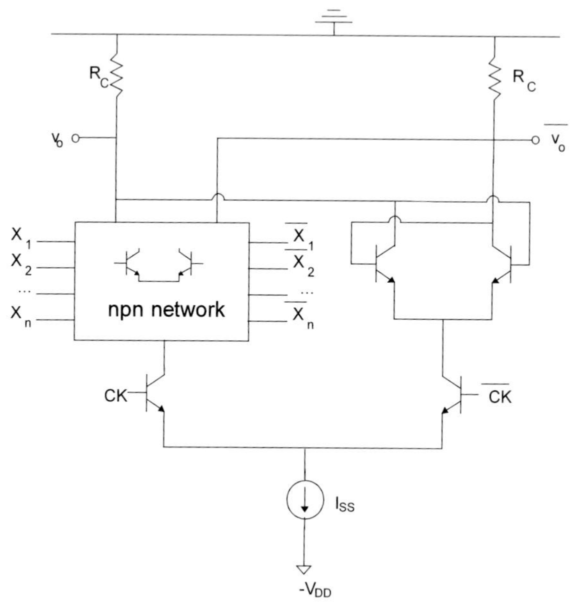

Fig. 3.3. Sequential series gate with embedded combinational logic.

3.1.2 Series-gate implementation of an assigned function $F(X_1...X_n)$

The unambiguous correspondence discussed in Section 3.1.1 between products of literals $x_1\cdot\ldots\cdot x_n$ and active paths from output node $\overline{v_o}$ (or v_o) to the current source provides a simple method to build an npn network which implements a given combinational function $F(X_1\ldots X_n)$. Indeed, each of the 2^n possible input values is associated with the unique product $x_1\cdot x_2\cdot\ldots\cdot x_n$ being equal to 1, in which each literal is complemented if the correspondent input bit is 0 (for example, input 0110 is associated with product $\overline{X_1}\cdot X_2\cdot X_3\cdot\overline{X_4}$). This product $x_1\cdot x_2\cdot\ldots\cdot x_n$ is in turn associated with a unique active path consisting of transistors driven by the correspondent literals. Therefore, an unambiguous correspondence between input values and stacked transistors' paths exists.

From the previous considerations, an arbitrary combinational function can be built by implementing an npn network having all transistor paths associated with the 2^n possible inputs (or, equivalently, the correspondent literal products) and then properly connecting each of the upper collector branch to $\overline{v_o}$ or v_o to set the output to the desired value. Assuming input X_j to be applied to the j-th level with j=1...n as in Fig. 3.1, such an npn network has a tree-like structure with an emitter coupled pair connected to each collector node of transistors lying at the lower level [CJ89], thereby doubling the number of transistors in a logic level compared to the lower level. For the sake of clarity, such a general npn network is shown in Fig. 3.4 for an arbitrary 3-variable function $F(X_1, X_2, X_3)$. In this figure, it is possible to identify all paths associated with the 8 possible input values. For example, the path Q7-Q3-Q1 is associated with (000), Q8-Q3-Q1 is associated with (100), and so on.

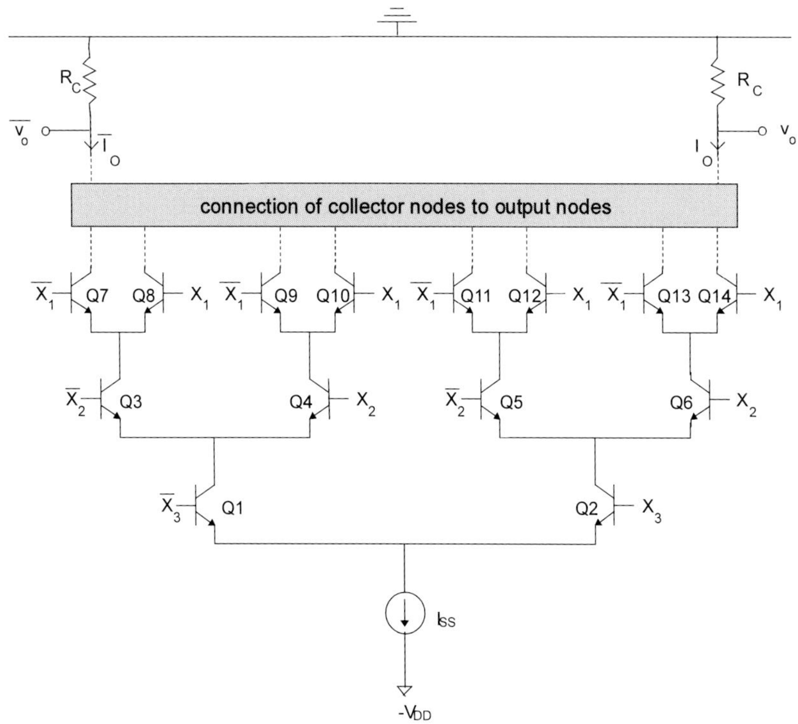

Fig. 3.4. General topology implementing an arbitrary 3-variable function.

To map the value of a boolean function F to the series-gate output voltage v_o (or equivalently its current $\overline{I_O}$), a proper choice of connections of collector nodes of the highest transistors to one of the two output nodes (represented by the gray box in Fig. 3.4) is needed. To understand this point, it is useful to express function $\overline{I_O}=V_O=F(X_1 \ldots X_n)$ in the standard sum-of-product form [C95], i.e. as the OR of products including all literals $x_1 \cdot x_2 \cdot \ldots \cdot x_n$ (usually referred to as *minterms*). Since, as shown in Section

3.1.1, $\overline{I_O}$ is equal to the OR of the products $x_1 \cdot x_2 \cdot \ldots \cdot x_n$ associated with all possible paths from $\overline{v_o}$ to the current source (i.e. the paths whose transistor at the highest level has its collector terminal connected to the output node $\overline{v_o}$), a given function $\overline{I_O}=F(X_1,\ldots, X_n)$ is simply obtained by identifying the transistor paths associated with minterms and then connecting the collector node of their highest transistor to the node $\overline{v_o}$. The collector terminal of the other transistors at the highest level must be connected to the output node v_o.

Obviously, the same series gate topology is obtained by considering the current I_O rather than $\overline{I_O}$, with the collectors of the highest transistors belonging to paths associated with minterms of I_O being connected to node v_o, and all other collectors being connected to node $\overline{v_o}$. In practical cases, it is convenient to implement the function having the lowest number of minterms (i.e. lower than or equal to 2^{n-1}) between I_O and $\overline{I_O}$.

As an example, let us consider the implementation of the 4-variable function $F(X_1, X_2, X_3, X_4)$ reported in the truth table in Table 3.1. From inspection of this table, the function $\overline{I_O}=F$ has 6 minterms, which is lower than $2^{4-1}=8$, thus it is convenient to implement current $\overline{I_O}$ rather than I_O.

According to Table 3.1, the standard sum-of-product form of function F is

$$F(X_1,X_2,X_3,X_4)=\overline{X_1}\,\overline{X_2}X_3\overline{X_4}+X_1\overline{X_2}\,\overline{X_3}X_4+X_1X_2\overline{X_3}\,\overline{X_4}+ \tag{3.4}$$
$$+X_1X_2\overline{X_3}X_4+X_1X_2X_3\overline{X_4}+X_1X_2X_3X_4$$

Implementation of this function starts from the general 4-variable npn network topology in Fig. 3.5a, where connections of collector terminals have to be specified. In this figure, the paths associated with the 6 minterms are highlighted with a gray line.

According to Table 3.1, or equivalently to relationship (3.4), function F is specified by connecting the collectors of transistors Q19, Q24, Q18, Q26, Q22 and Q30 to the output node $\overline{v_o}$, and the remaining collectors to the output node v_o. The resulting series gate is reported in Fig. 3.5b.

TABLE 3.1

$X_1\, X_2\, X_3\, X_4$	associated minterm	$F(X_1,X_2,X_3,X_4)$	$\overline{F(X_1,X_2,X_3,X_4)}$
0000	$\overline{X_1}\,\overline{X_2}\,\overline{X_3}\,\overline{X_4}$	0	1
0001	$\overline{X_1}\,\overline{X_2}\,\overline{X_3}X_4$	0	1
0010	$\overline{X_1}\,\overline{X_2}X_3\overline{X_4}$	1	0
0011	$\overline{X_1}\,\overline{X_2}X_3X_4$	0	1
0100	$\overline{X_1}X_2\overline{X_3}\,\overline{X_4}$	0	1
0101	$\overline{X_1}X_2\overline{X_3}X_4$	0	1
0110	$\overline{X_1}X_2X_3\overline{X_4}$	0	1
0111	$\overline{X_1}X_2X_3X_4$	0	1
1000	$X_1\overline{X_2}\,\overline{X_3}\,\overline{X_4}$	0	1
1001	$X_1\overline{X_2}\,\overline{X_3}X_4$	1	0
1010	$X_1\overline{X_2}X_3\overline{X_4}$	0	1
1011	$X_1\overline{X_2}X_3X_4$	0	1
1100	$X_1X_2\overline{X_3}\,\overline{X_4}$	1	0
1101	$X_1X_2\overline{X_3}X_4$	1	0
1110	$X_1X_2X_3\overline{X_4}$	1	0
1111	$X_1X_2X_3X_4$	1	0

Inspection of Fig. 3.5b shows that the resulting topology has many useless transistors, and thus can be significantly simplified to reduce the gate area and delay. For example, the emitter coupled pairs with short-circuited collectors Q15-Q16 and Q27-Q28 do not perform any logic operation, since they always steer the emitter current to the same output branch, regardless of their driving input value X_1. Accordingly, the two emitter coupled pairs can be deleted without affecting the gate operation. This also shows that in general a given boolean function can be implemented by different series-gate topologies.

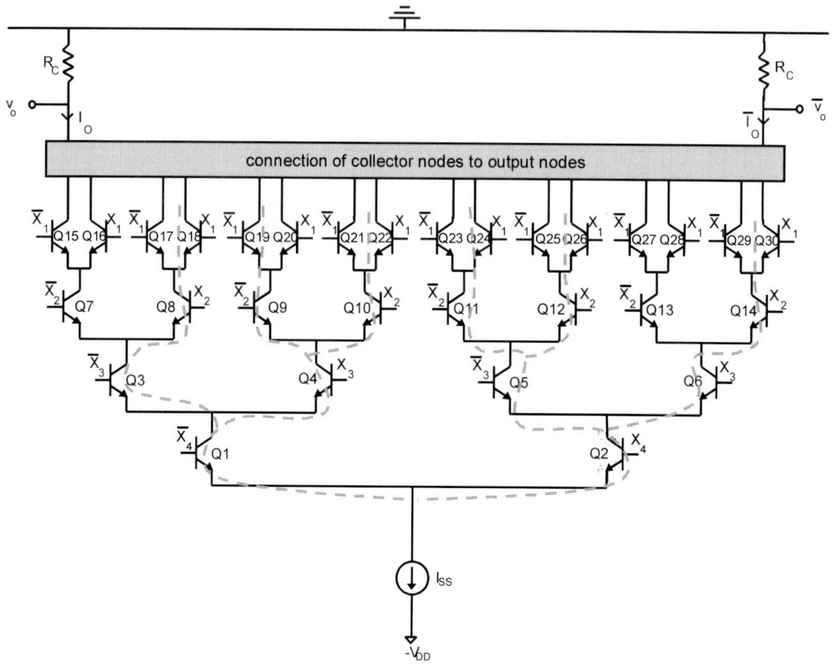

Fig. 3.5a. General topology implementing a 4-variable function.

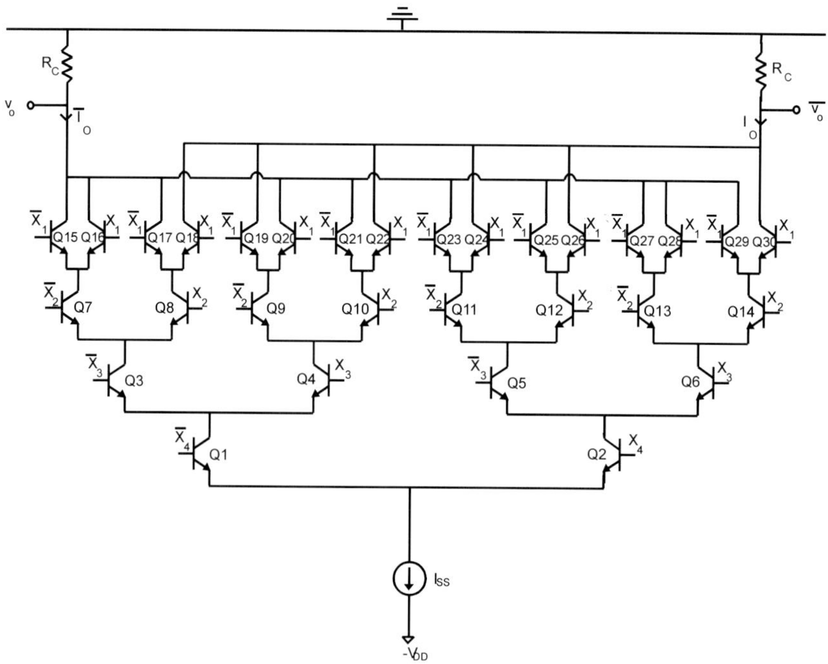

Fig. 3.5b. Series gate topology implementing function *F* in eq. (3.4).

3.1.3 Limitations of the general series-gate design approach

The series-gate design strategy discussed in the previous subsection is general and systematic, but is also very inefficient in terms of transistor count. Indeed, as a result, in an n-level series gate the npn network has an overall number of transistors given by

$$\sum_{j=1}^{n} 2^{j} = 2^{n+1} - 2 \approx 2^{n+1} \qquad (3.5)$$

which exponentially increases as increasing the number of logic levels. This rapid increase in the transistor count is highly undesirable in terms of silicon area and speed performance (due to the higher parasitic capacitances, as will be discussed in Chapter 4-6). Furthermore, as was shown in the example in Section 3.1.2, the design strategy presented leads to an npn network having an unnecessarily high number of transistors. Therefore, to avoid such transistor redundancy, it is essential to use alternative design strategies that are capable of implementing an arbitrary function in a series gate having a minimum transistor count. In the following, two different approaches proposed in the literature will be considered to minimize the series gate transistor count, a graphical procedure (Section 3.2), along with its analytical formulation (Section 3.3) and a VEM-based approach (Section 3.4). Moreover, in an n-level series gate there are many different possible input orderings, and their choice will be dealt with in Section 3.5 for different design goals.

3.2 A GRAPHICAL REDUCTION METHOD

Among the minimization techniques proposed in the literature, such as [CP86], [CJ89] and [MKA92], the graphical approach introduced in [CJ89] is discussed in this section.

3.2.1 Basic concepts on the graphical approach in [CJ89]

The design strategy in [CJ89] is based on the observation that redundant emitter-coupled pairs can often be found in the general series-gate npn network. To understand this point, consider the case where the general npn network topology contains two equal subcircuits N_1 and N_2 (connected to the same output nodes) being driven by two different currents i_1 and i_2, as depicted in Fig. 3.6a. It is apparent that the two subcircuits can be simplified into a single network N_{12} (equal to N_1 and N_2) driven by the current (i_1+i_2), as depicted in Fig. 3.6b.

The minimization method in [CJ89] consists of two formal ways to eliminate redundant emitter-coupled pairs in a general tree npn network topology (see Figs. 3.4-3.5a), and both of them are based on the circuit simplification described in Fig. 3.6b. To be more specific, let us consider an emitter-coupled pair Q1-Q2 driven by input X_j with the two collector terminals connected to two identical subcircuits N_1 and N_2, as depicted in Fig. 3.7a. Such an emitter-coupled pair is defined as a *complementary pair* in [CJ89].

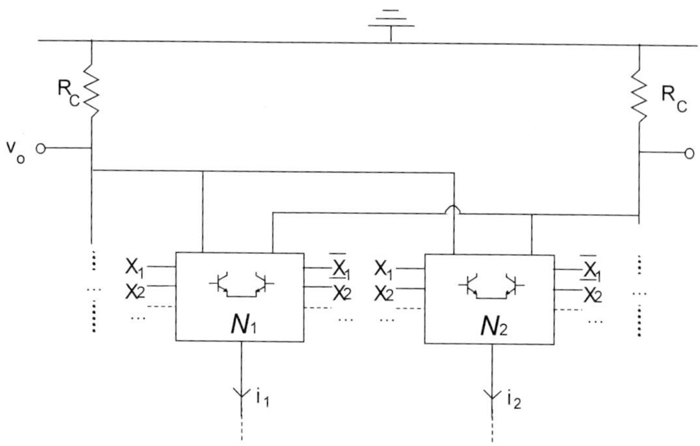

Fig. 3.6a. Equal subcircuits driven by two different currents.

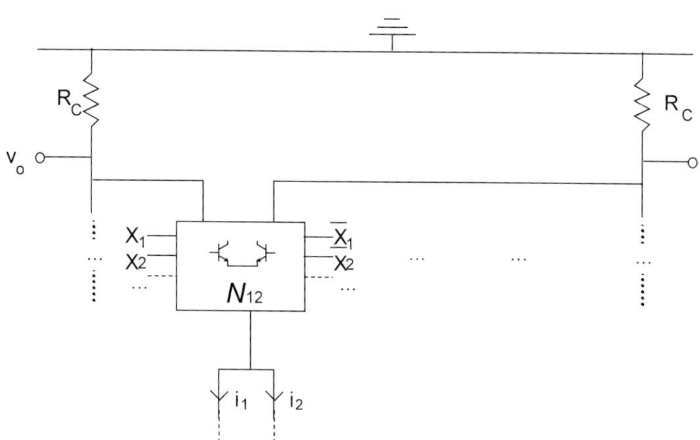

Fig. 3.6b. Simplification of two equal subcircuits into a single one.

By applying the simplification in Fig. 3.6, the two subcircuits can be lumped into the single network N_{12} driven by the emitter-coupled pair Q1-Q2, whose collector terminals are short-circuited, as in Fig. 3.7b. In this case, assuming $\alpha_F \approx 1$, the current $i_{C1}+i_{C2}$ provided to the network N_{12} is equal to the input current i_x of the transistor pair Q1-Q2, regardless of the value of the input X_j. As a consequence, the emitter-coupled pair Q1-Q2 can be eliminated, as shown in Fig. 3.7b. This simplification step is called CPE (Complementary Pair Elimination) and is performed for all transistor pairs belonging to the same level at a time, usually starting from the top level and ending at the lowest level. However, the same simplifications are obtained by starting from the latter one and ending at the top level.

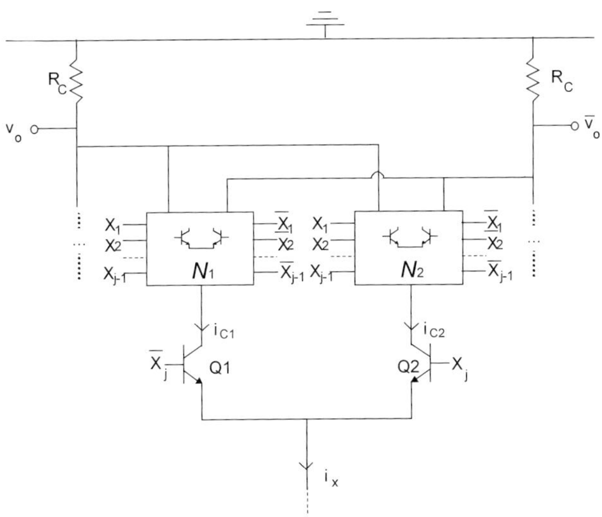

Fig. 3.7a. Emitter-coupled pair driving two identical subcircuits.

The second elimination step is the NPE (Normal Pair Elimination), and is applied to *normal pairs*, which are defined as emitter-coupled pairs lying in the same series-gate level. To understand this simplification, let us consider the two normal pairs Q1-Q2 and Q3-Q4 in Fig. 3.8a driving equal subcircuits N_1 and N_2, and being driven by the two different currents i_x and i_y, respectively. According to Fig. 3.6, the two normal pairs can be lumped into a single emitter-coupled pair driven by the sum of the currents i_x+i_y, as depicted in Fig. 3.8b. This elimination of normal pairs (NPE) is usually performed for all transistor pairs belonging to the same level at a time, starting from the top level and ending at the lowest level. However, the same simplifications are obtained by starting from the latter one and ending at the top level. Obviously, the NPE can be applied to all levels with the exception of the bottom level, since it has not subcircuits lying at a lower level.

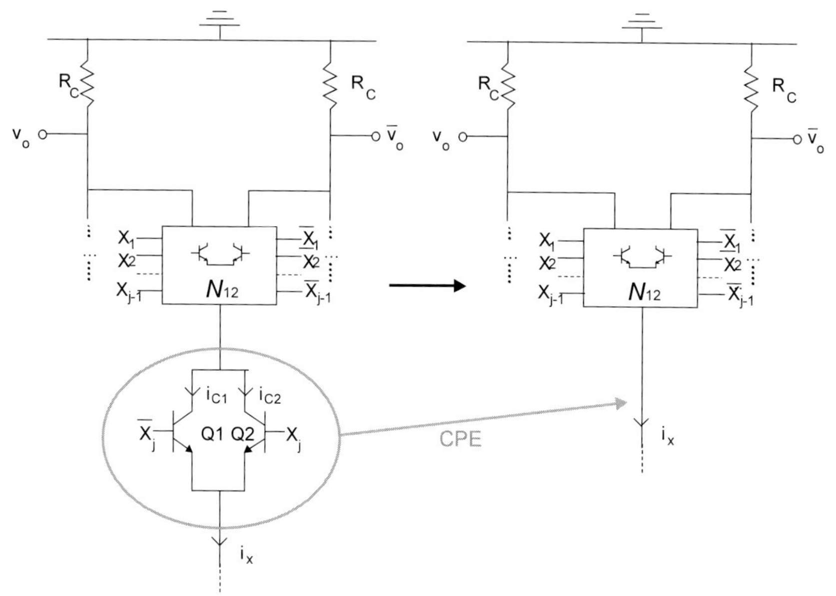

Fig. 3.7b. Complementary Pair Elimination (CPE) of Q1-Q2.

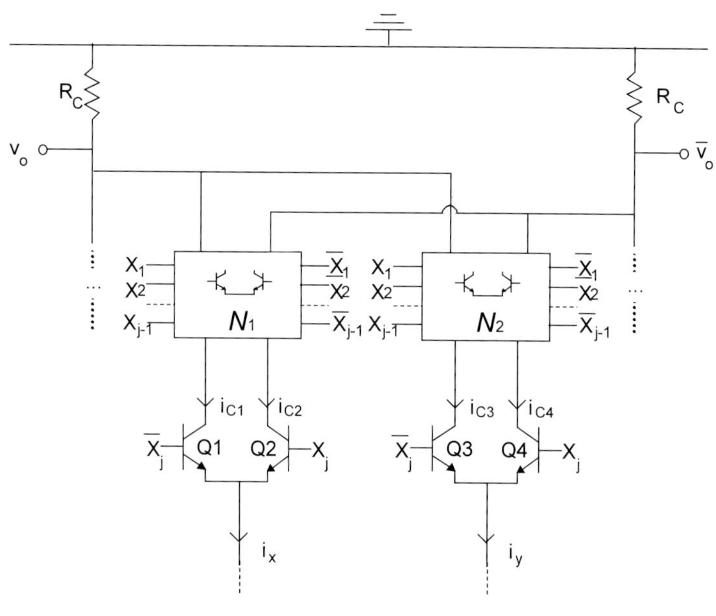

Fig. 3.8a. Normal pairs driving the same subcircuit.

This minimization strategy was developed by applying the CPE to each logic level starting from the top and ending at the lowest level, and then applying the NPE to the same levels. It is worth noting that the two simplifications may be applied in a different order without changing the topology of the resulting circuit. To better understand the procedure, a practical example is presented in the following subsection.

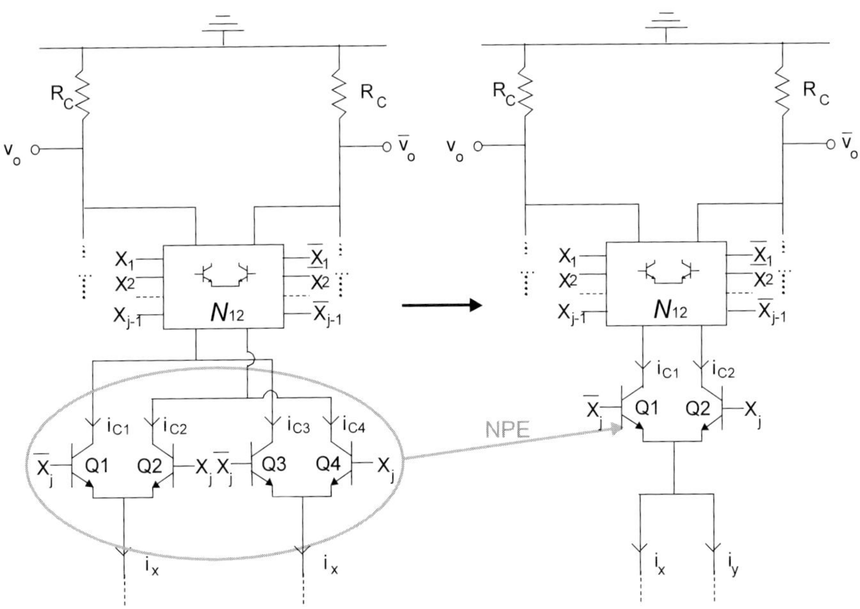

Fig. 3.8b. Normal Pair Elimination of Q3-Q4.

3.2.2 A design example

Let us consider the series-gate implementation of the function F expressed by eq. (3.4).

As a preliminary step, the general 4-variable npn network topology in Fig. 3.5a has to be built. Subsequently, the CPE must be applied to each transistor level. In the following, the CPE (NPE) applied to the j-th level will be referred to as CPEj (NPEj). When the CPE1 is applied, the two emitter-coupled pairs Q15-Q16 and Q27-Q28 with short-circuited collectors are substituted by a short circuit, according to Fig. 3.7b. The resulting circuit is depicted in Fig. 3.9a, where the CPE2 can be applied to eliminate the complementary pairs Q11-Q12 and Q25-Q26.

The simplified circuit after applying the CPE2 is depicted in Fig. 3.9b, where it can be noticed that no further simplification can be achieved by applying CPE3 and CPE4, since there are no other complementary pairs driving equal subcircuits.

Now, the NPE1 has to be applied to Fig. 3.9b, where normal pairs Q17-Q18, Q21-Q22, Q23-Q24 and Q29-Q30 drive the same subcircuit (in the specific case of the first level, it consists of the connections to the output nodes) and thus can be lumped into Q17-Q18, thereby connecting the collector terminals of the driving transistors Q8, Q10, Q5, Q14 to the emitter of Q17-Q18. The resulting simplified circuit is reported in Fig. 3.9c.

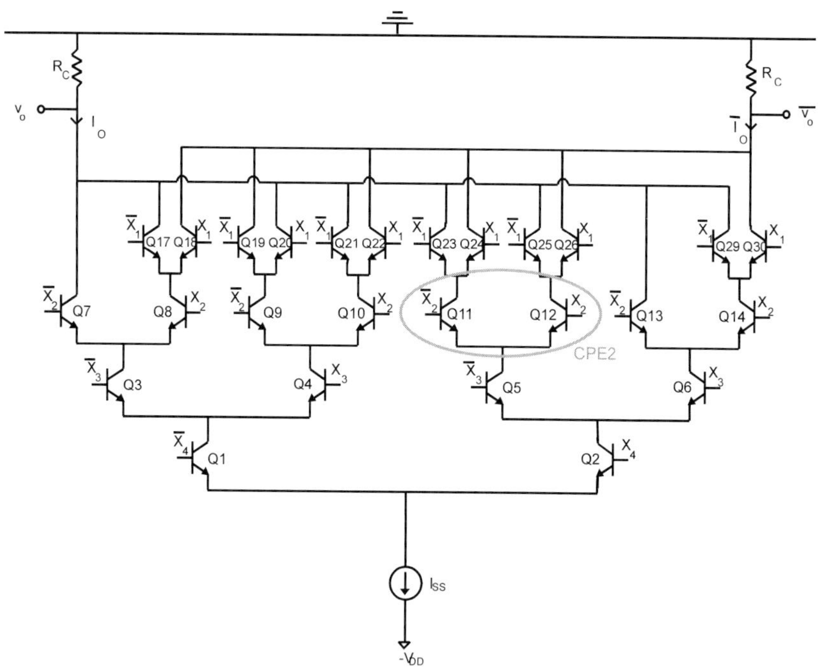

Fig. 3.9a. Series gate topology after CPE1.

The circuit in Fig. 3.9c can be further simplified by applying the NPE2, since normal pairs Q7-Q8 and Q13-Q14 drive equal subcircuits (the collector of the left-hand transistor directly drives the output node v_o, and the right-hand transistor drives the transistor pair Q17-Q18). Therefore, the two emitter-coupled pairs can be lumped into the transistor pair Q13-Q14, after connecting the collector of Q6 to that of Q3, as depicted in Fig. 3.9d. Since no normal pairs driving the same subcircuit exist, no further simplification is obtained by applying the NPE3, thus the circuit in Fig. 3.9d is the minimized series gate implementing the assigned function *F*, which is depicted in a more compact manner in Fig. 3.9e.

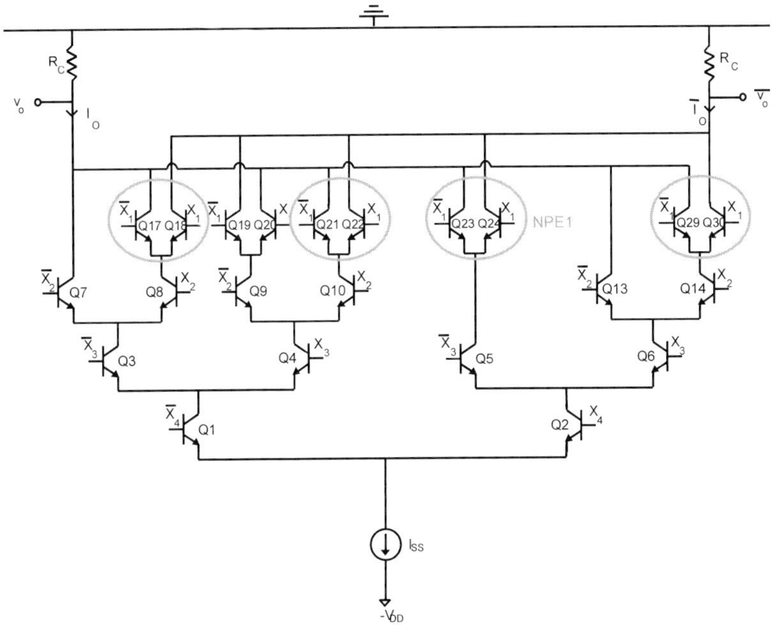

Fig. 3.9b. Series gate topology after CPE2.

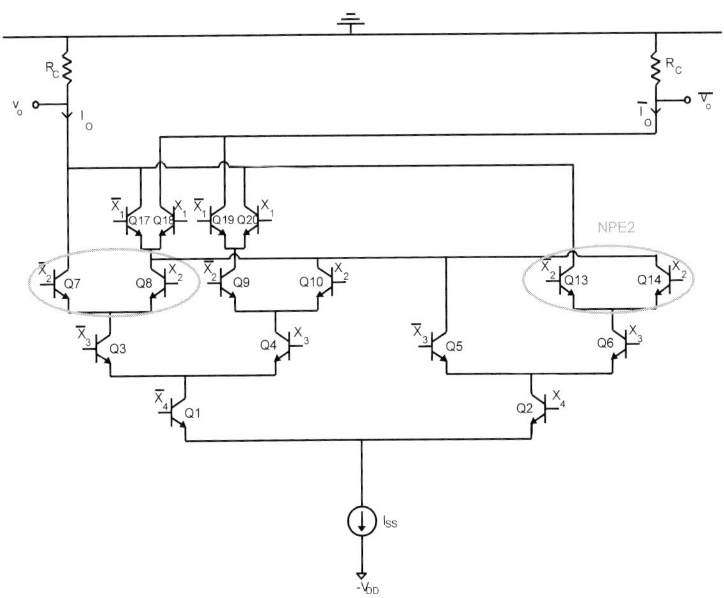

Fig. 3.9c. Series gate topology after NPE1.

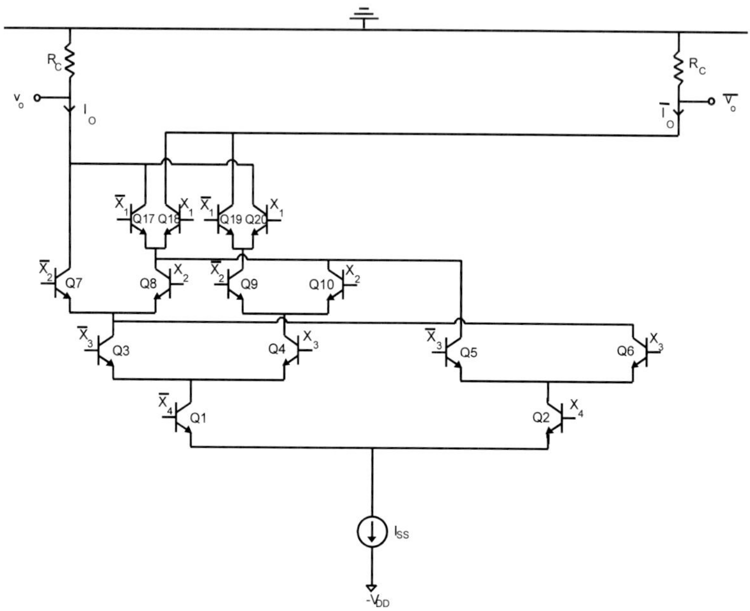

Fig. 3.9d. Series gate topology after NPE2.

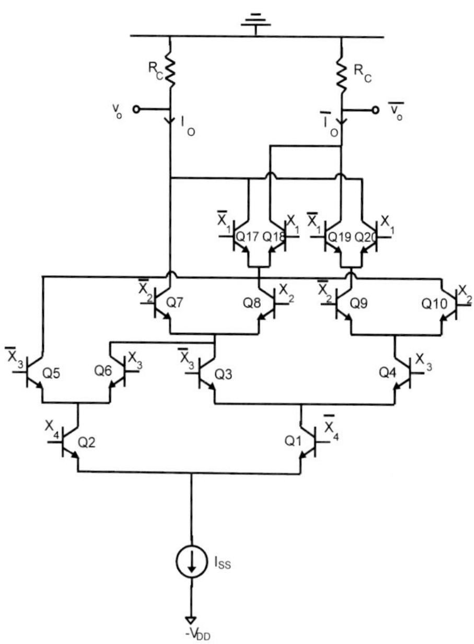

Fig. 3.9e. Final simplified circuit implementing function *F*.

Comparison of Figs. 3.5-3.9e shows that the transistor count has been reduced from 30 to 14 through the procedure described in Section 3.2.1, which is a significant advantage in terms of area, and is also beneficial in terms of delay.

3.3 AN ANALYTICAL FORMULATION OF THE DESIGN STRATEGY IN [CJ89]

The minimization strategy discussed in the previous subsection is based on a graphical approach leading to the elimination of redundant emitter-coupled pairs through the two basic steps CPE and NPE. However, analytical minimization procedures are usually preferred since they can be applied with less effort, and also afford a better understanding of the concepts on which the strategy is based. In the following, an interpretation of CPE and NPE simplification steps is discussed to derive an analytical formulation of the strategy presented in Section 3.2.

3.3.1 Analytical interpretation of CPE/NPE

An analytical interpretation of the CPE can be derived by exploiting the correspondence of each path from the output node $\overline{v_o}$ to the current source with a unique minterm. This minterm has the same literals of the inputs driving the transistors belonging to the path, as discussed in Section 3.1.2. Furthermore, it is useful to observe that the npn network implements both boolean functions F and $\overline{F}$, hence subcircuit N_1 in Fig. 3.7a implements some minterms of both functions. To be more specific, subcircuit N_1 implements all minterms that can be expressed as $g(X_1 \ldots X_{j-1}) \cdot \overline{X_j} \cdot i_x$, and all minterms of $\overline{F}$ which can be expressed as $\overline{g(X_1 \ldots X_{j-1})} \cdot \overline{X_j} \cdot i_x$ (being g a function of $X_1 \ldots X_{j-1}$, and i_x the product of the literals $x_{j+1} \ldots x_n$ associated with the path from the emitter of Q1-Q2 to the current source). Moreover, from Fig. 3.7a, N_2 implements all minterms of F and $\overline{F}$ which are expressed by $g(X_1 \ldots X_{j-1}) \cdot X_j \cdot i_x$ and $\overline{g(X_1 \ldots X_{j-1})} \cdot X_j \cdot i_x$, respectively. Therefore, lumping subcircuits N_1 and N_2 into N_{12} is analytically equivalent to collecting all pairs of minterms of F (as well as $\overline{F}$) containing the product $\overline{X_j} \cdot x_{j+1} \cdot \ldots \cdot x_n$ and $X_j \cdot x_{j+1} \cdot \ldots \cdot x_n$, respectively, into a single term

$$g(X_1 \ldots X_{j-1}) \cdot \left(\overline{X_j} + X_j\right) \cdot i_x = g(X_1 \ldots X_{j-1}) \cdot i_x \tag{3.6}$$

where simplification $\overline{X_j} + X_j = 1$ justifies the elimination of transistors Q1-Q2 in Fig. (3.7b). Obviously, the CPEj leads to a simplification only if N_1 and N_2 are equal, i.e. if for each term of F (or $\overline{F}$) including product $\overline{X_j} \cdot x_{j+1} \cdot \ldots \cdot x_n$ there is a corresponding one containing $X_j \cdot x_{j+1} \cdot \ldots \cdot x_n$. Instead, when at least a product without the corresponding one exists, the CPEj does not lead to any simplification in the circuit under design.

In regard to NPE, its analytical interpretation is easily found by inspecting Figs. 3.8aa-3.8b, in which transistor pairs Q1-Q2 and Q3-Q4 are lumped in a single transistor pair Q1-Q2, and equal subcircuits N_1-N_2 are lumped into a single subcircuit N_{12}. Subcircuit N_1 in Fig. 3.8a implements all minterms of F that can be expressed as $g(X_1 \ldots X_j) \cdot i_x$, as well as all minterms of $\overline{F}$ which can be expressed as $\overline{g(X_1 \ldots X_j)} \cdot i_x$. Analogously, subcircuit N_2 implements all minterms of F and $\overline{F}$ that can be expressed as $g(X_1 \ldots X_j) \cdot i_y$ and $\overline{g(X_1 \ldots X_j)} \cdot i_y$, respectively. Therefore, the NPEj simplification, which allows for lumping transistors Q1-Q2 and Q3-Q4 into Q1-Q2 and equal subcircuits N_1-N_2 into N_{12}, is analytically equivalent to collecting all products having equal literals $x_1, \ldots, x_j$ and different $x_{j+1}, \ldots, x_n$

$$g(X_1 \ldots X_j) \cdot i_x + g(X_1 \ldots X_j) \cdot i_y = g(X_1 \ldots X_j) \cdot (i_x + i_y) \tag{3.7}$$

As observed for the CPE, the NPEj leads to such a simplification only if N_1 and N_2 are equal, i.e. if for each term of F (or $\overline{F}$) in the form $g(X_1 \ldots X_j) \cdot i_x$ (being i_x a product of $x_{j+1} \ldots x_n$) there is a corresponding one equal to $g(X_1 \ldots X_j) \cdot i_y$ (being i_y a different product of $x_{j+1} \cdot \ldots \cdot x_n$). Equivalently, the simplification of two minterms $g(X_1 \ldots X_j) \cdot i_x$ and $g(X_1 \ldots X_j) \cdot i_y$ is possible only if there is not any other minterm having the same i_x (or, equivalently, i_y) but different literals $x_1 \ldots x_j$.

3.3.2 Analytical simplification through CPE/NPE: an example

In this subsection, function F in (3.4) is simplified through CPE and NPE in an analytical manner, according to their interpretation discussed in the previous subsection. For the sake of compactness, the AND operator "·" will be omitted.

When applying CPE1 to function F in (3.4), no simplification is obtained, since there are no minterms differing for only literal x_1. From a topological point of view, this means that it is not possible to simplify any transistor pair connected to $\overline{v_o}$, as confirmed in Figs. 3.5b and 3.9a. Instead, CPE2 allows for simplifying terms $X_1\overline{X_2}\,\overline{X_3}X_4$ and $X_1X_2\overline{X_3}X_4$, since they differ for only literal x_2 and there are no other terms including product $\overline{X_3}X_4$. According to relationship (3.6), after applying CPE2, function F becomes

$$F(X_1,X_2,X_3,X_4)=X_1\overline{X_3}X_4+\overline{X_1}\,\overline{X_2}X_3\overline{X_4}+X_1X_2\overline{X_3}\,\overline{X_4}$$
$$+X_1X_2X_3\overline{X_4}+X_1X_2X_3X_4 \tag{3.8}$$

This is equivalent to lumping transistor pairs Q23-Q24 and Q25-Q26 into Q23-Q24 and subsequently eliminating the complementary pair Q11-Q12, as in Fig. (3.9b). In regard to CPE3, it does not lead to any simplification into function F, since in (3.8) there are no pairs of products differing for only literal x_3. Even though there are two terms differing for x_4, i.e. $X_1X_2X_3\overline{X_4}$ and $X_1X_2X_3X_4$, CPE4 does not simplify (3.8), as there are other terms in the form $x_1x_2x_3X_4$ which do not have a corresponding term $x_1x_2x_3\overline{X_4}$. The resulting topology after CPE is that in Fig. 3.9b.

After simplifying F through CPEs, let us apply the NPE simplification. In particular, NPE1 applied to function F in (3.8) leads to

$$F(X_1,X_2,X_3,X_4)=X_1\left(\overline{X_3}X_4+X_2\overline{X_3}\,\overline{X_4}+X_2X_3\overline{X_4}\right.$$
$$\left.+X_2X_3X_4\right)+\overline{X_1}\left(\overline{X_2}X_3\overline{X_4}\right) \tag{3.9}$$

which is equivalent to lumping Q17-Q18, Q21-Q22, Q23-Q24 and Q29-Q30 into the transistor pair Q17-Q18 driven by transistors Q8, Q10, Q5 and Q14, as in Fig. 3.9c. Analogously, NPE2 allows for collecting terms having equal literals x_1-x_2 and different other literals. In particular, let us consider function F after NPE1 in (3.9), in which three terms including the same factor X_2 could potentially be collected, i.e. $X_1X_2\overline{X_3}\,\overline{X_4}$, $X_1X_2X_3X_4$ and $X_1X_2X_3\overline{X_4}$, even though only the first two can be collected to simplify the circuit. With referral to the previously introduced notation, this can be understood by observing that the three terms have the common factor $g(X_1,X_2)=X_1X_2$ and differ for other literals through products $i_x=\overline{X_3}\,\overline{X_4}$, $i_y=X_3X_4$ and $i_z=X_3\overline{X_4}$, respectively. The first two terms can be collected according to (3.7), while $X_1X_2X_3\overline{X_4}$ cannot be simplified since there is the other minterm $\overline{X_1}\,\overline{X_2}X_3\overline{X_4}$ having the same $i_z = X_3\overline{X_4}$ but different literals x_1x_2. Accordingly, after applying NPE2, (3.9) simplifies into

$$F(X_1,X_2,X_3,X_4) = X_1\left[X_2\left(\overline{X_3}\,\overline{X_4} + X_3X_4\right) + X_2X_3\overline{X_4} + \overline{X_3}X_4\right] + \overline{X_1}\left[\overline{X_2}\left(X_3\overline{X_4}\right)\right] \quad (3.10)$$

which is topologically equivalent to lumping transistor pairs Q7-Q8 and Q13-Q14 into the emitter-coupled pair Q7-Q8 driven by transistors Q3 and Q6, as reported in Fig. 3.9d. Since in (3.10) there are no other terms in brackets which include the same factor X_3 ($\overline{X_3}$), NPE3 does not lead to any further simplification.

Even though function F was manipulated through CPE-NPE in this example, it is apparent that the same topological simplifications are obtained by simplifying function $\overline{F}$ rather than F. Therefore, of the two functions, in practical cases it is convenient to manipulate that having the lowest number of minterms. For the sake of clarity, simplification of function $\overline{F}$ is also discussed in the following, by starting from its standard sum-of-product form obtained from Table 3.1

$$\begin{aligned}\overline{F}(X_1,X_2,X_3,X_4) = {} & \overline{X_1}\,\overline{X_2}\,\overline{X_3}\,\overline{X_4} + \overline{X_1}\,\overline{X_2}\,\overline{X_3}X_4 \\ & + \overline{X_1}\,\overline{X_2}X_3X_4 + \overline{X_1}X_2\overline{X_3}\,\overline{X_4} \\ & + \overline{X_1}X_2\overline{X_3}X_4 + \overline{X_1}X_2X_3\overline{X_4} \\ & + \overline{X_1}X_2X_3X_4 + X_1\overline{X_2}\,\overline{X_3}\,\overline{X_4} \\ & + X_1\overline{X_2}X_3\overline{X_4} + X_1\overline{X_2}X_3X_4\end{aligned} \quad (3.11)$$

From inspection of this expression, CPE1 simplifies minterms $X_1\overline{X_2}\,\overline{X_3}\,\overline{X_4}$ and $\overline{X_1}\,\overline{X_2}\,\overline{X_3}\,\overline{X_4}$ (since they only differ for literal x_1 and there are no other terms[2] including product $\overline{X}_2\overline{X}_3\overline{X}_4$), as well as $\overline{X}_1\overline{X}_2X_3X_4$ and $X_1\overline{X}_2X_3X_4$ (they also differ only for literal x_1 and there are no other terms including product $\overline{X}_2X_3X_4$)

$$
\begin{aligned}
\overline{F}(X_1,X_2,X_3,X_4)=\;&\overline{X}_2\overline{X}_3\overline{X}_4+\overline{X}_1X_2\overline{X}_3\overline{X}_4+\overline{X}_1\overline{X}_2\overline{X}_3X_4\\
&+\overline{X}_2X_3X_4+\overline{X}_1X_2\overline{X}_3X_4+\overline{X}_1X_2X_3\overline{X}_4 \qquad (3.12)\\
&+\overline{X}_1X_2X_3X_4+X_1\overline{X}_2X_3\overline{X}_4
\end{aligned}
$$

This is equivalent to eliminating transistor pairs Q15-Q16 and Q27-Q28, whose collectors are all connected to the output node v_o. When applying CPE2, terms $\overline{X}_1\overline{X}_2\overline{X}_3X_4$ and $\overline{X}_1X_2\overline{X}_3X_4$ differ for only literal x_2 and there are no other terms containing product $\overline{X}_3X_4$, thus they can be simplified according to (3.6) into the single term $\overline{X}_1\overline{X}_3X_4$

$$
\begin{aligned}
\overline{F}(X_1,X_2,X_3,X_4)=\;&\overline{X}_2\overline{X}_3\overline{X}_4+\overline{X}_2X_3X_4+\overline{X}_1\overline{X}_3X_4+\\
&+\overline{X}_1X_2\overline{X}_3\overline{X}_4+\overline{X}_1X_2X_3\overline{X}_4+ \qquad (3.13)\\
&+\overline{X}_1X_2X_3X_4+X_1\overline{X}_2X_3\overline{X}_4
\end{aligned}
$$

which is topologically equivalent to lumping transistor pairs Q23-Q24 and Q25-Q26 into Q23-Q24 and subsequently eliminating the complementary pair Q11-Q12, as in Fig. (3.9b). When CPE3 is applied to function $\overline{F}$, even though terms $\overline{X}_1X_2\overline{X}_3\overline{X}_4$ and $\overline{X}_1X_2X_3\overline{X}_4$ in (3.13) differ for only x_3, there is another term including $\overline{X}_4$ (i.e. $X_1\overline{X}_2X_3\overline{X}_4$) which does not have the correspondent term differing for only x_3 (i.e. $X_1\overline{X}_2\overline{X}_3\overline{X}_4$). Therefore CPE3 does not simplify such terms, and the same argument holds for CPE4 (indeed, even though terms $\overline{X}_1X_2X_3\overline{X}_4$ and $\overline{X}_1X_2X_3X_4$ differ for only literal x_4, there are several terms including X_4 which do not have a

[2] Actually, it is easy to realize that this condition is always satisfied when applying CPE to the top level.

corresponding term differing only for x_4). Detailed application of NPE is omitted and left to the reader.

The final step of the simplification procedure based on analytical CPE-NPE is the series-gate implementation of the simplified function, which is addressed in the following subsection.

3.3.3 Circuit implementation of the simplified function after CPE-NPE

After applying the analytical CPE-NPE, function F must be mapped into a series-gate topology by simply exploiting the correspondence of each minterm $x_1x_2\ldots x_n$ with the path connecting the output node $\overline{v_o}$ to the current source having transistors driven by the same literals. This is easily obtained by resorting to considerations in Section 3.1 and starting the design from the first level.

To better understand this step, the simplified function (3.10) of the example presented in Section 3.3.2 is implemented in the following. From (3.10), there are two terms proportional to literal x_1, thus the first level includes two transistor pairs Q1-Q2 and Q3-Q4, according to Fig. 3.10. It is worth noting that the term containing factor X_1 ($\overline{X_1}$) is implemented by connecting transistor Q2 (Q4) driven by the same signal to output $\overline{v_o}$ according to (3.1).

Now, let us consider the second level by implementing the factors multiplying X_1 and $\overline{X_1}$ in the first term. In particular, factor $\overline{X_2}\left(X_3\overline{X_4}\right)$ multiplying $\overline{X_1}$ contains one term including literal $\overline{X_2}$, which is implemented by the transistor pair Q5-Q6, where the collector of transistor Q5 driven by $\overline{X_2}$ is connected to the transistor pair Q3-Q4 driven by $\overline{X_1}$. Analogously, factor $X_2\left(\overline{X_3X_4}+X_3X_4\right)+X_2X_3\overline{X_4}+\overline{X_3}X_4$ multiplying X_1 contains three terms, which are implemented by the corresponding transistor pairs Q7-Q8, Q5-Q6 and Q9-Q10, as depicted in Fig. 3.10. It is worth noting that the third term does not contain literal x_2, thus it is implemented by the transistor pair Q9-Q10 lying at the third level (i.e., by skipping the second level).

In regard to the third level, the term $\overline{X_3X_4}+X_3X_4$ multiplying X_2 is implemented by the transistor pair Q11-Q12 and the existing Q9-Q10, with the collector of Q11 and Q10 being connected to the emitter of Q7-Q8. The fourth level consists of only the transistor pair Q13-Q14 since it is the lowest one. Once the insertion of transistor pairs has been completed, all floating collector nodes must be connected to the other output node, v_o, thereby leading to the final series gate topology in Fig. 3.10. As expected, this circuit is equal to that in Fig. 3.9e obtained by applying the graphical design strategy discussed in Section 3.2. It is apparent that an identical topology would have been obtained by implementing function $\overline{F}$, rather than F.

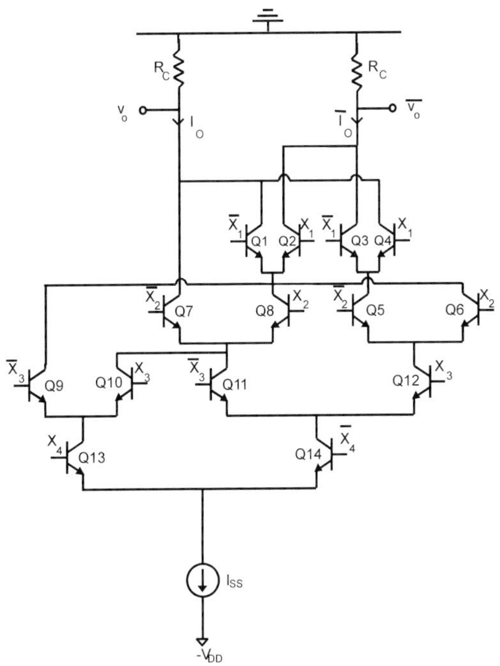

Fig. 3.10. Final series gate topology implementing simplified function F in (3.10).

3.4 A VEM-BASED REDUCTION METHOD

In this section, the tabular approach based on Variable-Entered Mapping (VEM) proposed in [MKA92] is presented. The Variable-Entered Mapping is a well-known technique to minimize multiplexer implementations of boolean functions, and can be exploited to minimize the number of transistors in a series gate. Indeed, the npn network of a series gate can be thought of as the connection of transistor pairs, each of which implements a 2:1 current multiplexer, as the current applied to their emitter is steered to one side according to the input value.

Let us consider an n-variable logic function $F(X_1 \ldots X_n)$ of input signals $X_1 \ldots X_n$, with X_1 being applied to transistors belonging to the first level, X_2

being applied to the second level, and so on. Function F can be minimized through the VEM technique [MKA92], [C95], which starts from its truth table with X_n placed into the leftmost column followed to the right by variables $X_{n-1}...X_1$. The simplification technique iteratively eliminates variable X_j with j=1...n by grouping contiguous rows in pairs and lumping each of them in a single row, thereby expressing in the output expression its dependence on X_j. Accordingly, at the j-th step the number of input variables of the truth table is reduced by 2^j through suppression of $X_1...X_j$. After n steps, the VEM technique leads to a single-row truth table, i.e. the expression of the minimized function F (and $\overline{F}$).

As an example, let us minimize function F in (3.4) by starting from its truth table in Table 3.1, which has to be rewritten in order to place variable X_n into the leftmost column and X_1 to the rightmost one, as reported in Table 3.2a.

TABLE 3.2a

$X_4\,X_3\,X_2\,X_1$	$F(X_1,X_2,X_3,X_4)$	$\overline{F(X_1,X_2,X_3,X_4)}$
0000	0	1
0001	0	1
0010	0	1
0011	1	0
0100	1	0
0101	0	1
0110	0	1
0111	1	0
1000	0	1
1001	1	0
1010	0	1
1011	1	0
1100	0	1
1101	0	1
1110	0	1
1111	1	0

The elimination of X_1 is achieved by grouping pairs of contiguous entries differing for only the value of X_1 into a single one. For example, the first two rows in Table 3.2a refer to input values 0000 and 0001, which differ for the value of X_1, which is equal to 0 and 1, respectively. Since output is equal to 0 in both cases, the two rows can be simplified into the first row of the reduced truth table (i.e. with only input variables X_4, X_3, X_2) in Table 3.2b, where output is equal to 0. The third and fourth row in Table 3.2a, which refer to input values 0010 and 0011 again differing for only X_1, can be simplified into the second row in Table 3.2b, where output has the same value of X_1 for both input values. By following the same reasoning, reduction of the input variable X_1 leads to the truth table in Table 3.2b, which also reports the simplification of function $\overline{F}$. Although in this case its results are exactly the complement of those obtained for function F, this is not true in general for the following steps, and hence the two functions will be minimized separately.

TABLE 3.2b

$X_4\,X_3\,X_2$	$F(X_2,X_3,X_4)$	$\overline{F(X_2,X_3,X_4)}$
000	0	1
001	X_1	$\overline{X_1}$
010	$\overline{X_1}$	X_1
011	X_1	$\overline{X_1}$
100	X_1	$\overline{X_1}$
101	X_1	$\overline{X_1}$
110	0	1
111	X_1	$\overline{X_1}$

Again, Table 3.2b can be reduced by eliminating input variable X_2 by grouping consecutive rows differing for only the value of X_2 into pairs and simplifying each of them into a single row. For example, the first two rows refer to input values 000 and 001, which differ for the value of X_2. In the two cases, output is equal to 1 only if both X_1 and X_2 are equal to 1, thus the two rows can be simplified into the first one in Table 3.2c, which also reports the complete truth table after elimination of X_2, as well as that of function $\overline{F}$, that is not the complement of function F[3].

[3] As observed before, these results are not exactly the complement of those obtained for function F. To be more specific, analytical results of $\overline{F}$ would be exactly the complement of those found for function F if property $A+\overline{A}B = A+B$ would be used. However, this simplification has not to be introduced in VEMs, since it no longer allows for minimizing the MUX-based (and hence series-gate) implementation.

The truth table of F and $\overline{F}$ obtained after eliminating variable X_3 from Table 3.2c is reported in Table 3.2d. Then, the expression of function F and $\overline{F}$ is derived from simple inspection of Table 3.2d

$$F = \left[X_1 X_2 \overline{X_3} + \left(\overline{X_1}\,\overline{X_2} + X_1 X_2\right) X_3\right] \overline{X_4} + \left[X_1 \overline{X_3} + X_1 X_2 X_3\right] X_4 \tag{3.14a}$$

$$\begin{aligned} \overline{F} = & \left[\left(\overline{X_1} X_2 + \overline{X_2}\right) \overline{X_3} + \left(X_1 \overline{X_2} + \overline{X_1} X_2\right) X_3\right] \overline{X_4} \\ & + \left[\overline{X_1}\,\overline{X_3} + \left(\overline{X_1} X_2 + \overline{X_2}\right) X_3\right] X_4 \end{aligned} \tag{3.14b}$$

TABLE 3.2c

$X_4 X_3$	$F(X_3, X_4)$	$\overline{F(X_3, X_4)}$
00	$X_1 X_2$	$\overline{X_1} X_2 + \overline{X_2}$
01	$\overline{X_1}\,\overline{X_2} + X_1 X_2$	$X_1 \overline{X_2} + \overline{X_1} X_2$
10	X_1	$\overline{X_1}$
11	$X_1 X_2$	$\overline{X_1} X_2 + \overline{X_2}$

TABLE 3.2d

X_4	$F(X_4)$	$\overline{F(X_4)}$
0	$X_1 X_2 \overline{X_3} + \left(\overline{X_1}\,\overline{X_2} + X_1 X_2\right) X_3$	$\left(\overline{X_1} X_2 + \overline{X_2}\right) \overline{X_3} + \left(X_1 \overline{X_2} + \overline{X_1} X_2\right) X_3$
1	$X_1 \overline{X_3} + X_1 X_2 X_3$	$\overline{X_1}\,\overline{X_3} + \left(\overline{X_1} X_2 + \overline{X_2}\right) X_3$

The series gate topology implementing function in (3.14) can be identified by following the procedure discussed in Section 3.3.3. It is worth noting that function (3.14) is written by first collecting terms including literal x_4, then x_3, etc., while in Section 3.3.3, function F (or $\overline{F}$) was written by first collecting terms including literal x_1, then x_2 and so on. Therefore, function (3.14) must be first expanded in a sum-of-product form and then rewritten by first collecting terms containing x_1, then x_2 and so on. In this example, manipulation of function F in (3.14a) leads to the same expression given in (3.10), thus the series gate topology obtained with the VEM minimization is equal to that obtained by applying CPE/NPE in Figs. 3.9e-3.10. It can also be shown that the same topology is obtained by minimizing

$\overline{F}$, rather than F. Actually, the equivalence of topologies obtained by resorting to CPE/NPE and VEM is general and is justified in the following.

The equivalence of the VEM-based minimization technique and that presented in Sections 3.3-3.4 can be understood by noting that, as was shown in the previous example, the simplified function F ($\overline{F}$) after VEM minimization is in the form

$$F = F(X_1, X_2, X_3, 0)\overline{X_4} + F(X_1, X_2, X_3, 1)X_4 \tag{3.15a}$$

where $F(X_1,X_2,X_3,0)$ (as well as $F(X_1,X_2,X_3,1)$) can be written again in the form

$$F(X_1, X_2, X_3, 0) = F(X_1, X_2, 0, 0)\overline{X_3} + F(X_1, X_2, 1, 0)X_3 \tag{3.15b}$$

Generalizing the result, all terms in each bracket are collected into two terms including factor X_j and $\overline{X_j}$, which is the same operation in (3.7) performed by NPEj. To be more specific, it is easy to verify that when applying NPE starting from the bottom level and ending at the top level, NPEj leads to the same analytical simplifications as the j-th step of the VEM-based minimization. Furthermore, CPEj is intrinsically performed at the j-th step of the VEM minimization, since terms differing for only literal x_j are lumped into a single term, which is equivalent[4] to perform the simplification in (3.6).

3.5 INPUT ORDERING VERSUS DESIGN GOAL

In the previous sections, minimization methods to reduce the transistor count in a series gate have been discussed by adopting a given input ordering, i.e. with input X_1 driving transistors belonging to the top level, X_2 driving those of the immediately lower level and so on. However, in general, in an n-variable function $F(X_1 \ldots X_n)$ there are $n!$ possible input orderings, which are reduced to $n!/k!$ if this function is symmetric in k variables [MKA92]. In the following, criteria to identify the most convenient input ordering for a design aiming at a high speed, minimum transistor count, low switching noise or reduced number of series-gate levels are discussed.

Let us consider the case of a high-speed design, such as that of a specific gate lying in the critical path. In this case, it is useful to observe that in

[4] For example, the minterms associated with the first two input values in Table 3.2a, which give F=0 regardless of the value of X_1, are simplified into the single minterm associated with the first row in Table 3.2b.

general the gate delay depends on the logic level which the switching input is applied to. Moreover, the gate input signals do not arrive at the same time, since they are generated by previous gates lying in different paths. To improve the speed performance, the gate output signal OUT must switch as soon as possible, therefore the latest arriving input signal X_{latest} must be applied to the gate input that exhibits minimum delay. In the specific case of Current-Mode circuits, as will be shown in Chapters 4 and 6, the delay is minimum for inputs applied to the top level of the series gate topology. Accordingly, the inputs $X_1 \ldots X_n$ must be reordered to connect the latest arriving signal to the top level and the progressively earlier signals to lower levels.

When minimizing the transistor count is the main target, it is useful to observe that it depends on the input ordering [MKA92]. Therefore, the most convenient input ordering is found by exhaustively minimizing function $F(X_1 \ldots X_n)$ for all possible different orderings, and then selecting that leading to the lowest number of transistors. In such cases, the comparison can be easily automated to reduce the design effort.

Another design option, namely the minimization of the switching noise, is an important target in practical applications involving high-resolution CMOS ICs, as discussed in Section 2.6.3. In these cases, the supply current can be kept almost constant by using Current-Mode gates, whose supply current is approximately their bias current I_{SS}. To understand the fundamental criteria to select the best input ordering which meets this goal, let us analyze how the switching noise is produced in CMOS Current-Mode circuits by considering the SCL inverter in Fig. 3.11, where capacitance C_{VDD1} (C_{VDD2}) is the sum of capacitances between the output node v_{o1} (v_{o2}) and the supply, whereas C_{GND1} (C_{GND2}) is the sum of capacitances between the output node v_{o1} (v_{o2}) and ground. Detailed calculation of parasitic capacitances will be developed in Chapter 5, and is not necessary to understand the following considerations.

Without loss of generality, let us assume an abrupt input switching of v_{in1} (v_{in2}) from low to high (high to low), which determines an output switching of v_{o1} (v_{o2}) from high to low (low to high). During the switching transient, the ground current changes from its steady-state value I_{SS} due to the additive current contributions i_{CGND1} and i_{CGND2} flowing through capacitances C_{GND1} and C_{GND2}. Being equal to the ground current, the supply current has a time-varying component equal to $i_{CGND1}+i_{CGND2}$, whose peak amplitude i_{VDD} must be kept within assigned bounds to avoid an excessive switching noise. As will be demonstrated in Chapter 5, output voltages v_{o1} and v_{o2} have an exponential waveform with a time constant $R_D \cdot C_{out}$, being C_{out} the overall capacitance at each node and R_D the equivalent PMOS resistance evaluated in Section 2.4.1. Since the overall capacitance at the output node v_{o1} and v_{o2}

is respectively (C_{VDD1}+C_{GND1}) and (C_{VDD2}+C_{GND2}), the two output voltages result as

$$v_{o1}(t) = V_{DD} - V_{SWING}\left(1 - e^{-\frac{t}{R_D(C_{VDD1}+C_{GND1})}}\right) \tag{3.16a}$$

$$v_{o2}(t) = V_{DD} - V_{SWING}e^{-\frac{t}{R_D(C_{VDD2}+C_{GND2})}} \tag{3.16b}$$

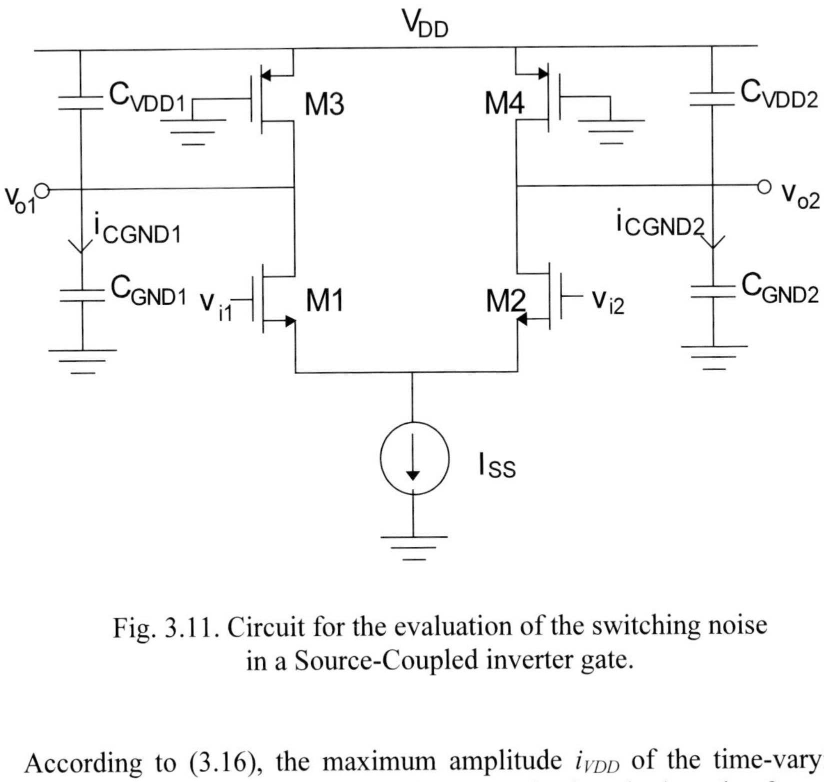

Fig. 3.11. Circuit for the evaluation of the switching noise in a Source-Coupled inverter gate.

According to (3.16), the maximum amplitude i_{VDD} of the time-varying contribution i_{CGND1}+i_{CGND2} of the supply current is given by its value for t=0

$$i_{VDD} = i_{CGND1}(0) + i_{CGND2}(0) = C_{GND1}\left.\frac{\partial v_{o1}}{\partial t}\right|_{t=0} + C_{GND2}\left.\frac{\partial v_{o2}}{\partial t}\right|_{t=0}$$

$$= \frac{V_{SWING}}{R_D}\left[\frac{C_{GND1}}{C_{VDD1}+C_{GND1}} - \frac{C_{GND2}}{C_{VDD2}+C_{GND2}}\right]$$

$$= I_{SS}\left[\frac{C_{GND1}}{C_{VDD1}+C_{GND1}} - \frac{C_{GND2}}{C_{VDD2}+C_{GND2}}\right] \tag{3.17}$$

which is proportional to the value of the bias current, and can be heavily reduced by matching the two capacitive terms with opposite sign in (3.17), i.e. by equalizing capacitances at the two output nodes. Except for the matching inaccuracy due to process tolerances, the capacitances at the two output nodes are equal in an SCL inverter gate. Indeed, it is symmetrical and drives symmetrical (differential) gates, thus parasitic transistor capacitances and gate input capacitances at the two output nodes are equal.

For more complex gates, the considerations introduced above still hold, and the switching noise is again minimized by matching the capacitances at the two output nodes. However, as opposite to the simple case of the inverter gate, in general the parasitic capacitances at the output nodes are different, since the number of transistors connected to an output node is different from the other one. For example, the circuit Fig. 3.9e has three transistors connected to output node v_o and two transistors to the other. Since the number of transistors connected to each output node depends on the input ordering, the switching noise in SCL gates is minimized by considering all possible input orderings and selecting that leading to the most balanced number of branches connected to the output nodes [MKA92], [ACK93].

Finally, it is interesting to observe that in some specific cases the number of series-gate levels in an n-input gate can be reduced with respect to n by properly selecting the input ordering, which allows for reducing the supply voltage (see Section 2.5.3). In particular, let us consider the transistor pairs belonging to two contiguous levels, the j-th level associated with input X_j and the (j+1)-th immediately lower one associated with X_{j+1}. In general, transistors lying at the (j+1)-th level cannot be driven by signals generated for the stacked j-th level, but a voltage level downshifting is needed to avoid operation in the triode region. However, if transistors driven by X_{j+1} are placed into the same series-gate level as transistors driven by X_j (i.e. they are both driven by signals generated for the j-th level), correct operation is still guaranteed in the particular case where transistors driven by X_j are never stacked to transistors driven by X_{j+1}. Equivalently, the number of series-gate levels can be reduced by unity if all possible active paths contain either transistors driven by X_j or transistors driven by X_{j+1}. From an analytical point of view, this occurs when all products of the minimized function $F(X_1 \dots X_n)$ written in the sum-of-product form alternatively contain either factor x_j or

x_{j+1}. As an example, the MUX gate in Fig. 2.15 has 3 inputs but only 2 series-gate levels, since in its expression

$$OUT(A,B,\Phi)=\Phi A+\overline{\Phi}B \tag{3.18}$$

each product contains either *A* or *B*.

Summarizing, if there is a pair of input variables which are never included into the same product in the minimized function $F(X_1,\ldots, X_n)$, a reduction in the number of levels is allowed, provided that the two input variables are associated with contiguous levels. Accordingly, in such cases an input ordering with contiguous inputs x_j and x_{j+1} must be adopted. Further reduction in the number of levels is allowed if there are various pairs of input variable with the requisites above discussed, and also in this case the two signals of each pair must be assigned to contiguous inputs.

Chapter 4

MODELING OF BIPOLAR CURRENT-MODE GATES

In this chapter, the delay modeling of CML and ECL gates is addressed. After briefly reviewing the previously proposed models, a more efficient approach is presented and applied to practical circuits.

4.1 INTRODUCTION TO MODELING METHODOLOGIES

Various approaches have been proposed in the literature to determine the delay expression of CML and ECL circuits. The most significant modeling approaches are based on the sensitivity analysis [TS79], [CBA88], [F90], [FBA90], the average branch current [KB92], [YC92], and the circuit linearization [SE94], [H95], [SE96]. Moreover, other strategies can be found in [GMO90].

The approach based on the sensitivity analysis was proposed in [TS79] in a simple form, and reused in other several works such as [CBA88] for the CML and ECL inverter, in [F90] for the ECL inverter and in [FBA90] for the ECL XOR gate with some approximations. By using the sensitivity analysis, the propagation delay of CML and ECL gates is represented as a weighted sum of all circuit time constants and transit time of transistors, according to

$$\tau_{PD} = \sum_{i=1}^{N}\sum_{j=1}^{M} k_{ij} R_i C_j + \sum_{l=1}^{P} k_l \tau_{F_l} \tag{4.1}$$

where N, M and P are the number of resistors, capacitors and transistors, respectively, that are included in the half circuit (due to the symmetry, only the half circuit can be used). The weighting factors k_{ij} in relationship (4.1) are the key parameters of the sensitivity analysis, and are determined by observing that

$$k_{ij} = \frac{\partial^2 \tau_{PD}}{\partial R_i \partial C_j} \tag{4.2}$$

where the derivative of delay τ_{PD} can be numerically evaluated by determining the delay change due to the increment of both R_i and C_j, through four circuit simulations.

The propagation delay is expressed as the sum of $(N \cdot M + P)$ terms. As an example, the simple case of a CML inverter has N=4 and M=6, thus leading to 24 terms [CBA88] and 96 simulations (one term was eliminated using circuit considerations).

The large number of simulations is the main drawback of the method. Besides, according to our experience and that of other authors, this approach is not independent of process parameters. Indeed, time constant factors k_{ij} and k_l have to be evaluated each time the technology changes (in fact, all papers have different coefficients). In addition, a model with such a high number of terms is not useful either to really understand the behavior of the gate or for an initial pencil-and-paper design.

To partially overcome these limitations, a simplification in the number of terms resulting from the sensitivity analysis was proposed in [FBA90]. The simplification is performed by neglecting the less significant terms by comparing the time constant values for the specific process used. Although this allows for reducing the number of terms to 10 for the case of the CML inverter gate, the procedure is not general, since it depends on the technology used. Moreover, the simplification in [FBA90] does not reduce the number of simulations required, since it must be performed after simulation.

The average branch current analysis was introduced in [KB92] and [YC92], and successively used in [SE96]. In this analysis strategy, the charge control transistor model is used to derive the differential equations describing the logic gates, that are solved by approximating currents with their mean value. This strategy was used to analytically model the propagation delay associated with the emitter follower in ECL circuits. By means of the average branch current analysis, a simple delay expression was derived in [KB92], and a quite similar expression was successively obtained in [SE96] by slightly different calculations. In [YC92] the delay is expressed by multiple expressions (whose validity depends on the logic swing value), that are much more complex than those in [KB92] and [SE96] and are thus useless. Although the emitter follower delay model in [KB92] and [SE96] is simple, it does not take into account the dependence of delay on the bias current, that is significant in practical current ranges, as will be shown in Section 4.2.

The approach to the delay modeling presented in [SE94], [H95], [SE96], is based on the linearization of the device models. In particular, [SE94] proposes a CML delay model written as the superposition of the delay contributions associated with capacitances, justifying it for the linear dependence of delay on the time constants which is demonstrated by the sensitivity analysis strategy. Each term is found by solving in the time domain the circuit including only a capacitor at a time, but no closed-form expression is found. Indeed, multiple expressions are derived for each delay contribution, whose validity depends on the input waveform rise time. This makes the procedure in [SE94] not so appealing for modeling and design purposes, especially for a pencil-and-paper approach. In addition, the model is not guaranteed to be process-independent, since an empiric correction is introduced in the final delay expression to improve fitting with simulations.

In [SE96], a similar approach is used to model the delay of CML and ECL series gates. However, some assumptions are not very reasonable. For example, small-signal expressions are used to model the transistor capacitances, and the delay associated with them is weighted in a different manner with respect to that of the load capacitance. This explains the low accuracy of the model, which may differ from simulation results by as high as 50%, according to the authors' experience.

The approach followed in [H95] starts from the linearization of the CML inverter gate by replacing transistors by their small-signal model. Successively, the delay is represented as the superposition of the delay contributions associated with each single capacitance, as justified by the Elmore's theory [E48]. The resulting expression is very simple, but there are some weak points in the delay evaluation strategy. For example, the base-emitter capacitance of the bipolar transistor is evaluated by means of its small-signal expression, in contrast with others [TS79], [SE96], [R96] that introduce a better linearized form. In addition, the base-collector capacitance is not split into the intrinsic and extrinsic contributions, as will be discussed below in regard to SPICE models. For these reasons, a high inaccuracy (as high as 60%) of the model by Harada can be shown by comparison with simulations (the real inaccuracy was not highlighted in the original paper since only two simulations were performed).

Other modeling methodologies were proposed in the literature [GMO90], [BK95a], but they are much more complex and are not expressed in a closed form, therefore they are less suitable for design than those above discussed.

It is worth noting that, among the approaches described above, only the sensitivity analysis is capable to model both the delay of CML and ECL gates. Indeed, the average branch current analysis was only used to evaluate the delay of the emitter follower in ECL gates, while the other methodologies can be only applied to CML gates. Moreover, none of the discussed modeling approaches is simple, general and accurate enough to be helpful in managing the design trade-offs in CML and ECL gates. For these reasons, the authors developed a novel strategy presented in the following that overcomes these limitations [AP99].

4.2 AN EFFICIENT APPROACH FOR CML GATES

The authors' approach starts from the observation that, in a bipolar Current-Mode gate, the transistors operate in the linear region, as saturation is avoided to allow high-speed performance. This suggests that transistors can be modeled by the small-signal model valid in the linear region, that has to be properly evaluated to account for a wide variation of currents and voltages. In the following, we consider the small-signal transistor SPICE model reported in Fig. 4.1.

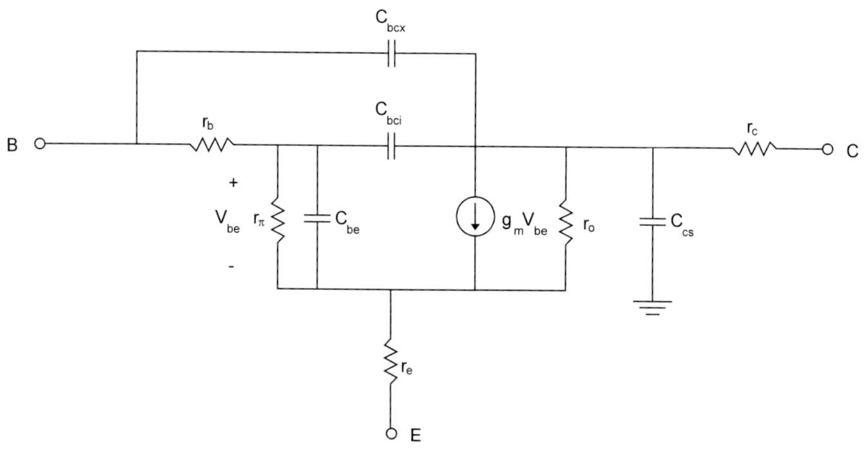

Fig. 4.1. SPICE small-signal model of the bipolar transistor.

The small-signal parameters g_m, r_π and r_o in Fig. 4.1 are the transconductance, the base-emitter resistance and the output resistance of the transistor. Due to the symmetry of the input-output DC characteristics, it is reasonable to linearize the circuit around the logic threshold v_d=0, or equivalently with the bias current equally divided into the two output branches. Under this assumption, the usual expressions $g_m=I_{SS}/2V_T$ and $r_\pi=2\beta_F V_T/I_{SS}$ hold for the transistors driven by the switching input. Moreover, the transistor output resistance r_o can be always neglected in practical cases, since it is much higher than the resistance R_C seen by the collector to ground.

The parasitic resistances r_b, r_e and r_c model the resistance associated with base, emitter and collector diffusion. The distributed base-collector junction capacitance is accounted for by two lumped capacitances connected to the outer and the inner base node, C_{bcx} and C_{bci}, that are the extrinsic and intrinsic contribution. To be more specific, the total base-collector capacitance C_{bc} is split into C_{bcx} and C_{bci} according to the model parameter *XCJC* that ranges between 0 and 1:

$$C_{bcx} = (1 - XCJC) \cdot C_{bc} \tag{4.3a}$$

$$C_{bci} = XCJC \cdot C_{bc} \,. \tag{4.3b}$$

The base-emitter capacitance C_{be} consists of the diffusion and junction contributions, C_D and C_{je}, while C_{cs} models the collector-substrate junction capacitance. It is worth noting that the well-known small-signal capacitance expressions are inadequate and the capacitances have to be properly linearized, since voltages move rapidly over a wide range.

Once the transistors are replaced by the model in Fig. 4.1, the equivalent linear circuit of a gate can be further simplified. Indeed, by observing that the circuit is generally symmetric and assuming differential signaling (as in practical cases, as discussed in Chapter 2), analysis can be limited to its half-circuit.

To simplify the delay evaluation, the linearized half circuit can be approximated by a single-pole system with time constant τ, that can be evaluated by means of the open-circuit time constant method [CG73], [MG87]. By assuming a step input waveform[1], the delay τ_{PD} can be expressed as

$$\tau_{PD} = 0.69\tau = 0.69 \sum_{i=1}^{M} R_{Ci} C_i \tag{4.4}$$

where R_{Ci} is the resistance seen by capacitance C_i when the others are considered open circuits. It is worth noting that the assumption of single-pole behavior has lead to a linear dependence of the delay on the time

---

1 The small delay dependence on the input rise time will be considered in Chapter 8.

constants, which is consistent with observations reported in the literature [CBA88], [F90], [FBA90], [SE94], [SE96].

From inspection of relationship (4.4), the approach followed makes it possible to represent propagation delay with very few terms, whose number is equal to that of circuit capacitances. Moreover, each term has and evident physical meaning, since it is the time constant associated with the related capacitance [H97].

The general delay expression of CML gates (4.4) provides a better insight into the dependence of delay on design parameters (i.e., the bias current I_{SS} and the load resistance R_C) and process parameters. This dependence will be made more explicit for various gates in the next sections. In addition, the dependence on the bias current will be carefully analyzed to develop design criteria to consciously manage the power-delay trade-off.

4.3 SIMPLE MODELING OF THE CML INVERTER

The CML inverter topology is shown in Fig. 4.2, where the capacitance C_L models the external load due to the input capacitance of driven gates (or a generic output load) and the wiring capacitance. Its delay can be modeled by applying the strategy described in the previous sections, by replacing each transistor by the linear circuit in Fig. 4.1 and limiting analysis to the half circuit. Therefore, the equivalent circuit in Fig. 4.3 of a CML inverter is obtained, where the transistor output resistance r_o is neglected since it is usually much higher than the load resistance R_C.

By linearizing the circuit around the logic threshold v_d=0, parameters g_m and r_π in Fig. 4.3 are evaluated as $I_{SS}/2V_T$ and $2\beta_F V_T/I_{SS}$, respectively. The junction capacitances C_{bcx}, C_{bci}, C_{je} and C_{cs} must be evaluated by using relationship (1.12)-(1.13), where voltages V_1 and V_2 are the minimum and maximum direct junction voltages, and are evaluated at the steady state when a high or a low input value is applied. For the sake of clarity, voltages V_1 and V_2 are analytically evaluated in Table 4.1 for the junction capacitances of the circuit in Fig. 4.2. For example, the maximum direct voltage V_2 seen by C_{bcx} occurs when the base voltage is high (i.e., equal to the ground voltage) and consequently the collector voltage is equal to -$R_C I_{SS}$=-$V_{SWING}/2$, thus yielding V_2=$V_{SWING}/2$. Instead, when the base voltage is low (i.e. equal to -$V_{SWING}/2$), the collector voltage is at the ground voltage, leading to V_1=-$V_{SWING}/2$. Maximum and minimum direct voltages across the other capacitances can be analogously evaluated.

In regard to the capacitance C_{je}, it has a maximum voltage equal to the base-emitter voltage $V_{BE,on}$ when the collector current is biased by the whole current I_{SS} (i.e. when a high input voltage is applied)

$$V_{BE,on} = V_T \ln \frac{I_{SS}}{\alpha_F I_{ES}} \tag{4.5}$$

where (1.17) was inverted and the other transistor was assumed to be in cut-off. In relationship (4.5), α_F is the common-base current gain in (1.19), and I_{ES} is the saturation current of the base-emitter junction, defined as the product $A_E \cdot J_{CS}$ in eq. (1.17). Due to the negligible dependence on the collector current, the maximum voltage $V_{BE,on}$ can be approximated to a constant value, which is evaluated for an intermediate bias current. The minimum voltage across the base-emitter junction results in ($V_{BE,on}$-$R_C I_{SS}$), since for a low input voltage the transistor base voltage is reduced by $R_C I_{SS}$ compared to the value for a high input, whereas the emitter voltage is the same as before (it is set to $-V_{BE,on}$ by the transistor in the ON state, regardless of the input voltage).

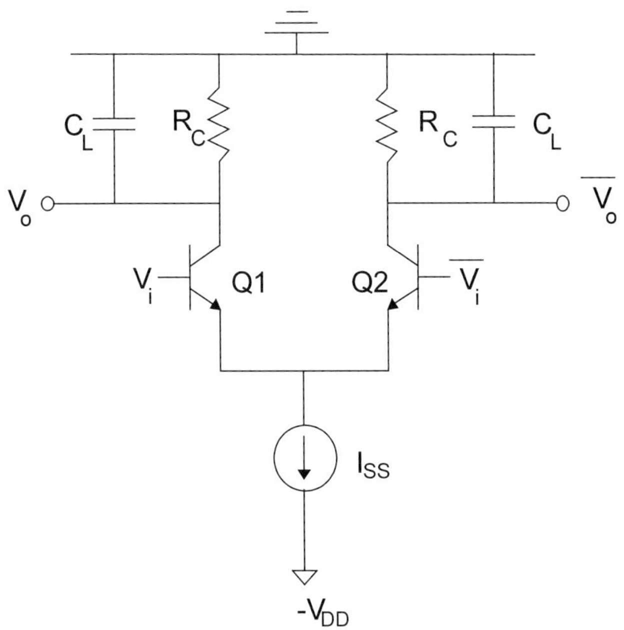

Fig. 4.2. CML inverter.

The base-emitter diffusion capacitance C_D is associated with the variation ΔQ_B of the base charge associated with minority carriers, that goes from $\tau_F I_{SS}$ (when I_{SS} is completely steered to the transistor) to about 0 (when I_{SS} is completely steered to the other transistor). As discussed for C_{je}, this charge variation is due to the base-emitter voltage variation V_2-V_1=$R_C I_{SS}$, therefore capacitance C_D can be linearized as the ratio $\Delta Q_B/\Delta V_{BE}=\tau_F/R_C$. However, as pointed out in [R96], this expression is affected by a significant error, and an empiric factor of two has to be introduced, leading to

$$C_D = 2\frac{\tau_F}{R_C}. \tag{4.6}$$

TABLE 4.1

capacitance	V_1	V_2
C_{bcx}	$-V_{SWING}/2$	$V_{SWING}/2$
C_{bci}	$-V_{SWING}/2$	$V_{SWING}/2$
C_{je}	$V_T \cdot \ln(I_{SS}/\alpha_F I_{ES})-V_{SWING}/2$	$V_T \cdot \ln(I_{SS}/\alpha_F I_{ES})$
C_{cs}	$-V_{DD}$	$-V_{DD}+V_{SWING}/2$

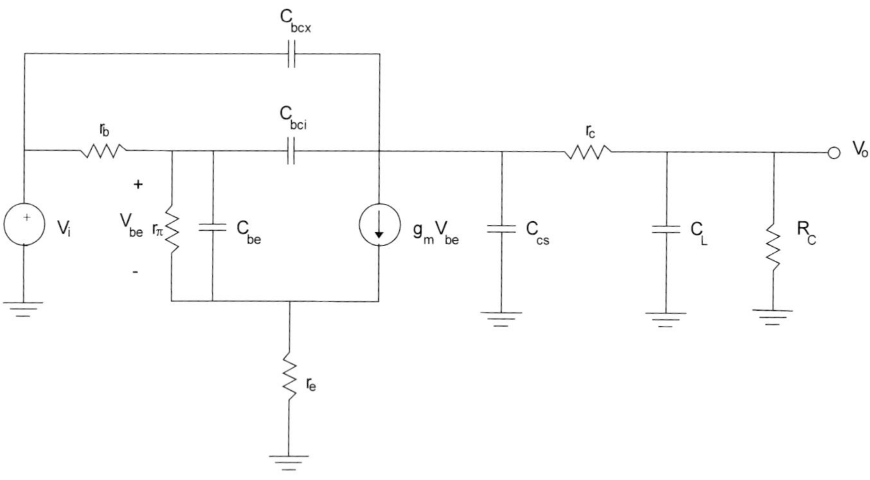

Fig. 4.3. Equivalent linear circuit of the CML inverter.

By approximating the equivalent circuit of the CML inverter in Fig. 4.3 to a single pole system with time constant τ, and evaluating it by the time constant method, the delay τ_{PD} can be expressed as

$$\tau_{PD} = 0.69\left\{\frac{r_e + r_b}{1 + g_m r_e}C_{be} + r_b C_{bci}\left[1 + \frac{g_m(r_c + R_C)}{1 + g_m r_e}\right]\right. \tag{4.7}$$

$$\left. + (r_c + R_C)(C_{bci} + C_{bcx} + C_{cs}) + R_C C_L \right\}$$

The propagation delay is the sum of four main terms, which have a simple circuit meaning and can be evaluated with pencil and paper. The first term is the contribution made by the base-emitter capacitance, the second is due to a Miller effect on the intrinsic base-collector capacitance, the third is a contribution which arises at the inner collector node (i.e., before parasitic resistance r_c) and the last one is due to the load capacitance at the output node. It is worth noting that the term $g_m/(1+g_m r_e)$ is the equivalent transconductance of the transistor having a series resistance r_e at the emitter node.

4.3.1 Accuracy of the CML simple model

To evaluate the accuracy of relationship (4.7), a comparison between the modeled propagation delay and SPICE simulations was carried out. To generalize the comparison, two different technologies were taken into consideration. The first is a BiCMOS technology whose npn bipolar transistor has a transition frequency equal to 6 GHz, the second is the HSB2 high-speed bipolar technology (by courtesy of ST Microelectronics) with an npn transition frequency equal to 20 GHz. The SPICE model of the 6-GHz and 20-GHz transistors considered is reported in Fig. 4.4a and 4.4b, respectively.

Some parameters extracted from the model in Fig. 4.4a and 4.4b are reported in Table 4.2, where $I_{\tau F}$ represents the bias current after which the transit time starts increasing due to high injection level effects, and was evaluated by a few DC simulations. In practical cases, the bias current should not significantly exceed $I_{\tau F}$ because the transit time degradation determines an increase in the diffusion capacitance (4.6), and in turn a speed degradation.

```
.MODEL QN
+ NPN
+ XTB=1.92          EG =1.15        XTI=3.84
+ IS =8.91E-18      BF =100             NF =1
+ VAF=100           IKF=0.0192      ISE=5.18E-18 NE =1.45
+ BR =0.0936        NR= 1
+ VAR=8             IKR=0.00289     ISC=5.46E-18 NC =1.07
+ RB =476           RBM=166         RE =9.53     RC =110
+ CJE=4.59E-14      VJE=1.07        MJE=0.5      FC =0.5
+ CJC=2.7E-14       VJC=0.646       MJC=0.35     XCJC=.146
+ CJS=7.28E-14      VJS=0.45        MJS=0.3
+ TF =2.64E-11      XTF=27.3        VTF=3        ITF=0.0201
+ PTF=30            TR =6.54E-08
```

Fig. 4.4a. SPICE model of the bipolar transistor (BiCMOS process).

```
.MODEL C12TYP
+ NPN
+ IS=7.40E-018 BF=1.00E+002 BR=1.00E+000 NF=1.00E+000
+ NR=1.00E+000 TF=6.00E-012 TR=1.00E-008 XTF=1.00E+001
+ VTF=1.50E+000 ITF=2.30E-002 PTF=3.75E+001 VAF=4.50E+001
+ VAR=3.00E+000 IKF=3.10E-002 IKR=3.80E-003 ISE=2.80E-016
+ NE=2.00E+000 ISC=1.50E-016 NC=1.50E+000 RE=5.26E+000
+ RB=5.58E+001 IRB=0.00E+000 RBM=1.55E+001 RC=8.09E+001
+ CJE=3.21E-014 VJE=1.05E+000 MJE=1.60E-001 CJC=2.37E-014
+ VJC=8.60E-001 MJC=3.40E-001 XCJC=2.30E-001 CJS=1.95E-014
+ VJS=8.20E-001 MJS=3.20E-001 EG=1.17E+000 XTB=1.70E+000
+ XTI=3.00E+000 KF=0.00E+000 AF=1.00E+000 FC=5.00E-001
```

Fig. 4.4b. SPICE model of the bipolar transistor (HSB2 process).

TABLE 4.2

	BiCMOS	HSB2
τ_F	26.4 ps	6 ps
r_b	476 Ω	56 Ω
r_c	110 Ω	81 Ω
r_e	10 Ω	5 Ω
$I_{\tau F}$	1.4 mA	2.4 mA

By using the parameters in Figs. 4.4a and 4.4b, assuming a logic swing of 500 mV and a supply voltage of 5 V, the resulting coefficients K_j used to evaluate the transistor junction capacitances are listed in Tables 4.3a and 4.3b, that refer to the BiCMOS and HSB2 process, respectively.

TABLE 4.3a
(BiCMOS process)

Capacitance	V_2 (V)	V_1 (V)	ϕ (V)	M	C_{j0} (fF)	K
C_{bcx}	0.250	-0.250	0.646	0.35	27	1.013
C_{bci}	0.250	-0.250	0.646	0.35	27	1.013
C_{je}	0.841	0.591	1.07	0.5	45.9	1.916
C_{cs}	-4.75	-5	0.45	0.3	72.8	0.477

TABLE 4.3b
(HSB2 process)

Capacitance	V_2 (V)	V_1 (V)	ϕ (V)	m	C_{J0} (fF)	K
C_{bcx}	0.250	-0.250	0.86	0.34	23.7	1.007
C_{bci}	0.250	-0.250	0.86	0.34	23.7	1.007
C_{je}	0.846	0.596	1.05	0.16	32.1	1.21
C_{cs}	-4.75	-5	0.82	0.32	23.7	0.538

After substituting these values, the delay predicted by relationship (4.7) versus the bias current is plotted in Figs. 4.5a and 4.5b for the two technologies, with a load capacitance C_L equal to 0, 100fF and 1pF.

The error of the model (4.7) with respect to SPICE simulations versus the bias current I_{SS} and with a load capacitance C_L equal to 0 F, 100 fF and 1 pF, is plotted in Figs. 4.6a and 4.6b for the BiCMOS and HSB2 technology, respectively.

The worst-case error is 18% and 42% in the unrealistic case of a zero load capacitance for the BiCMOS and HSB2 process, respectively. Moreover, outside the high-level injection region (i.e., for $I_{SS}>I_{\tau F}$) the mean error is equal to 5% and 17%. The error decreases as increasing the load capacitance. This is because the contribution due to the linear capacitance C_L becomes dominant, and the real circuit behavior is closer to that of a one-pole circuit whose time constant is equal to that resulting at the output node. Finally, it is worth noting that although the results for high-level injection bias currents are less significant, the error is still low.

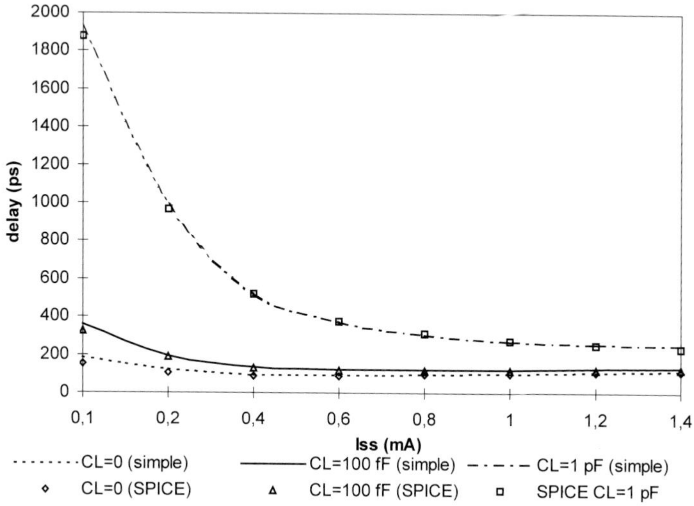

Fig. 4.5a. Predicted and simulated delay vs. I_{SS} for the BiCMOS process.

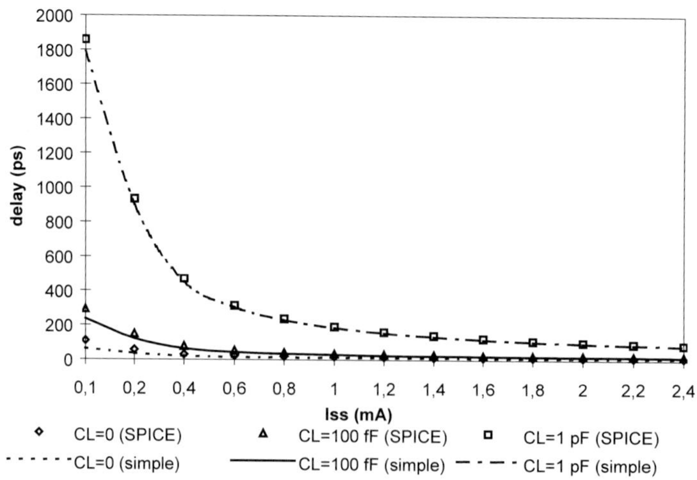

Fig. 4.5b. Predicted and simulated delay vs. I_{SS} for the HSB2 process.

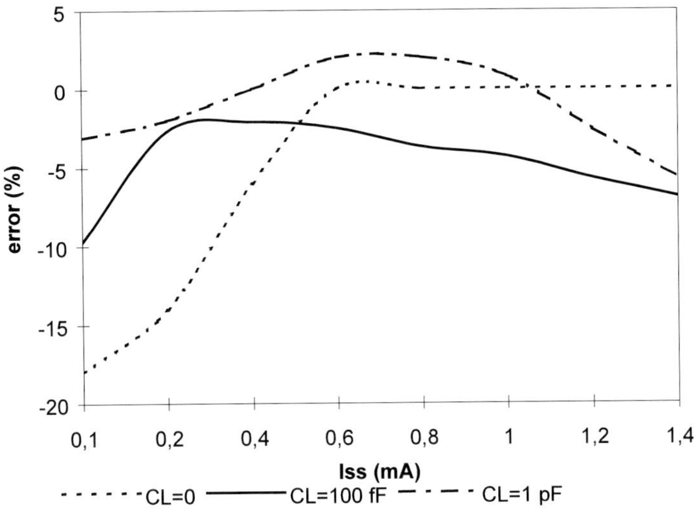

Fig. 4.6a. Error of (4.7) vs. I_{SS} for the BiCMOS process.

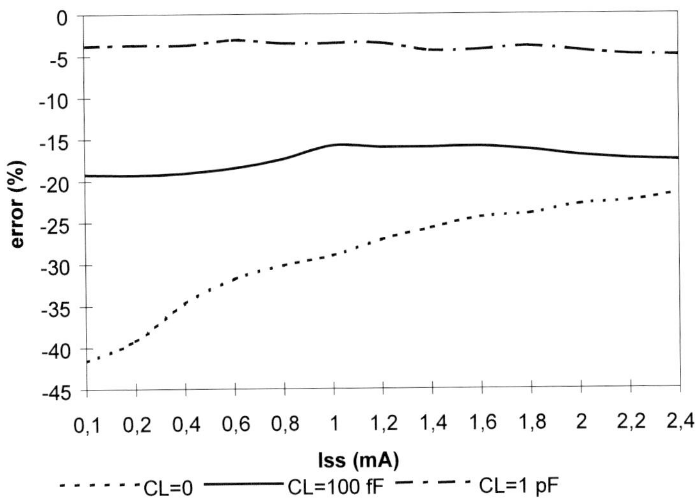

Fig. 4.6b. Error of (4.7) vs. I_{SS} for the HSB2 process.

4.4 ACCURATE MODELING OF THE CML INVERTER

The simple model discussed in the previous section has proved to be accurate enough for pencil and paper calculations. However, a greater accuracy would be desirable if the model were applied to a timing analyzer software, that is used to estimate the overall delay of complex logic networks in place of computationally expensive circuit simulators.

The main reason for the difference between the simulated and predicted delay concerns the non-linear behavior of the circuit. In particular, the terms affected by non-linearity in relationship (4.7) are the capacitances. Thus, the accuracy could be improved by introducing new values of coefficients K_j that multiply the capacitances, instead of those obtained from relationship (1.13). To be more specific, let us use the delay expression (4.7) by evaluating the capacitance coefficients from a few simulation runs by means of a numerical procedure. By representing each junction capacitance with its zero-bias value and introducing a corrective coefficient for each capacitance, from (4.7) we obtain the improved delay model

$$\tau_{PD} = 0.69\left\{\frac{r_e + r_b}{1+g_m r_e}\left(K_{be}C_{jeo} + K_D C_D\right) + r_b K_{bci} X_{cjc} C_{bco}\left[1+\frac{g_m\left(r_c + R_C\right)}{1+g_m r_e}\right]\right.$$
$$\left. + \left(r_c + R_C\right)\left(K_{bci}X_{cjc}C_{bco} + K_{bcx}\left(1-X_{cjc}\right)C_{bco} + K_{cs}C_{cso}\right) + R_C C_L\right) \qquad (4.8)$$

The five coefficients K_j associated with the capacitances have to be evaluated to fit simulation results of delay, by minimizing the error between analytical and simulated results. Among the many ways to obtain the coefficients K_j, the most efficient is that of minimizing the functional (4.9), since it requires few simulation runs

$$S\left(K_{bci}, K_{bcx}, K_{cs}, K_{be}, K_D\right) = \sum_{j=1}^{n}\left|\frac{\tau_{PDSpice}\left(I_{SSj}\right) - \tau_{PD}\left(I_{SSj}\right)}{\tau_{PDSpice}\left(I_{SSj}\right)}\right| \qquad (4.9)$$

where the least square formulation was not used to improve the convergence speed, since each term is much lower than unity. In (4.9), the number of simulations n is itself a variable, but, according to our experience, it can be set to the number of capacitances (i.e., n=5). Since we consider the load capacitance in the worst condition (i.e., C_L=0, as discussed in the previous section), simulations differ only for the bias current value. Minimization of functional S can be achieved with any of the numerical software tools such as Mathcad©, Matlab© or Mathematica©. Differently from traditional sensitivity analysis, our approach only needs a few Spice simulations, since only the circuit time constants associated with the capacitors are considered in the delay expression (4.8), rather than all possible products of resistances and capacitances in (4.1).

4.4.1 Accuracy of the CML accurate model

As an example, coefficients K were evaluated by using the bias current values (in mA) [0.1, 0.2, 0.6, 1, 1.4] and [0.1, 0.4, 1.2, 1.6, 2.2] to minimize the functional S in (4.9) for the BiCMOS and HSB2 technology, respectively. The coefficients K obtained are listed in Table 4.4 for both technologies. From the comparison with the values reported in Table 4.3a and 4.3b, it follows that the most evident difference between the simple and the improved model is due to the coefficients K_j of the capacitances C_{cs} and C_{je} for the BiCMOS process (which differ by a factor of two) and to the coefficients K_j of the capacitance C_{cs} for the HSB2 process (which differs by a factor higher than three).

TABLE 4.4

coefficients	**6 GHz process**	**20 GHz process**
K_{bcx}	0.816	0.998
K_{bci}	0.999	0.995
K_{cs}	0.942	1.789
K_{be}	1.009	0.984
K_D	1.076	1.100

The model error versus the bias current I_{SS} and with a load capacitance of 0 F, 100 fF and 1 pF, is plotted in Figs. 4.7a and 4.7b for the BiCMOS and HSB2 technology, respectively. It is worth noting that there is no difference in the accuracy of the model for the two technologies. The largest error is lower than 5% for both of them, and outside the high-level injection region the mean error is lower than 2%. This confirms that the improved model is well suited for the implementation of a timing analyzer.

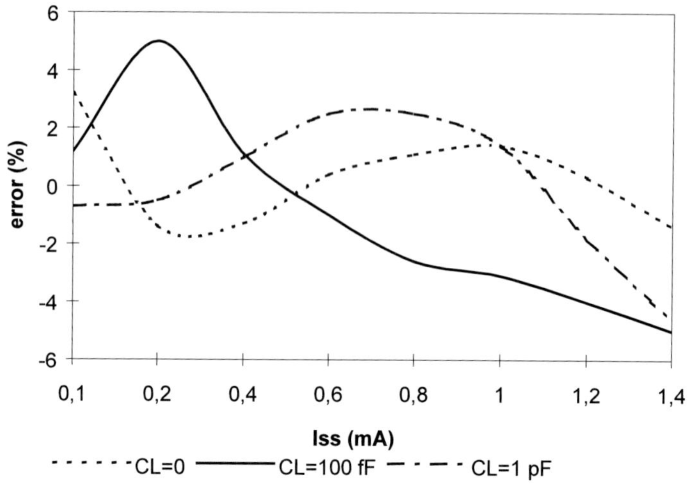

Fig. 4.7a. Error of (4.8) vs. I_{SS} for the BiCMOS process.

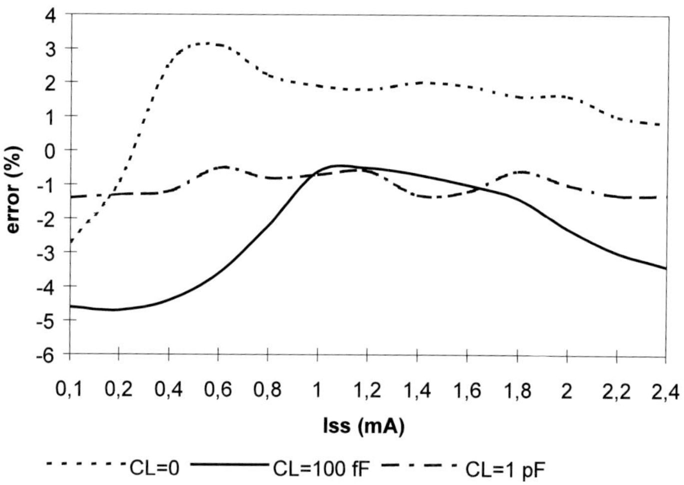

Fig. 4.7b. Error of (4.8) vs. I_{SS} for the HSB2 process.

4.5 SIMPLE AND ACCURATE MODELING OF THE ECL INVERTER

The modeling approach discussed in the previous sections can also be applied to ECL gates. To show this, let us consider an ECL inverter gate, depicted in Fig. 4.8, that is a symmetrical gate made up of a CML circuit followed by two common-collector (CC) stages with function of output buffers. Again, since transistors operate in the linear region (or at the boundary of the cut-off region), they can be replaced by their small-signal model in Fig. 4.1. Moreover, due to the symmetry and the assumption of differential signaling, the analysis can be again restricted to the half circuit.

Being the ECL gate made up of cascaded blocks (i.e., the CML circuit and the output buffer), we can assume the ECL propagation delay to be composed of two separated contributions: the CML propagation delay $\tau_{PD(CML)}$ and the common-collector propagation delay $\tau_{PD(CC)}$. To simplify the analysis, we can split the two contributions by applying the Thevenin theorem at the output of the CML stage. In particular, the intrinsic CML delay (i.e., without considering the CC stage) is represented by the delay of the Thevenin voltage with respect to the time when the input switches. The loading effect of the CC stage on the CML stage is accounted for by introducing the output impedance of the latter while driving the CC stage and evaluating its delay. Therefore, the CML propagation delay derives from that evaluated in the previous section by setting C_L=0, while the propagation delay of the CC stage must be evaluated by driving it with the Thevenin equivalent circuit seen by the output of the CML stage. To further simplify the circuit analysis, let us approximate the CML output impedance to its load resistance R_C.

From the above considerations, after using the transistor model in Fig. 4.1, the delay of the CC stage is evaluated by analyzing the equivalent circuit in Fig. 4.9. It is worth noting that in this circuit the base-collector parasitic is not split, since in this case it is not magnified by the Miller effect and makes a very small contribution.

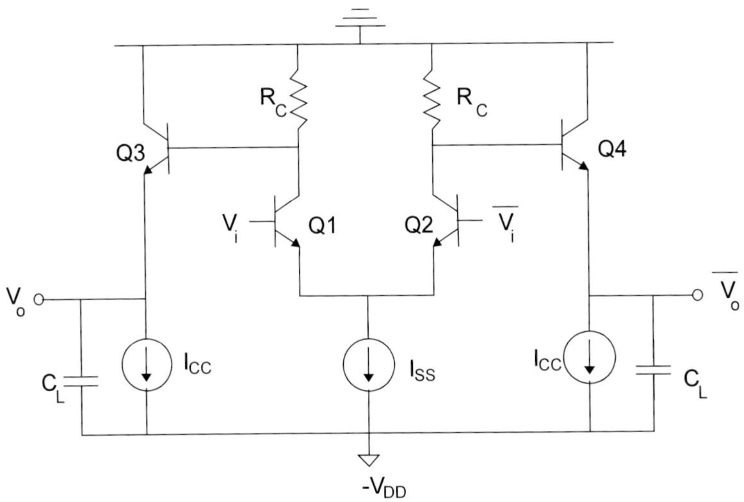

Fig. 4.8. ECL inverter.

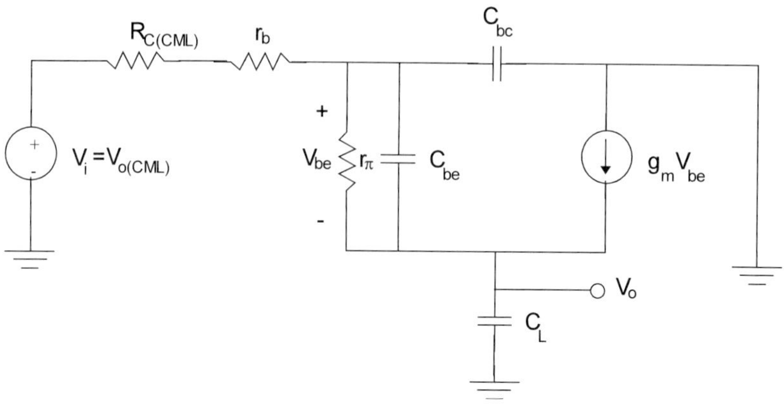

Fig. 4.9. Equivalent linear circuit of the CC stage driven by the Thevenin equivalent of the CML stage.

The transfer function of the circuit in Fig. 4.9 is given by

$$\frac{V_o}{V_{in}} \approx \frac{1+\frac{C_{be}}{g_m}s}{1+\left[\frac{R_C+r_b}{g_m r_\pi}C_L+\frac{C_{be}+C_L}{g_m}+(R_C+r_b)C_{bc}\right]s+\frac{R_C+r_b}{g_m}(C_{be}C_L)s^2} \tag{4.10}$$

where the product $C_{bc}(C_{be}+C_L)$ has been neglected with respect to $C_{be}C_L$ in the coefficient of s^2. Usually, (4.10) has two complex and conjugate poles which justify the well-known ringing behavior [LS94], [KB92]. By neglecting the zero of (4.10), which is at a frequency much higher than that of the poles, the normalized step response of (4.10) in the time domain is

$$y(t)=1-\frac{1}{\sqrt{1-\xi^2}}e^{-\xi\omega_n t}\sin\left(\sqrt{1-\xi^2}\,\omega_n t+arctg\frac{\sqrt{1-\xi^2}}{\xi}\right) \tag{4.11}$$

where the pole frequency, ω_n, and the damping factor, ξ, are found to be

$$\omega_n=\sqrt{\frac{g_m}{(R_C+r_b)C_{be}C_L}} \tag{4.12a}$$

$$\xi = \frac{1}{2}\left[\frac{1}{r_\pi}\sqrt{\frac{R_C + r_b}{g_m}}\sqrt{\frac{C_L}{C_{be}}} + \sqrt{\frac{1}{g_m(R_C + r_b)}\frac{(C_{be} + C_L)^2}{C_{be}C_L}}\right. \tag{4.12b}$$

$$\left. + \sqrt{g_m(R_C + r_b)\frac{C_{be}}{C_L}}\right]$$

The time, normalized to the inverse of the pole frequency, for which a fixed percentage of the steady-state output voltage is reached only depends on the damping factor. In particular the normalized propagation delay τ_{PDn} (i.e., $\tau_{PDn}=\tau_{PD}\cdot\omega_n$) can be found setting in (4.11) $y(t)$ equal to 0.5. The solution, plotted in Fig. 4.10, is quite linear and ranges from 1 to 1.6, for typical values of ξ from 0.1. to 0.8 [AP99].

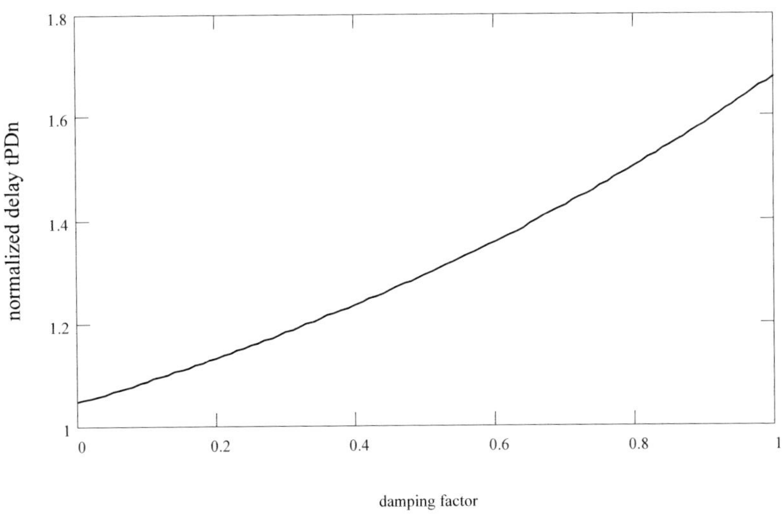

Fig. 4.10. Normalized delay of the CC stage τ_{PDn} vs. damping factor ξ.

Considering in the following the worst case of τ_{PDn}=1.6, and using relationship (4.12a) for the pole frequency, we get

$$\tau_{PD(cc)} = \frac{\tau_{PDn}}{\omega_n} \approx 1.6\sqrt{\frac{R_C + r_b}{g_m}C_{be}C_L} \; . \tag{4.13}$$

The use of relationship (4.13) allows evaluation of the common collector delay in a quite simple way, and takes into account the dependence of the delay on the bias current. By substituting the usual small-signal expressions of transconductance $g_m=I_{CC}/V_T$ and the base-emitter capacitance $C_{be}=C_{je}+C_D$ (where $C_D=g_m\tau_F$ and C_{je} are the diffusion and junction contributions), relationship (4.13) becomes

$$\tau_{PD(cc)}=\frac{\tau_{PDn}}{\omega_n}\approx 1.6\sqrt{\left(R_C+r_b\right)\left(\frac{C_{je}V_T}{I_{CC}}+\tau_F\right)C_L}\ . \tag{4.14}$$

Since the base-emitter voltage only slightly varies during a transition, the small-signal value of capacitances C_{je} and C_D can be used. From inspection of the CC stage delay model (4.14), its dependence on the bias current I_{CC} is small and becomes less significant as increasing I_{CC} since the diffusion capacitance tends to dominate over C_{je}. Hence the propagation delay of the ECL gate can be written as

$$\tau_{PD}=\tau_{PD(CML)}+\tau_{PD(CC)}$$

$$=0.69\left\{\frac{r_e+r_b}{1+g_{m1}r_e}C_{be1}+r_bC_{bci1}\left[1+\frac{g_{m1}\left(r_c+R_C\right)}{1+g_{m1}r_e}\right]\right. \tag{4.15}$$

$$\left.+\left(r_c+R_C\right)\left(C_{bci1}+C_{bcx1}+C_{cs1}\right)\right\}+1.6\sqrt{\frac{R_C+r_b}{g_{m2}}C_{be2}C_L}$$

where the indexes 1 and 2 refer to the CML and the CC stage, respectively. It is worth noting that, differently from the CML case, the minimum and maximum direct voltages across the base-collector junction are now $V_1=-V_{SWING}/2-V_{BE,on}$ and $V_2=V_{SWING}/2-V_{BE,on}$, where $V_{BE,on}$ is the base emitter voltage of the CC stage transistor, that is approximately independent of the bias current, as discussed in Section 4.3.

4.5.1 Validation and improvement of the ECL model

In order to evaluate the accuracy of the simple model proposed, the delay model (4.15) was compared to SPICE simulations for the bipolar technologies above discussed, by assuming the same conditions as in Section 4.3. The error found with the HSB2 technology is plotted in Figs. 4.11a and

4.11b for C_L equal 100 fF and 1 pF, respectively, by biasing the common collector with 0.4 mA, 1 mA and 1.8 mA. It is seen that the worst case percentage error is much lower than 20%. Analogous results are obtained for the BiCMOS process.

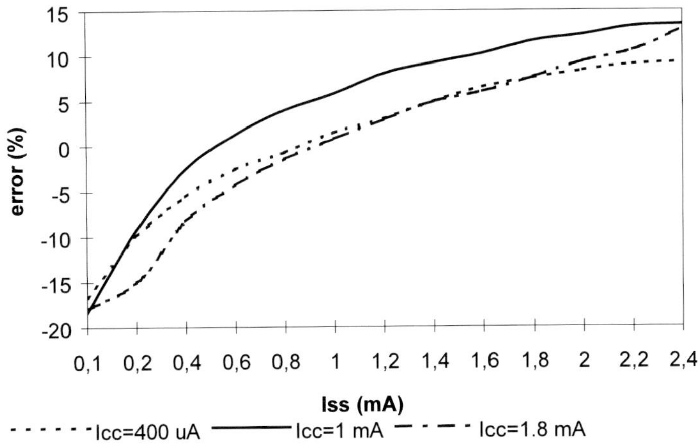

Fig. 4.11a. Error of the ECL simple model (4.15) vs. I_{SS} for the HSB2 process and C_L=100 fF.

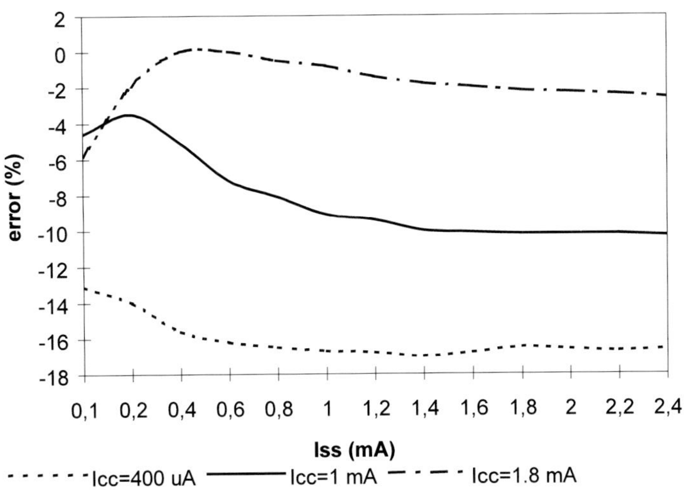

Fig. 4.11b. Error of the ECL simple model (4.15) vs. I_{SS} for the HSB2 process and C_L=1 pF.

The simple delay model discussed so far can be improved in terms of accuracy by adopting a numerical procedure similar to that used for the CML gate. More specifically, the propagation delay is represented as

$$\tau_{PD} = H_1 \tau_{PD(CML)} + H_2 \sqrt{\frac{g_{m2}\tau_F + C_{je2}}{g_{m2}} C_L^{\alpha_1} \left(R_C + r_b\right)^{\alpha_2}} \tag{4.16}$$

where $\tau_{PD(CML)}$ is given by (4.8) by setting C_L=0. The model has 9 unknown coefficients: α_1 and α_2 which regard only the CC terms, H_1 and H_2 which weight the CML and ECL contribution, and the 5 coefficients K inside $\tau_{PD(CML)}$ that were evaluated in the previous section.

Coefficients α_1, α_2, H_1 and H_2 in relationship (4.16) can be found by minimizing the functional in (4.17)

$$S_{ECL}\left(H_1, H_2, \alpha_1, \alpha_2\right) = \sum_{j=1}^{n} \left| \frac{\tau_{PDSpice}\left(I_{SSj}, I_{ccj}, C_L\right) - \tau_{PD}\left(I_{SSj}, I_{ccj}, C_L\right)}{\tau_{PDSpice}\left(I_{SSj}, I_{ccj}, C_L\right)} \right| \tag{4.17}$$

A good choice of the number of simulations n is 5. Moreover, the simulations should be run by setting the load capacitance to low and high values (100fF and 1pF, for the adopted processes), and properly distributing the bias currents I_{SS} and I_{CC} in the range outside high-level injection.

Coefficients α_1, α_2, H_1 and H_2 were evaluated for the BiCMOS and HSB2 process by running five simulations, uniformly covering the entire region of currents I_{SS} and I_{CC} in which the model is valid. The coefficients found are reported in Table 4.5.

TABLE 4.5

coefficients	**6 GHz process**	**20 GHz process**
α_1	0.486	0.506
α_2	0.490	0.597
H_1	0.486	0.650
H_2	1.000	1.040

In this model the main difference between the simple and the improved model is due to the values of parameters H_1 and H_2. Several simulations were run with different bias current and load capacitance values to evaluate the error of the accurate ECL model. The error found with the HSB2 process by biasing the common collector current with 0.4 mA, 1 mA and 1.8 mA is plotted in Figs. 4.12a and 4.12b, for C_L equal to 100 fF and 1pF, respectively. The worst-case error is lower than 10%, while the typical error is around 5%. Similar results were found for the BiCMOS process.

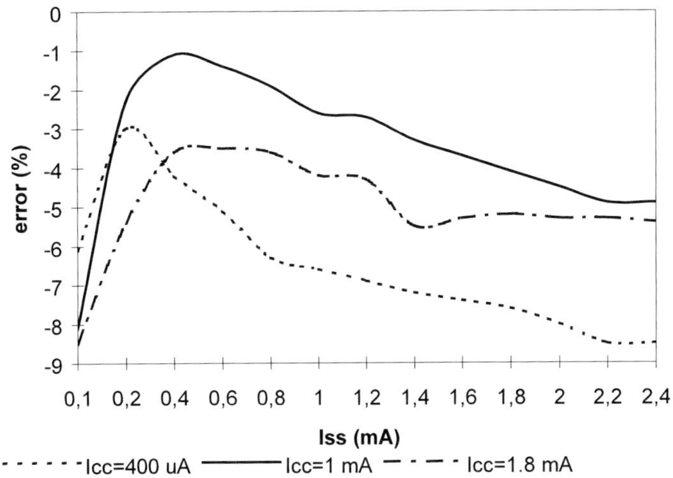

Fig. 4.12a. Error of the ECL accurate model (4.16) vs. I_{SS} for the HSB2 process and C_L=100 fF.

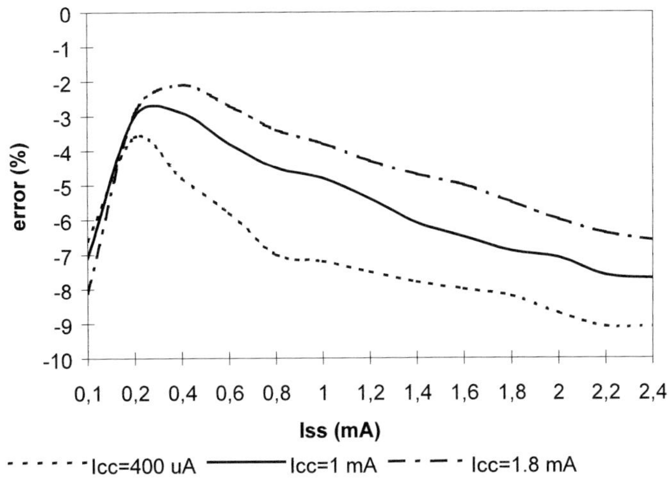

Fig. 4.12b. Error of the ECL accurate model (4.16) vs. I_{SS} for the HSB2 process and C_L=1 pF.

4.6 SIMPLE MODELING OF BIPOLAR CML MUX/XOR GATES

The CML MUX and XOR gates are shown in Figs. 4.13 and 4.14, respectively. Since these circuits are two-input logic gates, different delay values should be considered, depending on the switching input [AP00]. As a general consideration, the delay is greater for inputs associated with transistors that are at a lower level as stated in [FBA90] (i.e., Q1-Q2 in Figs. 4.13 and 4.14), as will be justified above.

Let us first consider the case where the input driving transistors Q1-Q2 switches, i.e., ϕ and B in the MUX and XOR gate, respectively, while the others are kept constant. Due to the symmetry, in the MUX and XOR gates without loss of generality since the opposite case is analogous, we assume input A to be high (and input B of the MUX to be low to have an output transition). As a consequence, transistors Q3 and Q6 are ON and can be replaced by the model in Fig. 4.1, while Q4 and Q5 are OFF and their effect on the circuit is accounted for by only junction capacitances (equal to those of Q3 and Q6).

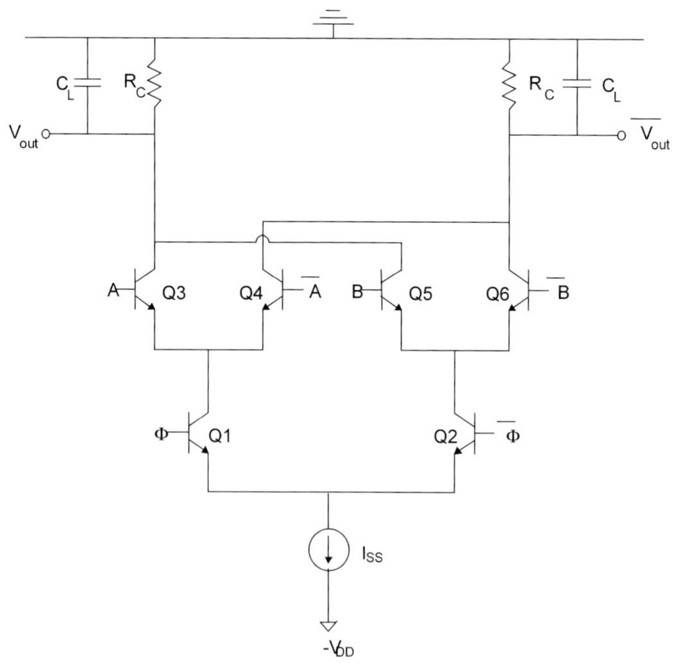

Fig. 4.13. CML MUX gate.

Since differential operation is assumed, the half-circuit model of both MUX and XOR gates in Fig. 4.15 is obtained, where the small-signal parameters g_m and r_π are evaluated around the logic threshold as for the inverter, thus resulting equal to $I_{SS}/2V_T$ and $2\beta_F V_T/I_{SS}$. The base-emitter diffusion capacitance of the switching pair Q1-Q2 is still given by (4.6), while that of transistors Q3 and Q6 is evaluated by the well-known small-signal expression $g_m \tau_F$, since the voltage variation at the emitter node of the upper transistors is small. For the same reason, small-signal values [GM93] for C_{cs1}, C_{je3}, C_{D3} and C_{je4} can be used. All the other junction capacitances are evaluated through relationship (1.12)-(1.13).

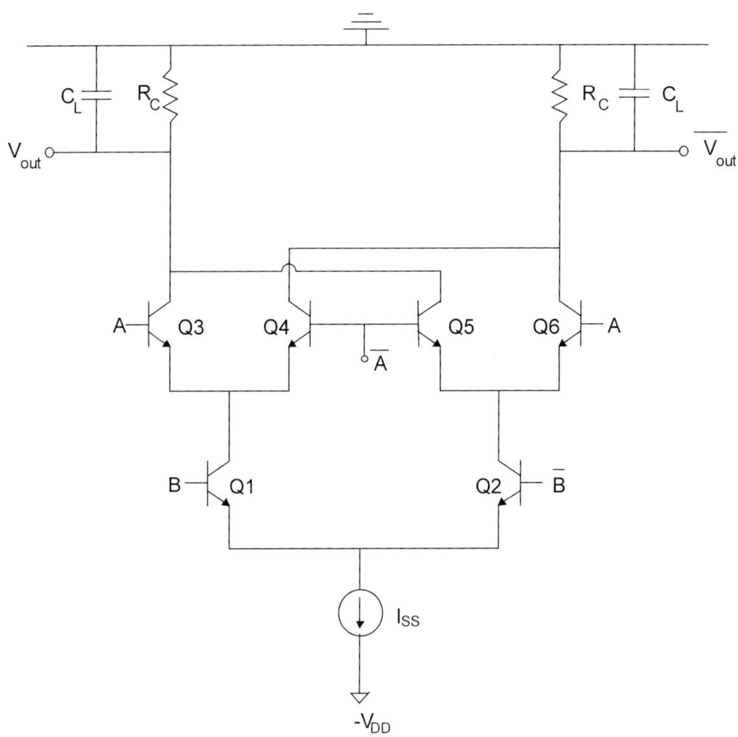

Fig. 4.14. CML XOR gate

Finally, it is worth noting that we are assuming a realistic case in which input A is driven by another CML gate. Hence, in the upper input we have to include the equivalent output resistance of the previous gate, represented by the two resistances R_C in series with r_{b3} and r_{b5}. However, this equivalent output resistance must not be considered in the switching lower input, because it is driven by an ideal voltage generator which represents the voltage variation at the input node. Loading effects at that node can be properly taken into account by evaluating the propagation delay of the previous gate.

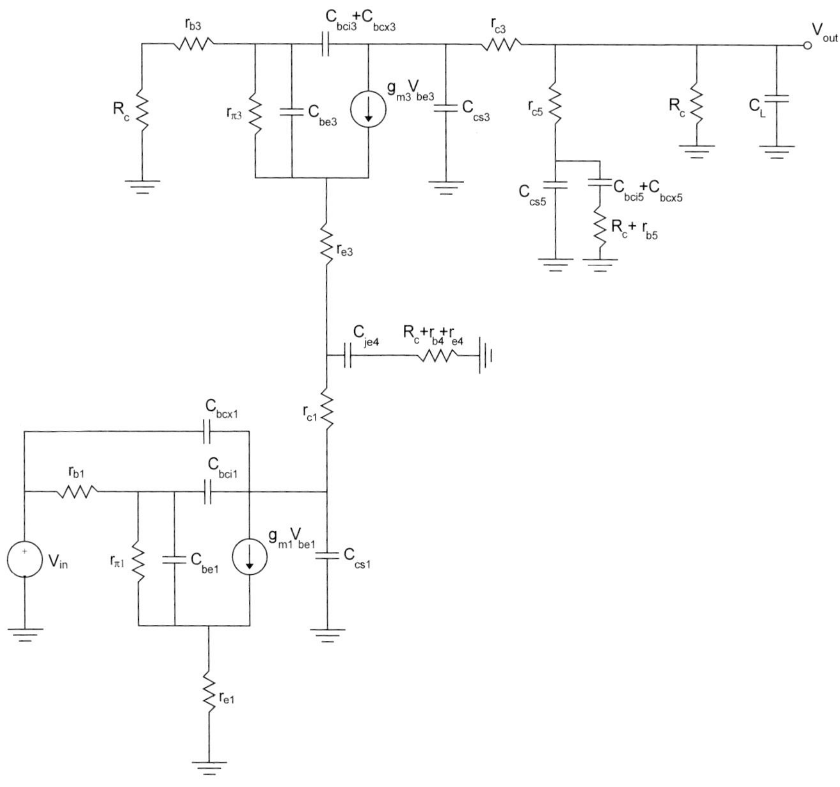

Fig. 4.15. Equivalent linear circuit of the CML MUX/XOR gate.

By assuming a dominant pole behavior, and thus using relationship (4.4), neglecting terms r_π with respect to terms $1/g_m$ and lumping C_{bcx3}, C_{bci3} and C_{bcx5}, C_{bci5} into C_{bc3} and C_{bc5}, respectively, the MUX/XOR equivalent circuit in Fig. 4.15 gives the following delay

$$\tau_{PD} = 0.69\left\{\frac{r_{el} + r_{bl}}{1 + g_{ml} r_{el}} C_{be1} + r_{bl} C_{bci1}\left[1 + \frac{g_{ml}\left(r_{cl} + R_{eq}\right)}{1 + g_{ml} r_{el}}\right] + \left(r_{cl} + R_{eq}\right)\left(C_{bc1} + C_{cs1}\right)\right.$$

$$+ \frac{1}{g_{mu}} C_{be3} + \left(2R_C + r_{cu} + r_{bu}\right)\left(C_{bc3} + C_{bc5}\right) \tag{4.18}$$

$$\left. + \left(R_C + r_{cu}\right)\left(C_{cs3} + C_{cs5}\right) + \left(R_C + r_{bu} + 2r_{eu}\right)C_{je4} + R_C C_L\right\}$$

where subscripts *l* and *u* refer to lower and upper transistors, respectively, and R_{eq} is the equivalent resistance at the emitter of Q3, given by

$$R_{eq} = r_{eu} + \frac{r_{\pi u} + r_{bu} + R_C}{1 + \beta_F} \approx \frac{1}{g_m} + \frac{R_C}{\beta_F} \tag{4.19}$$

From relationship (4.18), the propagation delay is the sum of eight main terms which have a simple circuit meaning. Three terms are due to the lower transistors, four to the upper transistors and one to the load.

4.6.1 Validation of the MUX/XOR model

The MUX/XOR delay in (4.18) was compared to SPICE simulations in the same conditions as in Section 4.3, that determine the junction capacitances given in Table 4.6a and Table 4.6b for the BiCMOS and HSB2 technology, respectively.

TABLE 4.6a
(BiCMOS process)

Capacitance	V_2 (V)	V_1 (V)	ϕ(V)	m	C_{j0} (fF)	K
C_{bcxl}	0.254	0	0.646	0.35	27	0.933
C_{bcil}	0.254	0	0.646	0.35	27	0.933
C_{csl}	-4.18^	-	0.45	0.3	72.8	0.4973
C_{jel}	0.841	0.591	1.07	0.5	45.9	1.467
C_{je4}	0.57^	-	1.07	0.5	45.9	1.467
C_{bc3}	0.25	-0.25	0.646	0.35	27	1.083
C_{je3}	0.82^	-	1.07	0.5	45.9	2.082
C_{cs3}, C_{cs5}	-4.75	-5	0.45	0.3	72.8	0.477
C_{bc5}	0.25	-0.25	0.646	0.35	27	0.942

^ this capacitance was evaluated using its small-signal expression.

$$\tau_{PD} = 0.69\left\{\frac{r_e + r_b}{1+g_m r_e} C_{be3,4} + r_b C_{bci3,4}\left[1+\frac{g_m(r_c + R_C)}{1+g_m r_e}\right]\right.$$

$$\left. + (r_c + R_C)(C_{bci3,4} + C_{bcx3,4} + C_{cs3,4}) + R_C(C_L + C_{eq})\right\} \tag{4.20}$$

where the capacitance C_{eq} accounts for the collector-substrate and base-collector contributions, C_{cs} and C_{bc},

$$C_{eq} = C_{bci5,6} + C_{bcx5,6} + C_{cs5,6} \tag{4.21}$$

that are equal to those of Q3-Q4. Obviously, capacitances of Q3-Q4 are evaluated as those of Q1-Q2 for the inverter in Table 4.3a-b. The error found with respect to SPICE simulations is equal to that of the inverter previously discussed, as expected.

4.7 ACCURATE MODELING OF BIPOLAR CML MUX/XOR GATES AND EXTENSION TO ECL GATES

To improve the accuracy of the CML MUX/XOR model (4.18) and (4.20), it is useful to apply the methodology presented in Section 4.4 (by introducing five coefficients to be evaluated by a few simulation runs with a numerical procedure). To be more specific, the model associated with lower transistors (4.18) becomes

$$\tau_{PD} = K_1\left\{\frac{r_{el} + r_{bl}}{1+g_{ml} r_{el}}(C_{je1} + K_2 C_{D1}) + r_{bl} C_{bci1}\left[1+\frac{g_{ml}(r_{cl} + R_{eq})}{1+g_{ml} r_{el}}\right]\right.$$

$$+ (r_{cl} + R_{eq})(C_{bc1} + C_{cs1}) + \frac{1}{g_{mu}} C_{be3}$$

$$+ (2R_C + r_{cu} + r_{bu})(K_3 C_{bco3} + K_5 C_{bco5})$$

$$\left. + K_4(R_C + r_{cu})(C_{cso3} + C_{cso5}) + (R_C + r_{bu} + 2r_{eu})C_{je4}\right\} + 0.69 R_C C_L \tag{4.22}$$

where the values for the base-collector and collector-substrate capacitances of Q3 and Q5 are those used in the small signal model (i.e., in zero-bias condition). It should be noted that, differently from the inverter, coefficients $K_1...K_5$ are not introduced for all the parasitic capacitances so that the procedure keeps its intrinsic simplicity. Thus, to minimize error without increasing the number of the parameters, four of the coefficients are associated to the most critical capacitances, while one is used for all the propagation delay terms (i.e. all capacitances) except that associated with the load capacitance. To evaluate the five coefficients, the functional (4.9) is to be minimized after performing five simulations.

The accurate model parameters are summarized in Table 4.7 for both technologies. The error found with respect to simulations is always lower than 5%, and is typically about 2%.

TABLE 4.7

	BiCMOS process	HSB2 process
K_1	0.27	0.452
K_2	3.633	1.895
K_3	1.224	1.471
K_4	4.509	2.169
K_5	1.034	2.171

Regarding the delay associated with upper transistors (4.20), that is equal to that of the inverter, the same considerations made in Section 4.4 still hold, and coefficients are again those reported in Table 4.4.

The ideas discussed until now about the MUX/XOR CML gates can be immediately extended to the ECL counterparts, as done for the inverter in Section 4.5. To be more explicit, the ECL delay is evaluated by using relationships (4.14)-(4.15) and substituting the CML inverter delay by the MUX/XOR CML delay. Comparison with SPICE simulations shows that this model provides results that are even better than those of the inverter model, since the maximum error reduces to 13% and 14% for the BiCMOS and HSB2 process, respectively.

The simple ECL delay model can be further improved by introducing the strategy in Section 4.5 based on relationship (4.16) and unknown coefficients. More specifically, coefficients $K_1...K_5$ in Table 4.7 are introduced in the inner CML gate and the four coefficient α_1, α_2, H_1 and H_2 are introduced in the overall delay expression (4.16). The resulting error is always lower than 4% and 7% for the BiCMOS and HSB2 process, respectively.

4.8 EVALUATION OF CML/ECL GATES INPUT CAPACITANCE

In the previous sections, the delay of a logic gate was evaluated by assuming a generic load capacitance C_L, that in practical cases consists of the wiring capacitance and the input capacitance of the following gates. Therefore, to evaluate the delay of cascaded gates, it is fundamental to evaluate the input capacitance of a CML/ECL gate. To this end, let us consider the emitter-coupled pair of a CML/ECL inverter, that can be simplified into the linear circuit in Fig. 4.3.

To simplify the evaluation of the input capacitance C_{in} of the circuit in Fig. 4.3, let us neglect the small parasitic resistance r_e and apply the Miller theorem to the intrinsic and extrinsic base-collector capacitance. As a result, the input impedance can be represented with the linear circuit in Fig. 4.17.

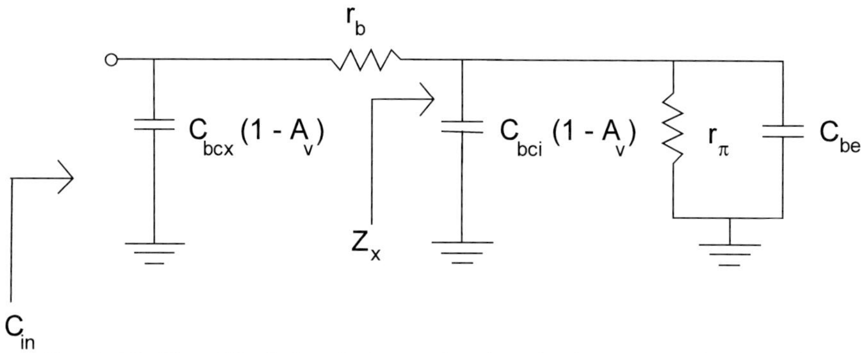

Fig. 4.17. Simplified circuit for evaluating the input capacitance of a CML/ECL gate.

In the circuit in Fig. 4.17, the input impedance is the parallel of that of $C_{bcx}(1\text{-}A_V)$ and the series of r_b and Z_x. If resistance r_b were equal to zero, the capacitive component of the input impedance would be the parallel of all capacitances $C_{bcx}(1\text{-}A_V)$, $C_{bci}(1\text{-}A_V)$ and C_{be} (in parallel to a resistance r_π). However, resistance r_b is far from being zero for current bipolar technologies, as was observed in regard to the delay model in Section 4.3, and cannot be neglected with respect to Z_x. Moreover, resistance r_b tends to significantly increase the impedance at the right of $C_{bcx}(1\text{-}A_V)$. As a consequence, as a first-order approximation, the input capacitance can be modeled by only the capacitance

$$C_{in} \approx C_{bcx}\left(1+g_m R_C\right)=C_{bcx}\left(1+\frac{I_{SS}R_C}{2V_T}\right) \tag{4.23}$$

which tends to underestimate the equivalent input capacitance, as the load effect on the previous gate is actually heavier, since the input impedance is actually lower than that of (4.23), due to the parallel of r_b+Z_x. It is worth noting that the same expression holds for the other CML/ECL logic gates.

Relationship (4.23) gives 138 fF and 109 fF for the BiCMOS and HSB2 technologies, respectively. The input capacitance of a CML/ECL gate can be evaluated from SPICE simulations by driving it through an inverter, and then evaluating the linear capacitance value that leads to the same delay. The simulated results obtained for the input capacitance of a CML/ECL gate are 120 fF and 100 fF for the BiCMOS and the HSB2 technologies, respectively. Therefore, relationship (4.23) is within 15 % of the input capacitance extracted from SPICE simulations. According to relationship (4.23), the simulated input capacitance of a CML/ECL gate was found to be essentially independent of the bias current for an assigned logic swing (or equivalently the product $R_C I_{SS}$).

The approximate relationship (4.23) was found by neglecting parasitic resistance r_e and approximating the input impedance to a single capacitance, i.e. the resistive contribution was neglected. According to the above considerations, this leads to a sufficient accuracy for practical purposes. However, a more accurate expression will be derived in Section 8.1.1, which is useful when a better accuracy is required.

4.9 BIPOLAR CURRENT-MODE D LATCH

The input capacitance evaluation discussed above allows for extending the model of MUX/XOR in Section 4.6 and 4.7 to the CML and ECL D latch gate, the first of which is depicted in Fig. 4.18 (the ECL gate is obtained by adding the output buffers). The only difference between the MUX/XOR gate and the D latch is due to the different load at the output nodes. Indeed, transistors Q5 and Q6, which are connected in a positive-feedback loop, load the output nodes with their equivalent input capacitance. Moreover, the effect of the positive feedback can be neglected since it is significant only for loop gain greater than unity, i.e. when the differential input voltage of Q5-Q6 (equal to the output voltage) crosses the gate logic threshold, after which the gate is assumed to have already switched by definition of gate delay. Since the feedback affects the circuit operation only at the end of the switching transient, we conclude that for most of the switching time the effect of feedback is negligible. As a consequence, the D latch delay is equal to that of the MUX/XOR gate loaded by a capacitance equal to the external capacitance C_L and the input capacitance (4.23) of the emitter-coupled pair Q5-Q6. Therefore, the D latch delay when the clock signal *CK* switches (often referred as the CK-Q latch delay [P01]) is given by relationship (4.18) where the load capacitance C_L is replaced by an equivalent capacitance C_L'

$$C_L' = C_L + C_{bcx5,6}\left(1+\frac{I_{SS}R_C}{2V_T}\right) \tag{4.24}$$

and the same observation holds for the delay associated with input D (often referred as D-Q latch delay), where the same substitution has to be carried out in relationship (4.20).

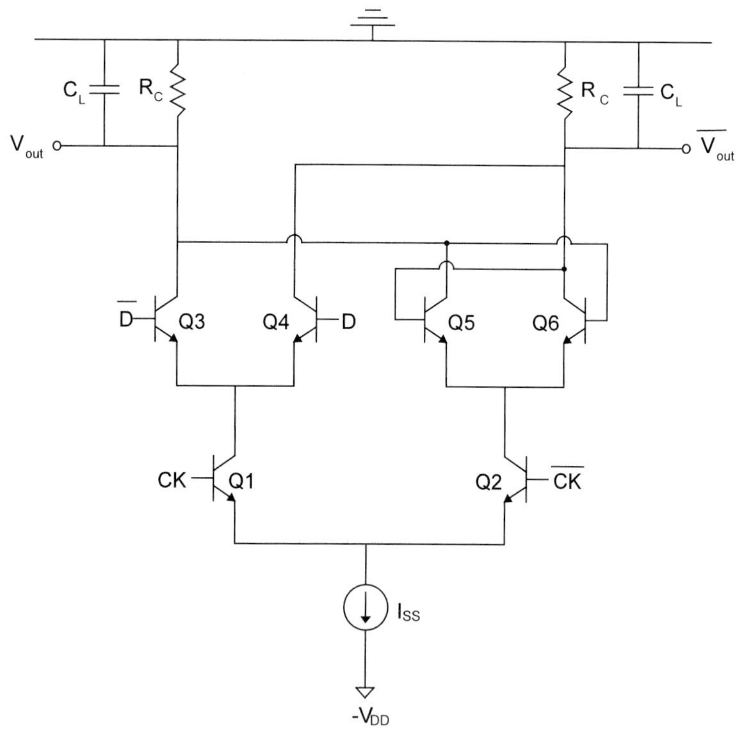

Fig. 4.18. CML D latch gate.

For the considered technologies and under the conditions above discussed, the transistor capacitances result equal to those of MUX/XOR gate in Table 4.6a-b. The only parameter which has to be modified in those Tables is the coefficient K_j of $C_{bc5,6}$, which is equal to 2.025 and 2.013 for the BiCMOS and HSB2 technology, respectively.

The simple CK-Q delay model (4.18) with (4.24) and the simulated delay of the D latch versus the bias current I_{SS} with a load capacitance C_L equal to 0 F, 100fF and 1pF, are plotted in Figs. 4.19a and 4.19b for the BiCMOS and HSB2 technologies, respectively.

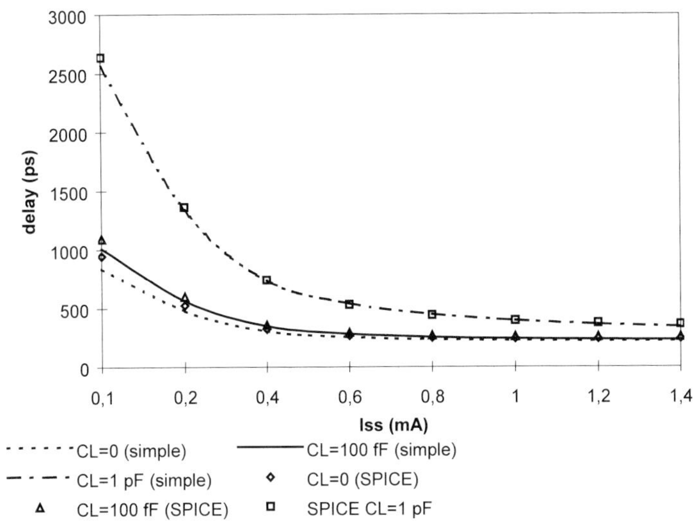

Fig. 4.19a. Simple CK-Q model (4.18) with (4.24) and simulated delay vs I_{SS} for the BiCMOS process.

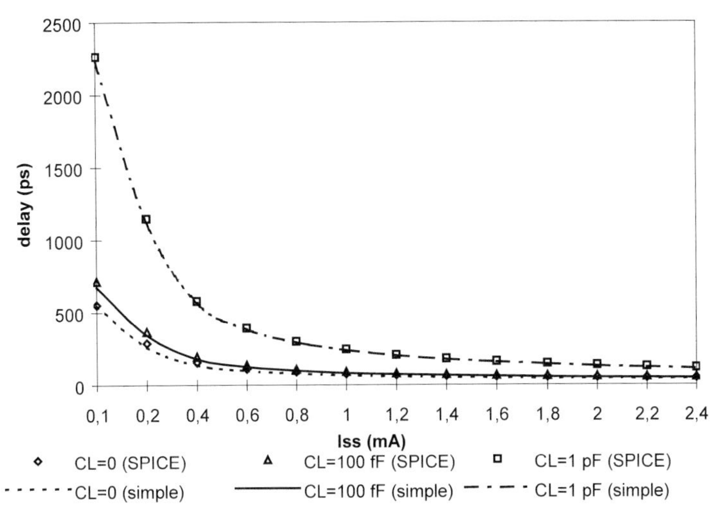

Fig. 4.19b. Simple CK-Q model in (4.18) and (4.24) and simulated delay vs. I_{SS} for the HSB2process.

The error, plotted in Figs. 4.20a and 4.20b, is always lower than 20%, and decreases as increasing the load capacitance, as for the MUX/XOR gate. Analogous results are obtained for the D-Q delay, for which the maximum error is always lower than 18%.

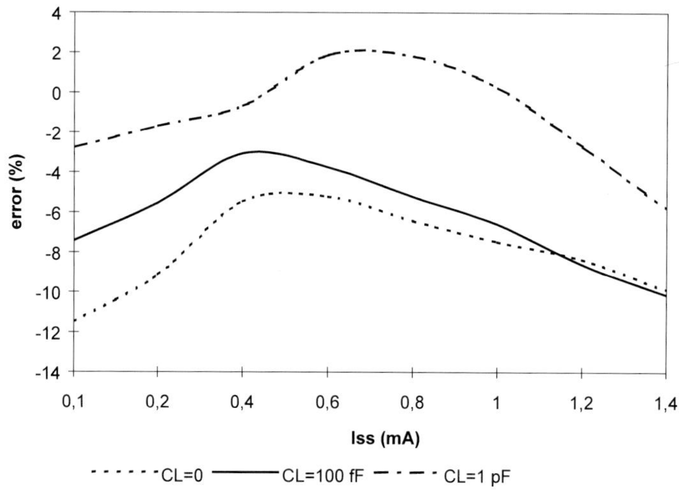

Fig. 4.20a. Error of the simple CK-Q delay model (4.18) with (4.24) vs. I_{SS} for the BiCMOS process.

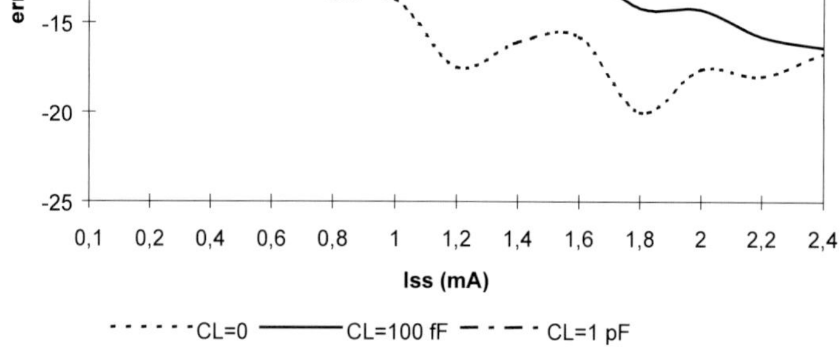

Fig. 4.20b. Error of the simple CK-Q delay model (4.18) with (4.24) vs. I_{SS} for the HSB2 process.

The accuracy of the simple model can be improved by introducing suitable unknown coefficients, as for the MUX/XOR gate. Since the loading effects of Q5 and Q6 are estimated in an approximated manner, the accurate model is obtained from the one developed for the MUX/XOR by adding a further term which accurately takes into account the loading effect associated with the input capacitance of transistors Q5 and Q6. According to relationships (4.22) and (4.24), in the model of the MUX/XOR gate we replace the load capacitance C_L by

$$C_L^{'} = C_L + C_6 \tag{4.25}$$

where the unknown capacitance C_6 is found by minimizing the functional (4.17). For the two technologies considered, the value of the unknown coefficients K_1...K_5 and C_6 is reported in Table 4.8.

TABLE 4.8

	BiCMOS process	HSB2 process
K_1	0.6	0.901
K_2	1.485	0.858
K_3	0.922	1.014
K_4	0.939	1.014
K_5	0.922	1.014
C_6	165E-15	53.7E-15

The error found for the accurate model of the CK-Q D latch delay, plotted in Figs. 4.21a and 4.21b for the two processes considered, is always lower than 5%, and is typically about 2%. Thus the same accuracy as the MUX/XOR is achieved. Analogous results are obtained for the accurate D-Q delay model.

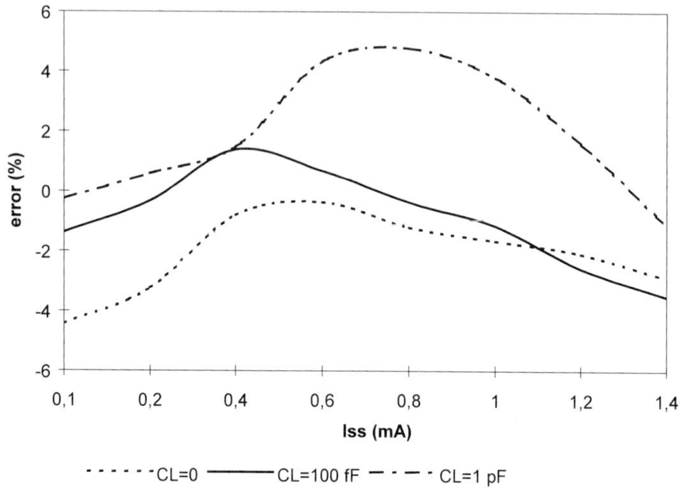

Fig. 4.21a. Error of the accurate CK-Q delay model (4.22) with (4.25) vs. I_{SS} for the BiCMOS process.

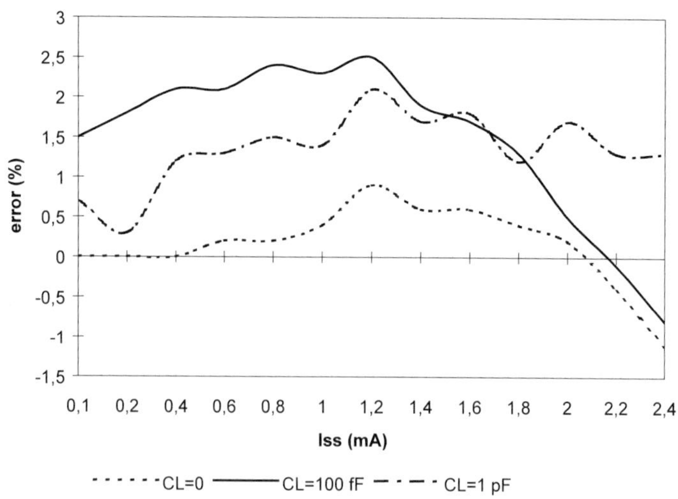

Fig. 4.21b. Error of the accurate CK-Q delay model (4.22) with (4.25) vs. I_{SS} for the HSB2 process.

Chapter 5

OPTIMIZED DESIGN OF BIPOLAR CURRENT-MODE GATES

In this chapter, a general methodology to maximize the speed performance and to manage the power-delay trade-off in CML and ECL gates is presented and applied to several fundamental gates.

5.1 INTRODUCTION TO OPTIMIZED METHODOLOGY IN CML GATES

The bias current of a bipolar Current-Mode gate determines its power consumption and significantly affects the delay, as shown in Chapter 4. Since the power consumption in bipolar Current-Mode gates is rather large in practical cases, an efficient strategy to manage the power-delay trade-off and to optimize the speed performance is highly desirable. Such a design strategy can be derived by analytically expressing the power-delay trade-off, or equivalently a delay expression as an explicit function of the bias current I_{SS}.

The delay expressions introduced in Chapter 4 depend on process and design parameters, that consist of the bias current I_{SS} and the load resistance R_C. In the design of bipolar gates, a suitable value of the logic swing has to be chosen from considerations on noise margin as discussed in Chapter 2, thus the product $R_C I_{SS}$ is constant and equal to $V_{SWING}/2$ from relationship (2.9). As a consequence, delay expressions reported in Chapter 4 can be expressed as a function of only I_{SS}, once the logic swing has been preliminarily set.

It will be shown in the following subchapters that the delay of a generic CML gate as a function of the bias current can be expressed as

$$\tau_{PD} = aI_{SS} + \frac{b}{I_{SS}} + c \tag{5.1}$$

where coefficients *a*, *b* and *c* depend on the logic swing, the power supply voltage and process parameters. We will show that coefficient *a* depends on the base-emitter diffusion capacitance of switching transistors, while coefficient *b* is due to an equivalent capacitance at the output node.

Delay (5.1) can be minimized by properly setting the bias current, as can be shown by differentiating (5.1) for I_{SS} and setting the result to zero. The minimum propagation delay, τ_{PDop}, is achieved by setting the bias current to

$$I_{SSop} = \sqrt{\frac{b}{a}} \tag{5.2}$$

and it is equal to

$$\tau_{PDop} = 2\sqrt{ab} + c \approx 2\sqrt{ab} \tag{5.3}$$

where the term *c* can be usually neglected, as will be discussed in the next subchapters.

Relationship (5.1) can also be used to express the power-delay product *PDP*, that quantitatively measures the efficiency of the power-delay trade-off [O99], [CBF01], giving

$$PDP = V_{DD} I_{SS} \tau_{PD} \approx V_{DD}\left(aI_{SS}^2 + b\right). \tag{5.4}$$

PDP always increases while increasing the bias current, which means that the power efficiency increases when the bias current is reduced. However, it is useful to observe that the power-delay product can be reduced only to some extent, since it tends to be constant for very low bias currents such that $aI_{SS}^2 << b$.

Reducing power dissipation is a key issue in modern digital circuits. Hence, a design strategy to achieve both high speed and reduced power dissipation is a target to pursue. A high-speed performance can be obtained by setting the bias current to its optimum value (5.2), while a reduction of power consumption can be only obtained by using a lower bias current [AP991], [AP00]. These opposite requirements can be accomplished by observing that, from relationship (5.4), the power-delay product decreases with the bias current through a square law, thus for a given bias current reduction a lower delay increase is expected. In other words, it is possible to reduce the power dissipation while paying for a smaller speed penalty with respect to the optimum case (5.2)-(5.3).

To evaluate the power-delay trade-off and quantitatively express the speed penalty due to the bias current reduction, we introduce the ratio T_{PD} between the propagation delay and its optimum value τ_{PDop} (i.e., $T_{PD}=\tau_{PD}/\tau_{PDop}$). We will also normalize the current to its optimum value I_{SSop} (i.e., $I_N = I_{SS}/I_{SSop}$), and from relationships (5.1) and (5.3) we get

$$T_{PD} = \frac{\sqrt{ab}\left(I_N + \dfrac{1}{I_N}\right) + c}{2\sqrt{ab} + c} \approx \frac{1}{2}\left(I_N + \frac{1}{I_N}\right) \tag{5.5}$$

where the coefficient c was assumed to be lower than $\sqrt{ab}$, as occurs in practical cases. It is worth noting that, even in the unrealistic case $c=\sqrt{ab}$, the error on the approximated expression of (5.5) with respect to (5.1) for values of I_N typically greater than 0.5 is lower than 7%.

The expression (5.5) of T_{PD} is independent of the circuit and process parameters. Moreover, inspection of its plot, depicted in Fig. 5.1, shows that a reduction in the bias current around its optimum value (i.e., I_N lower than 1) determines only a small increase in the resulting propagation delay, as expected.

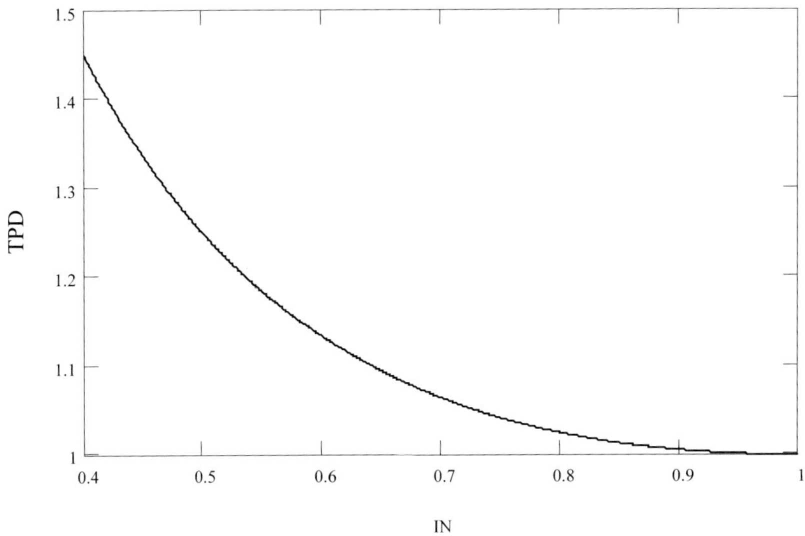

Fig. 5.1. T_{PD} versus I_N through relationship (5.5).

As an example, Fig. 5.1 shows that a 50% bias current reduction with respect to the optimum value (i.e., I_N=0.5) leads to a delay increase by only 25%. Moreover, a further bias current reduction leads to a greater delay increase that is not usually acceptable, as can be seen for a 60% bias current reduction which gives a 45 % increase in the propagation delay with respect to the optimum case. This is because for I_N<0.5 the term $1/I_N$ in relationship (5.5) tends to dominate over I_N, yielding

$$T_{PD} \approx \frac{1}{2}\frac{1}{I_N} \tag{5.6}$$

which shows that, for a given reduction of the bias current by a factor $1/I_N$ with respect to I_{SSop}, the related increase of the delay factor is about halved. From a power-delay trade-off point of view, this is easily explained by observing that for I_N<<0.5 (i.e. $aI_{SS}^2 << 0.5 \cdot b$) the power-delay product (5.4) no longer decreases as decreasing the bias current, but tends to be constant.

In our opinion, a good trade-off between power and speed, which will be used in the following, is achieved by setting I_N=0.6 (i.e., a bias current equal to 60% of the optimum value). This choice determines a propagation delay which is only 10% worse than the optimum value, while it reduces the power dissipation by 40% with respect to that needed in an optimum design [AP991].

5.2 OPTIMIZED DESIGN OF THE CML INVERTER

In the previous section, a general design methodology of CML circuits has been discussed without referral to the specific gate considered. Now, the inverter gate is analyzed from a design point of view, by resorting to the concepts and notations above introduced. Since in general the transistor emitter area is a design parameter, the case of minimum-area transistors is first dealt with, and then extended to the more general case of a larger area.

5.2.1 Design with minimum transistor area

Consider a CML inverter gate, whose design parameters are the bias current I_{SS}, the load resistance R_C and the transistor emitter area A_E. The latter parameter A_E defines the transistor resistance and capacitance parasitics, and, for the sake of simplicity, is assumed to be equal to the minimum value allowed by the adopted process, therefore in the design is a constant. Moreover, as clarified in Section 5.1, the load resistance R_C has to be set equal to $V_{SWING}/2I_{SS}$ to achieve the desired logic swing preliminarily chosen (typically around 400-500 mV), therefore the only design parameter is the bias current.

The explicit delay dependence on the bias current can be found by properly manipulating the delay model (4.7). In particular, by assuming $g_m r_e$<<1 as in practical cases[1] and substituting $R_C=V_{SWING}/2I_{SS}$, the delay expression (4.7) can be written in the form of relationship (5.1) with coefficients a, b and c being equal to

$$a = 0.69\left(4\frac{r_e + r_b}{V_{SWING}}\tau_F + r_b C_{bci}\frac{r_c}{2V_T}\right) \approx 2.76\frac{r_e + r_b}{V_{SWING}}\tau_F \tag{5.7a}$$

$$b = 0.35 \cdot V_{SWING}\left(C_{bci} + C_{bcx} + C_{cs} + C_L\right) \tag{5.7b}$$

$$c = 0.69\left[r_b C_{bci}\left(1 + \frac{V_{SWING}}{4V_T}\right) + r_c\left(C_{bc} + C_{cs}\right) + \left(r_e + r_b\right)C_{je}\right] \tag{5.7c}$$

where coefficient a (responsible for the delay increase for $I_{SS}>I_{SSop}$) is essentially due to the base-emitter diffusion capacitance (4.6) through transit time τ_F, while coefficient b is due to the total capacitance at the output node (represented by the term in brackets in (5.7b)).

From relationship (5.7) and (5.2)-(5.3), the resulting values of optimum bias current I_{SSop} and the correspondent minimum delay τ_{PDop} are

$$I_{SSop} = \sqrt{\frac{b}{a}} = \frac{V_{SWING}}{2}\sqrt{\frac{C_{bc} + C_{cs} + C_L}{2\left(r_e + r_b\right)\tau_F}} \tag{5.8}$$

$$\tau_{PDop} \approx 2\sqrt{ab} \approx 2\sqrt{\left(r_e + r_b\right)\tau_F\left(C_{bci} + C_{bcx} + C_{cs} + C_L\right)} \tag{5.9}$$

Relationship (5.8) shows that the optimum bias current, and hence the power consumption, proportionally increases as increasing the logic swing, whereas the minimum delay achievable (5.9) is independent of the logic swing. Thus, the logic swing should be kept as low as possible, in order to exploit the speed potential of the technology used without wasting an excessive power.

---

[1] In actual cases, $g_m r_e$ tends to reduce the voltage gain $A_V=g_m R_C/(1+g_m r_e)$. Since CML gates have a small logic swing and thus a low noise margin, a decrease of A_V is not unacceptable, and is avoided by satisfying the condition $g_m r_e$<<1.

Relationships (5.8) and (5.9) clarify the usual belief that CML circuits' speed is due to their small logic swing. Indeed, from (5.9) this assumption fails when the optimum bias current (5.8) is used, while is correct when the delay dependence on V_{SWING} is evaluated with an assigned bias current. This can be understood by observing that for $I_{SS}<I_{SSop}$ the dominant term in delay (5.1) is b/I_{SS}, that from (5.7b) can be lowered by decreasing the logic swing.

Relationship (5.9) can also be used to identify the main speed limits while optimizing a bipolar process. Indeed, by assuming the inverter to be driving other CML gates modeled by their input capacitance (4.23) (i.e., C_L proportional to C_{in}), the main capacitive contributions are the diffusion capacitance (through the transit time τ_F), as well as the sum of the extrinsic base-collector capacitance and the collector-substrate capacitance. Finally, the main resistive contributions are the base and the emitter resistance, of which the first is usually the dominant term.

Of course, even though the optimum case has been discussed so far, the same design strategy to manage the power-delay trade-off as in Section 5.1 can be also used to reduce the power consumption without significantly degrading the speed performance.

5.2.2 Design with non-minimum transistor area

When non-minimum transistors are used, they are (or can be assumed to be) made up of the parallel of N unitary transistors, each of which being characterized by junction capacitances C_{bci}, C_{bcx}, C_{cs} and C_{je}, as well as parasitic resistances r_b, r_e and r_c. These unitary transistors connected in parallel can be modeled as a single transistor with each junction capacitance equal to that of the minimum-area transistor multiplied by a factor N, and parasitic resistances equal to those of the minimum-transistor divided by the factor N [AM88]. As a consequence, from inspection of (5.7), substituting the parasitic capacitances and resistances, relationship (5.1) must be rewritten as

$$\tau_{PD} = a\frac{I_{SS}}{N} + \left(b_1 + \frac{b_0}{N}\right)\frac{N}{I_{SS}} + c \tag{5.10}$$

where a and c are again given by (5.7a) and (5.7c), while b_0 and b_1 are

$$b_0 = 0.35 \cdot V_{SWING} C_L \tag{5.11a}$$

$$b_1 = 0.35 \cdot V_{SWING}\left(C_{bci} + C_{bcx} + C_{cs}\right) \tag{5.11b}$$

The delay expression (5.10) is minimized by setting the bias current to

$$I_{SSop}(N) = \sqrt{\frac{b_1 + \frac{b_0}{N}}{a}} = N\sqrt{\frac{1+\frac{b_0}{Nb_1}}{1+\frac{b_0}{b_1}}} I_{SSop}(1) \tag{5.12}$$

that was found by equating to zero the derivative of (5.10) with respect to I_{SS}, and defining $I_{SSop}(1)$ as the optimum current (5.8) found for minimum area (i.e, for N=1). The resulting minimum delay as a function of N, found by substituting (5.12) into (5.10), results in

$$\tau_{PDop}(N) = 2\sqrt{a\left(b_1 + \frac{b_0}{N}\right)} + c \approx \sqrt{\frac{1+\frac{b_0}{Nb_1}}{1+\frac{b_0}{b_1}}}\tau_{PDop}(1) \tag{5.13}$$

where $\tau_{PDop}(1)$ is the minimum delay (5.9) obtained for N=1. By observing that the term in the square root of (5.13) is lower than unity, the delay of the CML gate can be reduced by increasing the transistor emitter area, but this is obtained at the cost of a greater power dissipation from (5.12). To be more specific, when the load capacitance C_L is much greater than the parasitic capacitances at the output node, from (5.11) the ratio b_0/b_1 is much higher than unity. As a consequence, the delay reduction and the increase in the bias current are both proportional to $\sqrt{N}$ from relationships (5.12) and (5.13), which means that the power-delay product is equal to that already found assuming minimum-area transistors. In this case, the power efficiency of the CML is equal to that achieved by optimizing with a unitary transistor, since a given delay reduction is achieved at the price of an equal increase of the power consumption.

On the other hand, with a small capacitive load, the delay reduction is low compared to the increase in bias current, which is proportional to N from (5.12). Therefore, an inconvenient amount of power dissipation is needed.

Unless under very high speed constraints, the above approach is seldom useful, since the other fundamental target is usually to minimize the power dissipation, and a bias current greater than $I_{SSop}(1)$ is rarely tolerable. Thus, in practical cases a different approach should be followed, as explained in the following. In particular, let us consider relationship (5.10) for the fixed bias current value $I_{SSop}(1)$ and optimize it for parameter N (i.e., differentiate it for N and, setting the result to zero, solve for an integer N). The resulting optimum value of N is

$$N' = \mathrm{int}\left(\sqrt{\frac{a}{b_1}} I_{SSop}(1)\right) = \mathrm{int}\left(\frac{I_{SSop}(1)}{I_{SSop}(1)\big|_{C_L=0}}\right) \tag{5.14}$$

where the function $int(x)$ gives the maximum integer value that is not greater than the argument x.

From (5.14), the number of unitary transistors needed to reduce the propagation delay for $I_{SS}=I_{SSop}(1)$ is equal to the ratio of the optimum current with a unitary area and the optimum current with a unitary area evaluated without a load capacitance. Using these design criteria, after substituting (5.14) and (5.8) into (5.10), the resulting delay is

$$\tau_{PD}\left(I_{SSop}(1), N'\right) = 2\sqrt{ab_1} + b_0\sqrt{\frac{a}{b_1 + b_0}} + c \tag{5.15}$$

By comparing relationship (5.15) with that evaluated under unitary area (5.9), we see that the emitter area optimization with the bias current set to $I_{SSop}(1)$ allows for a delay reduction at the cost of an increased area occupation.

The increase in the transistor area can also be helpful when the optimum current for the unitary case, $I_{SSop}(1)$, leads to non-negligible high-injection level effects, that determine an increase in the transistor transit time and thus reduce the transistor speed performance [AM88]. However, it is worth noting that the increase with respect to the unitary area leads to a proportional increase in the input capacitance of the CML gate, from (4.23). Thus, if the gate under design is the main load of a previous gate, it may not be convenient to design a transistor with a non-unitary area.

Obviously, the same design criteria highlighted in Section 5.1 using a current which is 40% lower than the optimum value can be applied here to reduce the power consumption with respect to the optimum case, without significantly degrade the speed performance.

5.2.3 Design examples

To better understand the design procedure previously described, let us consider the design of a CML inverter gate, by using both the BiCMOS and the HSB2 technology discussed in Chapter 4, and assuming fan-out values equal to 1 and 10. By assuming the gates to be powered with 5 V and with a logic swing V_{SWING}=500 mV, we obtain the data summarized in Tables 4.3-4.4.

As already discussed in Section 4.8, for the BiCMOS technology the two load conditions are equivalent to 120 fF and 1.2 pF load capacitance, respectively. The resulting optimum bias currents from (5.8) are 0.7 mA and 1.8 mA, which lead to a propagation delay of 129 ps and 277 ps (the values are very close to those given by SPICE simulations, which are 126 ps and 276 ps). For a fan-out of 10 the current is slightly higher than the high-injection level, but there is no appreciable degradation in the propagation delay. However, since the optimum current evaluated with C_L=0 is 0.4 mA, according to (5.14) the delay can be further reduced to 166 ps by setting the number of unitary transistors equal to 4. In this case, the simulated delay is 15% higher than that predicted analytically.

For the HSB2 technology, the two fan-out conditions above considered are equivalent to 100 fF and 1 pF load capacitance, respectively. The optimum bias currents are 3.4 mA and 9.4 mA, which lead to a nominal propagation delay of 19 ps and 42 ps. While for the former case the analytical results are very close to the SPICE simulation, which gives 21 ps, with the bias current of 9 mA the transistors suffers from strong high-injection level effects, and the analytical propagation delay obtained is not realistic. According to (5.14), we set the number of unitary transistors to 3. Thus the delay is reduced to 30 ps, which differs from the simulated value by 10%, and avoids the strong condition of high-injection level.

A reduction of power dissipation can be achieved by decreasing the bias current by 40% with respect to the optimum value, that only leads to a small increase in the propagation delay, as discussed in Section 5.1.

5.3 OPTIMIZED DESIGN OF THE ECL INVERTER

The delay of an ECL gate given by (4.15) consists of two main terms (see Section 4.5), one associated with the inner CML gate and the other due to the output buffer, thus we split the optimized design into two steps.

First the bias current I_{SS} of the CML stage has to be computed to minimize its delay $\tau_{PD(CML)}$. As discussed in Section 4.5, the delay contribution $\tau_{PD(CML)}$ of the inner CML gate is evaluated by setting its load capacitance to zero, thus its optimum bias current is found by following the same strategy as in Section 5.2 by setting C_L=0 into (5.8). From (5.2) the optimum bias current results to be

$$I_{SSop} = \frac{V_{SWING}}{2}\sqrt{\frac{C_{bc1}+C_{cs1}}{2(r_e+r_b)\tau_F}} \tag{5.16}$$

The delay contribution of the output buffer $\tau_{PD(CC)}$ in (4.15) should also be reduced to minimize the ECL inverter delay. However, relationship (4.15) tends asymptotically to a minimum value for the output buffer bias current I_{CC} approaching infinity. Therefore, a design strategy to keep (4.15) as low as possible with reasonable values of I_{CC} is required. To this end, observe that, during the falling output signal transition, if the emitter voltage cannot follow the base voltage at the same speed, the CC stage cuts off, and this results in a much larger delay with respect to the value resulting from relationship (4.15), that was derived by assuming the CC stage to be working in the linear region. As a consequence, we have to guarantee that the CC stage works almost in the linear region most of the time.

A simple design strategy allowing the output buffer transistor to work in the linear region was proposed in [KB92]. This strategy is based on the consideration that the medium current value which discharges the load capacitance is $C_L \Delta V_O/\Delta t$, where ΔV_O is the voltage swing at each output node (equal to $V_{SWING}/2$) and Δt is the time required for the output transition (approximately equal to $2\tau_{PD(CC)}$, since the delay $\tau_{PD(CC)}$ is needed for a half output transition). Thus, in [KB92] the bias current I_{CC} is set equal to the average discharge current (i.e., $I_{CC}=C_L V_{SWING}/2\tau_{PD(CC)}$), and a bias current greater than this critical value would not substantially reduce the delay. However, this choice has two drawbacks. The first is the dependence of $\tau_{PD(CC)}$ on I_{CC}, that from the expression $I_{CC}=C_L V_{SWING}/2\tau_{PD(CC)}$ leads to a dependence of I_{CC} on itself. The second is the linear dependence of I_{CC} on the load capacitance, which can determine excessive power dissipation for the practical ranges of C_L. To overcome these drawbacks, an alternative approach is outlined below.

The delay of the common collector in (4.15) can be written as

$$\tau_{PD(cc)} = 1.6\sqrt{(R_C + r_b)\left(\tau_F + \frac{V_T}{I_{CC}} C_{je2}\right) C_L}$$

$$= 1.6\sqrt{\left(\frac{\tau_{PD(cc)\min}}{1.6}\right)^2 + (R_C + r_b)\frac{V_T}{I_{CC}} C_{je2} C_L} \qquad (5.17)$$

where $\tau_{PD(CC)min}$ is the asymptotic value of $\tau_{PD(CC)}$ reached for $I_{CC}\rightarrow\infty$ (i.e. $\tau_{PD(CC)\min} = 1.6\sqrt{(R_C + r_b)C_L\tau_F}$). Therefore, after defining the output buffer delay normalized to its asymptotic minimum τ_N (i.e., $\tau_N=\tau_{PD(CC)}/\tau_{PD(CC)min}$), relationship (5.17) can be inverted to express the bias current as

$$I_{CC} = \frac{C_{je2}}{\left(\tau_n^2 - 1\right)\tau_F} V_T \tag{5.18}$$

that allows for evaluating the bias current I_{CC} needed to achieve a given value of τ_n. From simulations, it was found that the resulting I_{CC} maintains the CC stage in the linear region most of the time if τ_n<1.4. Therefore, under this condition, the delay evaluation in Section 4.5 and thus relationship (5.18) are valid.

As a rule of thumb to achieve a low CC propagation delay (i.e., close to the asymptotic value) with a low power dissipation, it is suggested to set τ_n=1.1, that leads to a CC delay worse by 10% with respect to the asymptotic value. It is worth noting that, for practical load capacitance values, (5.18) gives a lower value of the bias current, compared to that resulting from [KB92]. This can be seen by comparing relationship (5.18) with that proposed in [KB92] (i.e., $I_{CC}=C_L V_{SWING}/2\tau_{PD(CC)}$). Indeed, straightforward calculations show that the latter is lower for load capacitance values such that

$$C_L < \frac{10}{\left(\tau_n^2 - 1\right)^2} \left(\frac{V_T}{V_{SWING}}\right)^2 \frac{C_{je2}^2}{C_{D2}} \approx \frac{C_{je2}^2}{C_{D2}} \tag{5.19}$$

where the right-hand side represents a capacitance much lower than the input capacitance of a bipolar current-mode gate (4.23), since C_{je2}/C_{D2} is surely lower than unity and C_{je2} is lower than the input capacitance (4.23). Therefore, in practical cases where at least a unitary fan-out is considered, condition (5.19) is certainly satisfied.

By following the approach discussed, the power dissipation of the optimized ECL inverter is found to be independent of the load capacitance. It is, in fact, composed of the CML and CC stage contributions which are constant with respect to C_L. In order to reduce the power dissipation, the bias current I_{SS} can be conveniently chosen lower than the optimum value (5.16) by 40%, as discussed in Section 5.1.

To better understand the design criteria introduced for the bias current of the output buffers, let us consider the HSB2 technology and design the ECL inverter gate for a fan-out of 1 and 10. The optimum bias currents are independent of the load, and their values are I_{SSop}=1.7 mA and I_{CC}=0.8 mA. The delay of the differential stage is independent of the load, and its value is 12 ps. Thus the resulting propagation delays are 30 ps and 72 ps for a load capacitance of 0.1 pF and 1 pF, respectively. These values are very close to 28 ps and 86 ps which are those given by SPICE.

Regarding the transistor area, the transistors of the inner CML gate must be minimum sized. Indeed, the CML delay is evaluated for C_L=0, thus coefficient b_0 in (5.11a) is equal to zero and, from relationship (5.13), the delay cannot be reduced by increasing the emitter area, or equivalently the factor *N*.

From simulations, it can be seen that there exists an optimum emitter area of the output buffers transistors that minimizes the delay. This is because the area increase leads to a reduction of resistance r_b in the buffer delay (4.14) (even though this is not significantly beneficial, since $R_C \gg r_b$) and at the same time to an increase in the capacitance C_{bc} (that was neglected in (4.14)). Therefore, due to the two opposite effects of the buffer delay, an optimum exists. However, since the buffer delay is only slightly affected by the emitter area, this optimization leads to a very low delay reduction, that is in the order of a few percentage points, as found from simulations.

From these considerations, it is clear that the output buffer transistors' area should be increased only to avoid high-injection level effects. To be more specific, the high-injection condition must be avoided at the average current that flows in the buffer transistors, that is evaluated in the following according to the method in [KB92]. Since the buffer delay is defined as the time needed by each output node to cross half of its overall voltage swing (that is equal to $V_{SWING}/2$), the average value of the current flowing in the load capacitance during the charge transition is

$$\bar{i}_{C_L} = \frac{V_{SWING}}{4}\frac{C_L}{\tau_{PD(CC)}} \tag{5.20}$$

where $\bar{i}_{C_L}$ is equal to the difference between the average emitter current, $\bar{i}_E$, and the bias current I_{CC}. As a result, by substituting the optimized bias current (5.18) into (5.20), the average emitter current is equal to

$$\bar{i}_E = I_{CC} + \frac{V_{SWING}}{4}\frac{C_L}{\tau_{PD(CC)}} = \frac{C_{je2}}{(\tau_n^2 - 1)\tau_F}V_T + \frac{V_{SWING}}{4}\frac{C_L}{\tau_{PD(CC)}} \tag{5.21}$$

To avoid high-injection level effects, the emitter area of the buffer transistors has to be set to the ratio of the average emitter current (5.21) and the maximum current after which the high-injection level occurs in a minimum-area device, that is a characteristic parameter of the process used. For both the BiCMOS and HSB2 technologies, the emitter area increase is not required, because of the high value of the current after which the high-injection level occurs.

5.4 COMPARISON BETWEEN THE CML AND ECL INVERTER

It is a common belief that ECL gates exhibit superior speed performance compared to CML gates. However, this does not appear so evident by comparing these gates with an equal bias current per gate. Indeed, the CML bias current is entirely spent to provide current I_{SS}, while in the ECL gate the current is divided between I_{SS} and the bias current I_{CC} of each of the two output buffers. Therefore, in the ECL gate, the delay of the inner CML may be lower than that of the CML gate (due to the lower capacitive load), but the output buffer introduces a further delay (4.14). As a consequence, a quantitative analysis is required to better understand the performance of CML and ECL gates in terms of delay, power consumption and their trade-off.

To compare the speed performance of CML and ECL gates designed for minimum delay, the optimum CML delay (5.3) is rewritten using (5.11) as

$$\tau_{PD(CML)op} = 2\sqrt{a\left(b_1 + 0.35V_{SWING}C_L\right)} + c \tag{5.22}$$

Note that term b_1 represents term b without the contribution due to the load capacitance

$$b_1 = b - 0.35V_{SWING}C_L \tag{5.23}$$

Thus the ECL delay can be rearranged as

$$\tau_{PD(CML)op} = 2\sqrt{ab_1} + c + 1.6\sqrt{\left(R_C + r_b\right)\tau_F C_L} \tag{5.24}$$

By solving the inequality $\tau_{PD(CML)op}<\tau_{PD(ECL)op}$, it is found that when $R_C>2.4r_e+1.3r_b$, which is generally true for high-speed technologies, $\tau_{PD(CML)}$ is always lower than $\tau_{PD(ECL)}$ [AP991]. On the other hand, in the infrequent case $R_C<2.4r_e+1.3r_b$, the inequality $\tau_{PD(CML)op}<\tau_{PD(ECL)op}$ is true for load capacitance values such that

$$C_L < 15\frac{\left(r_e + r_b\right)\left(R_C + r_b\right)}{\left(2.4r_e + 1.3r_b - R_C\right)^2}\left(C_{be} + C_{cs}\right) \tag{5.25}$$

where the value given by the ratio is usually greater than one. The right-hand side of relationship (5.25) represents an impractically high load capacitance in high-speed circuits, thus in actual cases the CML gate is again faster than the ECL. Summarizing, the CML gates have always a better speed performance than ECL gates with modern technologies, and as opposite to the traditional assumption, this property also holds for a high capacitive loads. To further illustrate this, the delay of optimized CML and ECL gate are plotted versus the load capacitance for the BiCMOS and HSB2 technologies in Figs. 5.2a and 5.2b, where a minimum-delay design under the conditions previously used is assumed.

The speed advantage of the CML gate is achieved at the cost of a higher power consumption, in some cases. Indeed, the optimized CML and ECL gates require a bias current equal to

$$I_{CML} = I_{SSop} = \frac{V_{SWING}}{2}\sqrt{\frac{C_{bc}+C_{cs}+C_L}{2(r_e+r_b)\tau_F}} \tag{5.26}$$

$$I_{ECL} = I_{CMLop} + 2I_{CC} = \left[\sqrt{\frac{C_{bc1}+C_{cs1}}{2(r_e+r_b)\tau_F}}\frac{V_{SWING}}{2} + \frac{2C_{je2}}{\left[\tau_n^2-1\right]\tau_F}V_T\right] \tag{5.27}$$

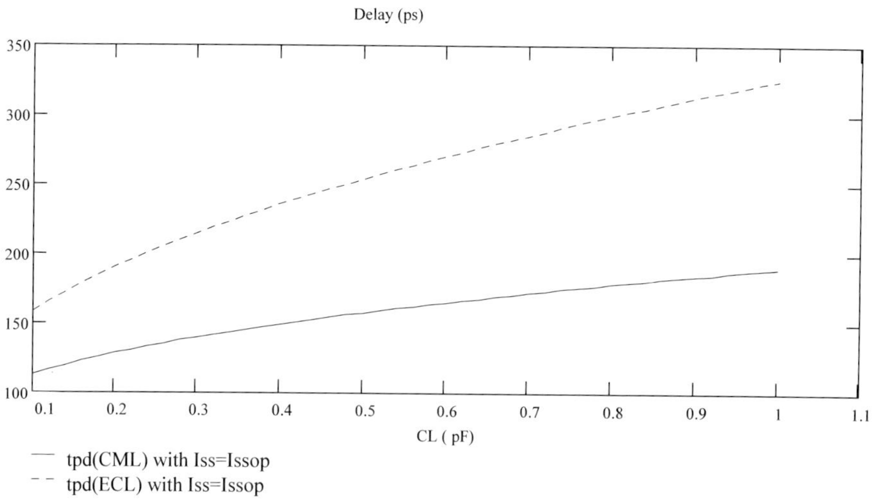

Fig. 5.2a. CML and ECL delay versus C_L for the optimized design (BiCMOS technology).

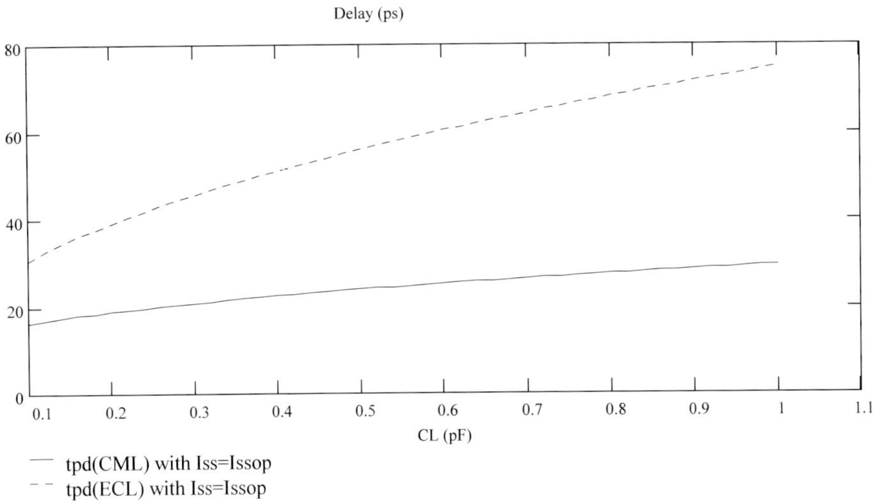

Fig. 5.2b. CML and ECL delay versus C_L for the optimized design (HSB2 technology).

From a straightforward analysis, it can be seen that the current value given by relationship (5.26) is lower than that given by (5.27) for load capacitance values such that

$$C_L < \frac{1}{\tau_n^2 - 1}\left[4\frac{(r_e + r_b)C_{je2}}{(\tau_n^2 - 1)\tau_F}\frac{V_T}{V_{SWING}} + \sqrt{\frac{2(r_e + r_b)(C_{bc1} + C_{cs1})}{\tau_F}}\right]\frac{V_T}{V_{SWING}} C_{je2} \tag{5.28}$$

The right-hand side of relationship (5.28) results equal to 0.6 pF and 0.1 pF, for the BiCMOS and HSB2 process, as illustrated in detail in Figs. 5.3a and 5.3b, where the bias current required by the optimized CML and ECL gate is plotted versus the load capacitance.

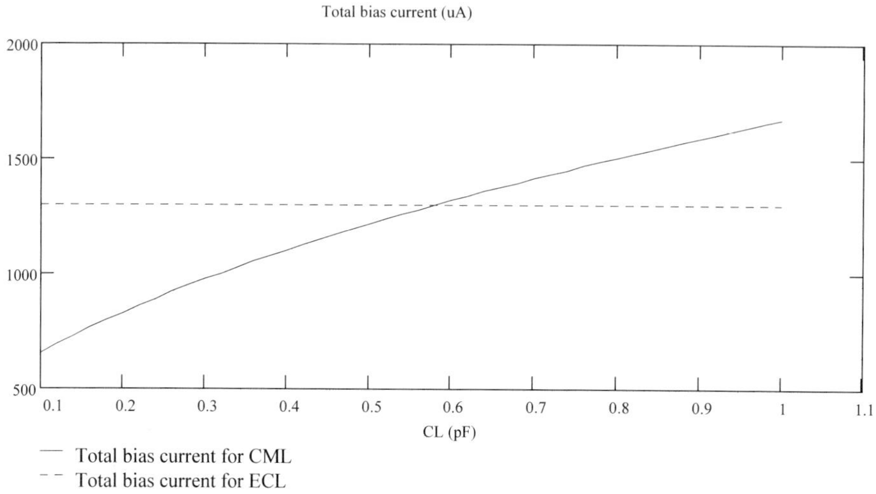

Fig. 5.3a. Total bias current of the CML and ECL inverters versus C_L for optimized design (BiCMOS process).

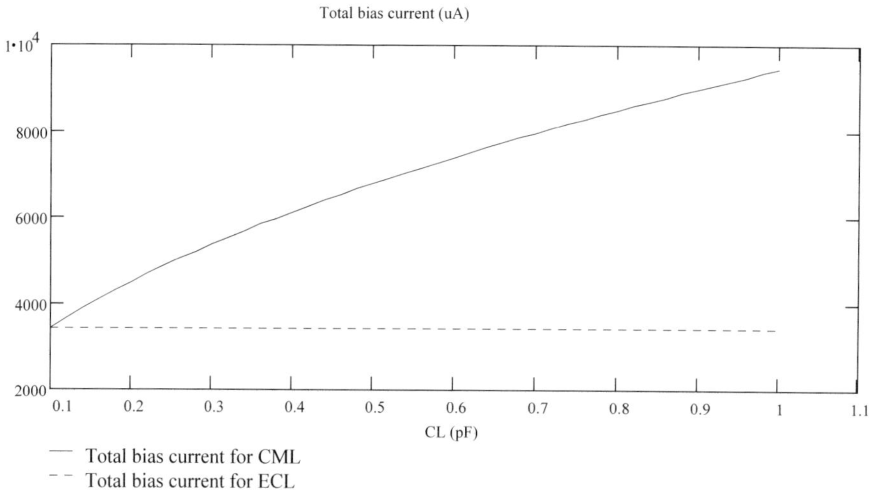

Fig. 5.3b. Total bias current of the CML and ECL inverters versus C_L for optimized design (HSB2 process).

From inspection of Figs. 5.3a and 5.3b, as predicted by relationship (5.28), the CML inverter implemented with the BiCMOS technology needs less power than the ECL gate for a load capacitance lower than 0.6 pF, i.e. a fan out of 5 as discussed in Section 4.8. Analogously, for the HSB2 technology the CML needs more power for a load capacitance higher than 0.1 pF, that corresponds to a unity fan-out. As a consequence, in practical cases where the fan-out is not lower than one, for the HSB2 technology the CML inverter always exhibits a greater dissipation than the ECL counterpart.

Finally, in order to analyze the trade-off between power and speed of the CML and ECL gate, their power-delay product is depicted in Figs. 5.4a and 5.4b. From inspection of these figures, it is apparent that the power-delay product of the CML gate is always better than that of the ECL gate. In the same way, it can be seen that the same consideration holds for the CML and ECL gates designed with a bias current 40% lower than the optimum value.

In summary, a comparison of the CML and ECL inverter designed for minimum delay has been carried out. The analytical results demonstrate that the CML inverter can achieve the lowest delay, even though this advantage is sometimes obtained at the cost of a greater power dissipation. Moreover, the power efficiency of the CML gate is always better than the ECL gate, which means that the speed advantage of the former is greater than its power increase.

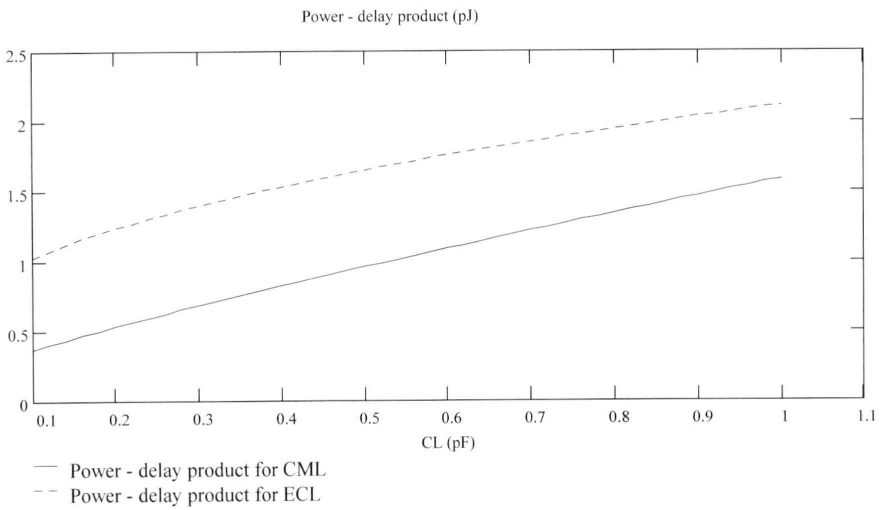

Fig. 5.4a. Power-delay product of the CML and ECL inverter versus C_L for optimized design (BiCMOS process).

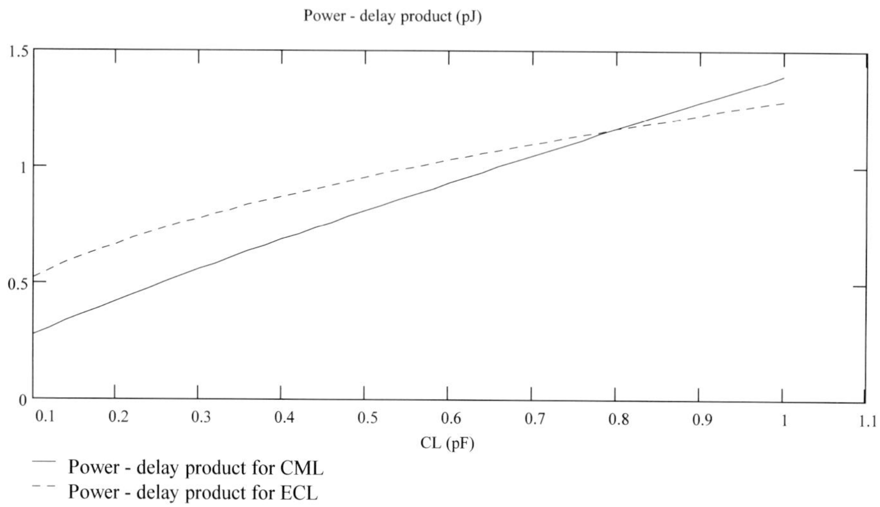

Fig. 5.4b. Power-delay product of the CML and ECL inverter versus C_L for optimized design (HSB2 process).

5.5 OPTIMIZED DESIGN OF BIPOLAR CURRENT-MODE MUX/XOR AND D LATCH

In this section, the methodology presented in Section 5.2 is applied to the design of the MUX/XOR and D latch gates for best speed or an efficient power-delay trade-off.

5.5.1 Design of MUX/XOR CML gates with minimum transistor area

Let us consider the MUX and XOR CML gates depicted in Fig. 5.5 and 5.6, respectively. For the sake of simplicity, let us assume the transistor emitter area to be minimum, while the more general case of a larger area will be dealt in the next subsection.

As in the case of a CML inverter in Section 5.4, under the assumption of minimum-area transistors, the only design parameter is the bias current I_{SS}, since the product $R_C I_{SS}$ is constant and is determined from the logic swing preliminarily assigned. The delay of a MUX/XOR as a function of I_{SS} can be easily obtained from analytical manipulation of the delay model discussed in Section 4.6. Since the delay of a MUX/XOR depends on whether the switching input is applied to the lower or the upper level transistors, the two cases will be discussed separately.

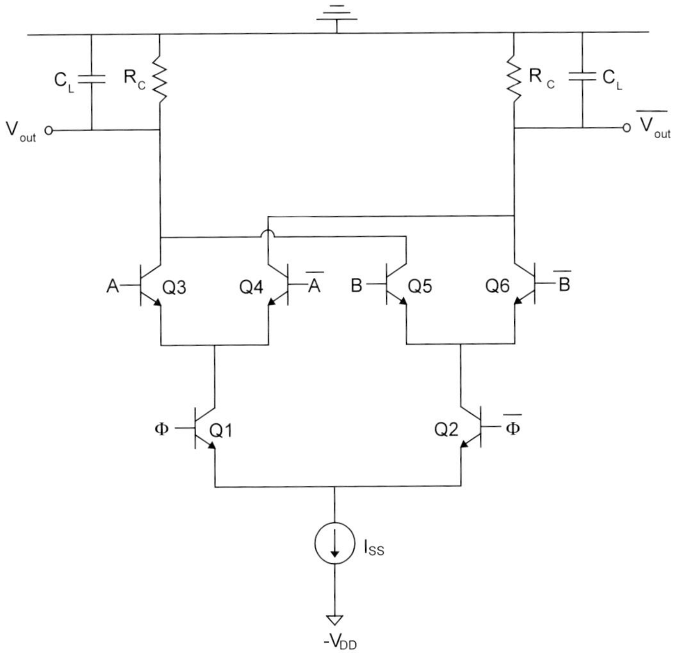

Fig. 5.5. CML MUX gate.

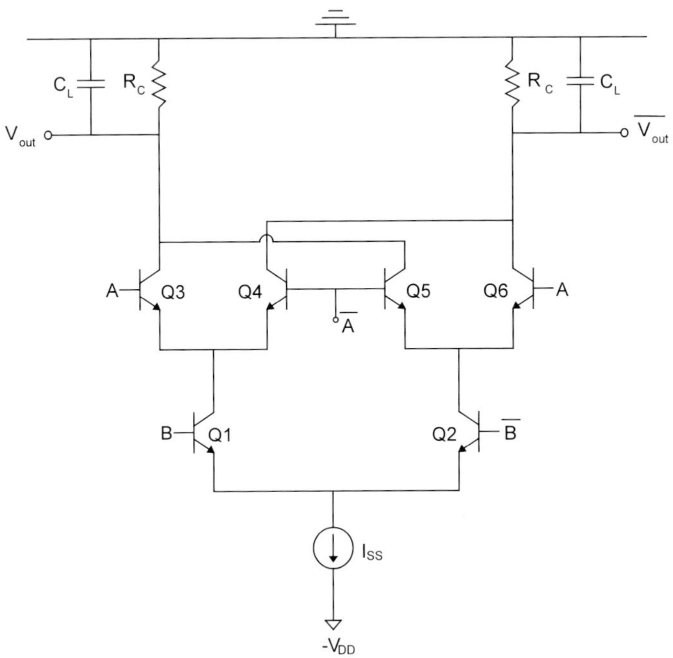

Fig. 5.6. CML XOR gate.

Let us first consider the case where the switching input is applied to the lower level transistors Q1-Q2, for which the correspondent delay is modeled by relationship (4.18). As discussed for the inverter, by assuming $g_m r_e \ll 1$ and substituting $R_C = V_{SWING}/2I_{SS}$, the delay expression (4.18) can be rewritten in the form of relationship (5.1) with coefficients a, b and c being equal to [AP00]

$$a = 0.69\left(4\frac{r_{el}+r_{bl}}{V_{SWING}}\tau_F + r_{bl}C_{bci1}\frac{r_{cl}+r_{eu}+\left(\dfrac{r_{bu}}{\beta+1}\right)}{2V_T}\right) \approx 2.76\frac{r_{el}+r_{bl}}{V_{SWING}}\tau_F \qquad (5.29a)$$

$$b = 0.69\left[\frac{V_{SWING}}{2}\left(2C_{bc3}+C_{cs3}+C_{je4}+2C_{bc5}+C_{cs5}+C_L\right)+2V_T C_{je3}\right.$$
$$\left.+\left(C_{bc1}+C_{cs1}\right)\left(2V_T+\frac{V_{SWING}}{2\beta}\right)\right] \qquad (5.29b)$$
$$\approx 0.35V_{SWING}\left[2C_{bc3}+C_{cs3}+C_{je4}+2C_{bc5}+C_{cs5}+C_L\right.$$
$$\left.+4\frac{V_T}{V_{SWING}}\left(C_{bc1}+C_{cs1}+C_{je3}\right)\right]$$

$$c = 0.69\left[r_{bl}C_{bci1}\left(2+\frac{V_{SWING}}{4V_T(\beta+1)}\right)+\left(r_{cl}+r_{eu}+\frac{r_{bu}}{\beta+1}\right)\left(C_{bc1}+C_{cs1}\right)+\tau_F\right.$$
$$\left.+\left(r_{el}+r_{bl}\right)C_{je1}+\left(r_{cu}+r_{bu}\right)\left(C_{bc3}+C_{bc5}\right)+r_{cu}\left(C_{cs3}+C_{cs5}\right)+\left(r_{bu}+2r_{eu}\right)C_{je4}\right] \qquad (5.29c)$$
$$\approx 0.69\left[2r_{bl}C_{bci1}+r_{cl}\left(C_{bc1}+C_{cs1}\right)+\left(r_{el}+r_{bl}\right)C_{je1}+\tau_F\right.$$
$$\left.+\left(r_{cu}+r_{bu}\right)\left(C_{bc3}+C_{bc5}\right)+r_{cu}\left(C_{cs3}+C_{cs5}\right)+\left(r_{bu}+2r_{eu}\right)C_{je4}\right]$$

where coefficient a (responsible for the delay increase for $I_{SS}>I_{SSop}$) is essentially due to the base-emitter diffusion capacitance, and is equal to that of an inverter (5.7a). Coefficient b consists of two terms, one proportional to

V_{SWING} that accounts for the total capacitance at the output node, and the other proportional to V_T that takes into account the total capacitance at the emitter node of Q3-Q6 (their contribution is lowered by a factor $4V_T/V_{SWING}$ since they see the resistance (4.19) approximately equal to $1/g_m$, that is lower than that seen by the output node, R_C, by the same factor). In analogy to the inverter gate, the weighted sum of capacitances in (5.29b) in brackets can be interpreted as an equivalent capacitance at the output node, i.e. it introduces the same delay contribution as if it was connected directly to the output node. By comparison with relationship (5.7b), this equivalent capacitance (and thus coefficient b) is roughly four times as that of an inverter for $C_L \to 0$ and is equal when the external load is dominant (i.e., for $C_L \to \infty$).

From (5.2) and (5.29), the optimum bias current for the MUX/XOR CML gate is

$$I_{SSop} = \frac{V_{SWING}}{2} \cdot \sqrt{\frac{2C_{bc3} + C_{cs3} + C_{je4} + 2C_{bc5} + C_{cs5} + C_L + 4\frac{V_T}{V_{SWING}}\left(C_{bc1} + C_{cs1} + C_{je3}\right)}{2\left(r_{el} + r_{bl}\right)\tau_F}} \tag{5.30}$$

that is essentially proportional to V_{SWING}, as for the inverter gate. Indeed, the last addend in the numerator under the square root in (5.30) is significantly lower than the other, because it has a lower number of capacitances and is multiplied by the term $4V_T/V_{SWING}$ that is lower than unity for practical values of the logic swing. By comparison of relationship (5.30) with (5.8), the MUX/XOR optimum bias current for a negligible load capacitance is more than twice as that of a CML inverter, but tends to the same optimum bias current for high values of the load capacitance.

Regarding the minimum delay, from (5.3) and (5.29), its expression results to

$$\tau_{PDop} = 2\sqrt{\left(r_{el} + r_{bl}\right)\tau_F} \cdot \sqrt{2C_{bc3} + C_{cs3} + C_{je4} + 2C_{bc5} + C_{cs5} + C_L + 4\frac{V_T}{V_{SWING}}\left(C_{bc1} + C_{cs1} + C_{je3}\right)} \tag{5.31}$$

$$\approx 2\sqrt{\left(r_{el} + r_{bl}\right)\tau_F} \cdot \sqrt{2C_{bc3} + C_{cs3} + C_{je4} + 2C_{bc5} + C_{cs5} + C_L}$$

It is worth noting that, for the optimum bias current, the optimum propagation delay (5.31) is roughly twice that of an inverter for low values of the load capacitance, and tends to it for high values of C_L. As observed for relationship (5.30), the first addend under square root is significantly lower than the other, thus the minimum delay is basically independent of the logic swing. As a result, according to observations made for the inverter gate, the logic swing should be kept as low as possible to reduce the power consumption, since it does not give any benefit on speed performance.

Regarding the delay of a MUX/XOR CML gate associated with the switching of an input that is applied to transistors at the upper level (Q3-Q4 or Q5-Q6), its expression (4.20) can be rewritten in the form (5.1) with coefficients given by

$$a = 0.69\left(4\frac{r_e + r_b}{V_{SWING}}\tau_F + r_b C_{bci3,4}\frac{r_c}{2V_T}\right) \approx 2.76\frac{r_e + r_b}{V_{SWING}}\tau_F \tag{5.32a}$$

$$\begin{aligned} b &= 0.35 \cdot V_{SWING}\left(C_{bc3,4} + C_{cs3,4} + C_{bc5,6} + C_{cs5,6} + C_L\right) \\ &= 0.35 \cdot V_{SWING}\left(2C_{bc3,4} + 2C_{cs3,4} + C_L\right) \end{aligned} \tag{5.32b}$$

$$c = 0.69\left[r_b C_{bci3,4}\left(1 + \frac{V_{SWING}}{4V_T}\right) + r_c\left(C_{bc3,4} + C_{cs3,4}\right) + \left(r_e + r_b\right)C_{je3,4}\right] \tag{5.32c}$$

From inspection of relationships (5.32), coefficient a is equal to that of an inverter in (5.7a), while coefficient b is roughly twice as that of an inverter for $C_L \to 0$ and equal to it for $C_L \to \infty$. As a consequence, the optimum bias current

$$I_{SSop} = \sqrt{\frac{b}{a}} = \frac{V_{SWING}}{2}\sqrt{\frac{2\left(C_{bc3,4} + C_{cs3,4}\right) + C_L}{2\left(r_e + r_b\right)\tau_F}} \tag{5.33}$$

is roughly greater than that of an inverter by a factor of $\sqrt{2}$ for $C_L \to 0$ and equal to it for $C_L \to \infty$. The same consideration holds for the optimum MUX/XOR delay associated with an input that drives the transistors at the upper level

$$\tau_{PDop} \approx 2\sqrt{ab} \approx 2\sqrt{\left(r_e + r_b\right)\tau_F\left(2C_{bc3,4} + 2C_{cs3,4} + C_L\right)} \tag{5.34}$$

5.5.2 Design of MUX/XOR CML gates with non-minimum transistor area and examples

In Section 5.5.1, the assumption of minimum-area transistors was made to simplify the analysis. Now let us consider the more general case where the emitter area of the transistors Q1-Q2 at the lower level is N_1 times the minimum area, while that of transistors at the upper level Q3-Q6 is N_2 times the minimum area.

First, let us consider a MUX/XOR gate with switching inputs applied to transistors Q1-Q2, whose delay can be evaluated by remembering that the minimum-area transistor capacitances have to be multiplied by the factor N_1 (N_2) for lower transistors (upper transistors), whereas resistances are divided by the same factor. Therefore, by following the same procedure as in Section 5.2.2 for the inverter gate, delay can be written as

$$\tau_{PD} = a\frac{I_{SS}}{N_1} + \left(b_0 + b_1 N_1 + b_2 N_2\right)\frac{1}{I_{SS}} + c \tag{5.35}$$

where a and c are still given by (5.29a) and (5.29c) that were derived under the assumption of minimum area, while b_0, b_1 and b_2 are

$$b_0 = 0.35 V_{SWING} C_L \tag{5.36a}$$

$$b_1 = 1.38 V_T \left(C_{bc1} + C_{cs1} + C_{je3}\right) \tag{5.36b}$$

$$b_2 = 0.35 V_{SWING} \left(2C_{bc3} + C_{cs3} + C_{je4} + 2C_{bc5} + C_{cs5}\right) \tag{5.36c}$$

where b_0 is due to the load capacitance, b_1 is due to the capacitance at the emitter node of Q3-Q6, while b_2 is due to the capacitance at the output node, according to considerations presented in Section 5.5.1.

From relationship (5.35), the delay increases as increasing the emitter area of transistors at the upper level, according to [SE96]. As a result, parameter N_2 has to be set to its minimum value (as close to unity as possible) that avoids high-injection effects.

As far as the area of transistors at the lower level is concerned, once N_2 is set on the basis of the considerations reported above, parameter N_1 can be optimized along with the bias current to reduce the delay (5.35). To be more specific, the optimum bias current that minimizes the delay as a function of N_1 is

$$I_{SSop}(N_1)=\sqrt{\frac{b_0+b_1N_1+b_2N_2}{\dfrac{a}{N_1}}}=N_1\sqrt{\frac{1+\dfrac{b_0+b_2N_2}{b_1N_1}}{1+\dfrac{b_0+b_2}{b_1}}}I_{SSop}(1) \qquad (5.37)$$

while the optimum delay versus factor N_1 becomes

$$\tau_{PDop}(N_1)=2\sqrt{\frac{a}{N_1}\left(b_0+b_1N_1+b_2N_2\right)}=\sqrt{\frac{1+\dfrac{b_0+b_2N_2}{b_1N_1}}{1+\dfrac{b_0+b_2}{b_1}}}\tau_{PDop}(1) \qquad (5.38)$$

where $I_{SSop}(1)$ and $\tau_{PDop}(1)$ refer to the case with minimum transistors Q1-Q2 (i.e., N_1=1). As already observed for the inverter gate in Section 5.3.2, the term in the square root of (5.38) is lower than unity, hence a delay reduction is allowed by increasing the emitter area of transistors Q1-Q2. This speed improvement is obtained at the cost of a greater power dissipation from (5.37). To better understand this point, it is worth noting that N_1 is always much lower than $(b_0+b_2)/b_1$ in relationship (5.37)-(5.38). This can be easily verified by analytically expressing $(b_0+b_2)/b_1$

$$\frac{b_0+b_2}{b_1}=\frac{V_{SWING}}{4V_T}\frac{C_L+2C_{bc3}+C_{cs3}+C_{je4}+2C_{bc5}+C_{cs5}}{C_{bc1}+C_{cs1}+C_{je3}} \qquad (5.39)$$

that is very high with respect to practical values of N_1. Indeed, even in the pessimistic case of C_L=0, the ratio of capacitances in (5.39) is greater than 2, while term $V_{SWING}/4V_T$ is greater than 4 since the logic swing is at least 400 mV. From these considerations, $(b_0+b_2)/b_1N_1$ results much greater than unity, hence the delay reduction and the bias current increase are both proportional to $\sqrt{N_1}$, from (5.37) and (5.39). As for the case of the inverter gate, the power efficiency is equal to that achieved by optimizing with a unitary transistor.

In general, except for the case of very heavy speed constraints, the approach discussed above is seldom useful, since a bias current greater than $I_{SSop}(1)$ is rarely tolerable. Thus, the alternative approach to the area optimization developed for the inverter gate in Section 5.3.2 has to be used. To be more explicit, let us minimize delay (5.35) for parameter N_1 with the fixed bias current value $I_{SSop}(1)$, thus leading to the following optimum value of N_1

$$N_1^{'} = \text{int}\left(\sqrt{\frac{a}{b_1}} I_{SSop}(1)\right) \tag{5.40}$$

where the function int(x) was defined in Section 5.2.2. For the design criteria (5.40), by substituting (5.30) into (5.35), the resulting delay is

$$\tau_{PD}\left(I_{SSop}(1), N_1^{'}\right) = 2\sqrt{ab_1} + \left(b_0 + b_2 N_2\right)\sqrt{\frac{a}{b_0 + b_1 + b_2 N_2}} + c \tag{5.41}$$

By comparing relationship (5.41) to the delay with a unitary area (5.31), the emitter area optimization with the bias current set to $I_{SSop}(1)$ allows for a delay reduction at the cost of an increased area. As already discussed for the inverter gate, this area increase leads to a proportional increase in the input capacitance (4.23) of the CML gate, which slows down the response of the previous gate. As a consequence, the area increase is helpful only when the optimum current for the unitary case, $I_{SSop}(1)$, leads to non-negligible high-injection level effects.

Regarding the MUX/XOR gate with switching inputs applied to transistors Q3-Q6, results are equal to those obtained for an inverter gate. Indeed, as clarified in Section 4.6, in this case delay is equal to that of an inverter with properly modified capacitances. Therefore, exactly the same considerations as in Section 5.2.2 hold by substituting factor N_2 to parameter N, as well as using unitary-area coefficients (5.32) instead of (5.7).

5.5.3 Design of the CML D latch

The results presented in the previous subsections can be easily extended to the case of a CML D latch. Indeed, as observed in Section 4.9, the only difference between the XOR and the D latch is an additive capacitance (4.23) associated with cross-coupled transistors Q5-Q6. As it will be demonstrated above, this leads to an increase in coefficient b with respect to the MUX/XOR gate, while the other coefficients in (5.29a) and (5.29c) do not change. Therefore, from (5.2) and (5.3), for negligible load capacitances, the increased load due to transistors Q5-Q6 leads to an optimum bias current and delay worse than those found for XOR and the inverter gates. However, the difference is strongly reduced for load conditions in which the load capacitance is dominant. In the following a more detailed analysis is presented.

When the delay is evaluated for an input applied to the lower transistors and for a minimum-area design, coefficient b becomes

$$b \approx 0.35V_{SWING}\Big[2C_{bc3}+C_{cs3}+C_{je4}+2C_{bc5}+C_{cs5}+C_{L} + C_{bcx5,6}\left(1+\frac{V_{SWING}}{4V_T}\right)+4\frac{V_T}{V_{SWING}}\left(C_{bc1}+C_{cs1}+C_{je3}\right)\Big] \tag{5.42}$$

where, as usual, it was assumed $g_m r_e$<<1 and substituted $R_C=V_{SWING}/2I_{SS}$. Hence, the optimum value of the bias current results to

$$I_{SSop}=\frac{V_{SWING}}{2}\cdot\sqrt{\frac{2C_{bc3}+C_{cs3}+C_{je4}+2C_{bc5}+C_{cs5}+C_{L}+C_{bcx5,6}\left(1+\frac{V_{SWING}}{4V_T}\right)+4\frac{V_T}{V_{SWING}}\left(C_{bc1}+C_{cs1}+C_{je3}\right)}{2\left(r_{el}+r_{bl}\right)\tau_F}} \tag{5.43}$$

$$\tau_{PDop}=2\sqrt{\left(r_{el}+r_{bl}\right)\tau_F}\cdot\sqrt{2C_{bc3}+C_{cs3}+C_{je4}+2C_{bc5}+C_{cs5}+C_{L}+C_{bcx5,6}\left(1+\frac{V_{SWING}}{4V_T}\right)+4\frac{V_T}{V_{SWING}}\left(C_{bc1}+C_{cs1}+C_{je3}\right)} \tag{5.44}$$

When transistors Q1-Q2 are allowed to have an emitter area greater than the minimum one by a factor N_1, as well as transistors Q3-Q6 by a factor N_2, the only modification to coefficients (5.36) of the MUX/XOR gate delay (5.35) with an input applied to lower transistors Q1-Q2 has to be introduced in b_2, that becomes

$$b_2=0.35V_{SWING}\left[2C_{bc3}+C_{cs3}+C_{je4}+2C_{bc5}+C_{cs5}+C_{bcx5,6}\left(1+\frac{V_{SWING}}{4V_T}\right)\right] \tag{5.45}$$

therefore the considerations on the sizing of N_1 and N_2 for the MUX/XOR gate in Section 5.5.2 are again extended to the D latch.

For the delay associated with an input driving transistors Q3-Q4 and minimum-area design, coefficient b is obtained by substituting relationship (4.23) into (5.32b)

$$b = 0.35 \cdot V_{SWING}\left[2C_{bc3,4} + 2C_{cs3,4} + C_L + C_{bcx5,6}\left(1 + \frac{V_{SWING}}{4V_T}\right)\right] \tag{5.46}$$

whereas the other coefficients are given by (5.32a) and (5.32c). Once again, coefficient a is equal to that of an inverter (5.7a), while coefficient b is more than four times larger than that of an inverter for $C_L \rightarrow 0$ (more explicitly, by a factor 6.3 and 5 for the BiCMOS and the HSB2 process, respectively) and equal to it for $C_L \rightarrow \infty$.

As a consequence, the optimum bias current

$$I_{SSop} = \sqrt{\frac{b}{a}} = \frac{V_{SWING}}{2}\sqrt{\frac{2\left(C_{bc3,4} + C_{cs3,4}\right) + C_L + C_{bcx5,6}\left(1 + \frac{V_{SWING}}{4V_T}\right)}{2\left(r_e + r_b\right)\tau_F}} \tag{5.47}$$

that is roughly greater than that of an inverter by more than twice (more explicitly, by a factor 2.5 and 2.2 for the BiCMOS and HSB2 process, respectively) for $C_L \rightarrow 0$ and equal to it for $C_L \rightarrow \infty$. The same increase is observed in the optimum delay, whose expression is

$$\tau_{PDop} \approx 2\sqrt{\left(r_e + r_b\right)\tau_F\left[2C_{bc3,4} + 2C_{cs3,4} + C_L + C_{bcx5,6}\left(1 + \frac{V_{SWING}}{4V_T}\right)\right]} \tag{5.48}$$

When area factor N_2 of transistors Q3-Q4 is also a design parameter, the same criteria as in Section 5.5.2 can be used, by observing that the only difference with respect to the MUX/XOR gate is in coefficient b_2, that becomes

$$b_2 = 0.35V_{SWING}\left[2C_{bc3} + C_{cs3} + C_{je4} + 2C_{bc5} + C_{cs5} + C_{bcx5,6}\left(1 + \frac{V_{SWING}}{4V_T}\right)\right] \tag{5.49}$$

whereas the other coefficients are given by relationship (5.29a), (5.29c), (5.36a) and (5.36b).

In regard to the ECL implementation of MUX/XOR gates, the design criteria for an inverter discussed in Section 5.3 are again valid by simply substituting the expression of coefficients a, b and c of MUX/XOR gates to those of the inverter.

5.5.4 Design examples

To illustrate the proposed procedure in detail, the MUX/XOR gate with both the BiCMOS and the HSB2 technologies were designed assuming a supply voltage of 5 V, V_{SWING}=500 mV, thus Tables 4.2-4.3 can be used. Very similar results can be obtained for the D latch.

The fan-out was assumed to be equal to 1 and 10, equivalent to a load capacitance of 120 fF and 1.2 pF for the BiCMOS process, and a capacitance of 100 fF and 1 pF for the HSB2 technology, respectively.

For the BiCMOS process and applying the input to lower transistors, the two load conditions lead to an optimum bias current of 1 mA and 1.9 mA respectively, which lead to a delay of 243 ps and 338ps. The values are very close to those given by SPICE simulations which yield 234 ps and 351 ps. Even though the current is slightly higher than 1.4 mA for a fan-out of 10, the transistors work marginally at the high-injection level and there is no appreciable degradation in the propagation delay.

For the HSB2 process and the two fan-out conditions, the optimum bias current is 4.8 mA and 10 mA respectively, which leads to a high power dissipation and is significantly greater than the high-injection level. Hence, we can use the 60% optimum bias current, i.e. a bias current of 2.8 mA and 6 mA for the two fan-out conditions, respectively. In the former case minimum-size transistors can be still used and the resulting delay is 45 ps which has an error of 15% with respect to the value predicted. In the second case the bias current is excessively higher than the high-injection level and we have to use a transistor with an area 3 times the minimum size. Under this condition a simulated delay of 69 ps is obtained, that differs by 13% from the theoretical value.

5.6 SUMMARY AND REMARKS

In this chapter, a design methodology for CML and ECL gates has been discussed. This strategy allows for sizing the bias current, the load resistance and the transistor area to efficiently manage the power-delay trade-off, that is crucial in this kind of logic.

In regard to CML gates, their general delay expression (5.1) introduced permits to use the same design approach regardless of the specific logic gate considered. Indeed, although the general delay expression (5.1) was validated for the case of the inverter, MUX, XOR and D latch gates, it can be immediately extended to the other CML gates. This is because delay can always be written as the sum of time constants (4.4), thus for more complex gates, and even more series-gating levels, the capacitances and the resistances seen by them have the same dependence on the bias current as the gate considered. As a result, analysis based on relationship (5.1) is extremely general, and the only difference among the various gates is in the expression of coefficients *a*, *b* and *c* (the latter of which is negligible and does not affect the optimization).

In the design approach discussed, coefficient *a* is due to the base-emitter diffusion capacitance, whose contribution to the delay increases as increasing the bias current. Its expression is always the same, regardless of the considered gate. Coefficient *b* is due to the other capacitances, that can be represented with an equivalent capacitance at the output node (i.e. as the gate were a simple inverter), and this contribution to delay decreases as increasing the bias current. This explains why delay can always be minimized for the bias current by relationship (5.2), that makes the two contributions equal.

By following this approach, the power-delay trade-off was shown to be represented in the general form (5.5), that holds regardless of the specific logic gates considered. From this representation, it was shown that the power-delay product increases as increasing the bias current. As a consequence, the power consumption can be reduced without significantly affecting the speed performance by reducing the bias current with respect to the optimum value. The rule of thumb proposed to achieve a favorable power efficiency is to set the bias current equal to 60% of the optimum value, that allows for a 40% power saving and only a 10% delay increase with respect to the optimum case.

Various criteria to size the transistors' emitter area were discussed. However, unless heavy speed constraints are imposed, the area must eventually be increased only to the minimum value that avoids the high-injection level effects. When a very high-speed performance is required, the suggested approach is to use the optimum bias current evaluated for minimum emitter area, and then to properly increase the emitter area.

A design strategy was also proposed for ECL gates to obtain nearly the minimum achievable delay. The approach discussed is independent of the specific gate considered and allows a suitable output buffer bias current to be found. The bias current obtained is demonstrated to be independent of the load capacitance for high-speed design cases.

A comparison of the CML and ECL inverter designed for high-speed design was also carried out. In contrast with the usual belief, results demonstrate that the CML inverter can achieve the lowest delay, even though this advantage is sometimes obtained at the cost of a greater power dissipation. Moreover, the power efficiency of CML gates is always better than ECL gates, i.e. the speed advantage of CML gates is greater than their power increase with respect to ECL gates.

Chapter 6

MODELING OF MOS CURRENT-MODE GATES

In Chapter 4, a strategy to model the delay of bipolar Current-Mode gates has been discussed and applied to several gates. In this chapter, the same subject is investigated for CMOS Current-Mode logic circuits by extending the fundamental ideas introduced in Chapter 4.

6.1 INTRODUCTION TO THE DELAY MODELING OF MOS CURRENT-MODE GATES

The bipolar CML/ECL delay model can be extended to MOS SCL gates by properly simplifying the equations of the BSIM3v3 model currently used for submicron MOS transistors [CH99], whose complex equations are unsuitable for formulating an efficient model and understanding the delay dependence on design and process parameters. To simplify the general analysis of CMOS Current-Mode circuits, the PMOS active load included in every CMOS SCL gate is approximated as an equivalent resistor, as discussed in Section 2.4.1. Parasitic capacitances of MOS transistors will also be evaluated and linearized starting from the exact relationships of the BSIM3v3 model. Moreover, the relationship between the current and the gate-source voltage of NMOS transistors is simplified by introducing an equivalent linear transconductance. Once NMOS and PMOS transistors are properly linearized, the approach previously used for bipolar gates can be applied to CMOS SCL gates.

In this chapter, a methodology to evaluate the delay of MOS Current-Mode gates is developed and applied to the inverter, MUX, XOR and D latch. Gates both with and without an output buffer are considered, and the effect of the input waveform rise time on the delay is also treated. To the best authors' knowledge, except for works written by themselves [AP01], [AP02], no other approaches to the delay modeling of CMOS Current-Mode gates have been proposed in the literature.

6.2 MODELING OF THE SOURCE-COUPLED INVERTER

Let us consider the SCL inverter gate in Fig. 6.1, where I_{SS} is the bias current, V_{DD} is the supply voltage and C_L is the external load capacitance that accounts for the wiring capacitance and the input capacitance of the driven gates. It is worth noting that, unlike CML gates, a negative supply is rarely used because SCL gates are typically used in mixed-signal circuits, where the same positive supply voltage is used for analog blocks.

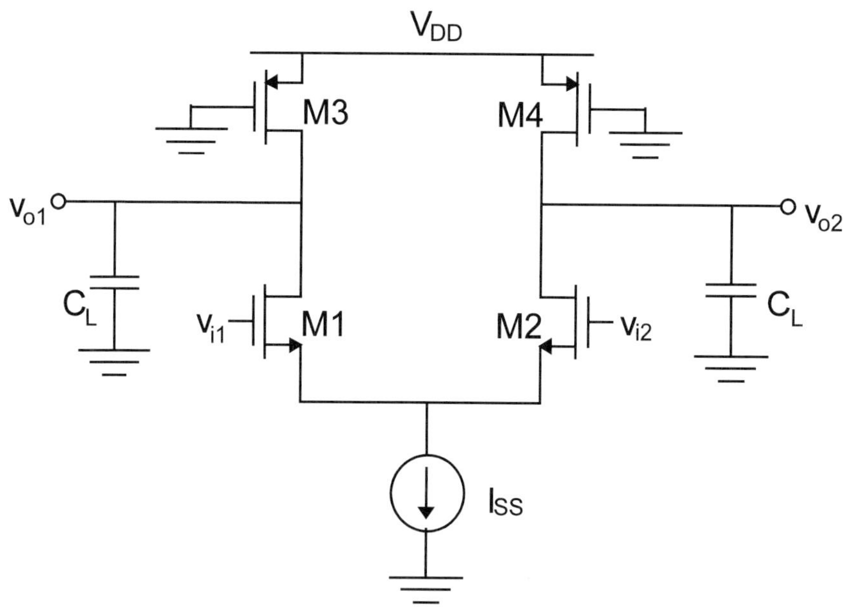

Fig. 6.1. Source-Coupled inverter gate.

In this section, the delay τ_{PD} of the SCL inverter gate in Fig. 6.1 is evaluated when the differential input $v_i = v_{i1} - v_{i2}$ switches. To simplify analysis, the PMOS active load is modeled as an RC circuit in Section 6.2.1. The timing analysis of the circuit in Fig. 6.1 is carried out in Subsection 6.2.2 by assuming a step input for the sake of simplicity, and is then extended to more general input waveforms in Section 6.2.3.

6.2.1 Circuit model of the PMOS active load

To simplify the circuit analysis of Fig. 6.1 and evaluate the SCL inverter delay, PMOS transistors M3-M4 can be approximated as a linear resistance R_D given by relationship (2.38) in a DC condition. Moreover, to accurately model the transient behavior of the gate, the PMOS drain-bulk, $C_{db,p}$, and gate-drain capacitance, $C_{gd,p}$, must also be accounted for.

In the BSIM3v3 model, the PMOS drain-bulk capacitance is associated with the depletion region of the drain-bulk junction, and consists of the two contributions in (1.55a) and (1.55b), the bottom capacitance and the side-wall capacitance [R96], [CH99]. Due to a wide variation of their voltage during switching, the two contributions must be evaluated by multiplying the zero-bias capacitances C_j and C_{jsw} by coefficients $K_{j,p}$ and $K_{jsw,p}$ respectively, according to relationship (1.13). Voltages V_1 and V_2 in (1.13) are evaluated as explained in Section 4.3 for the bipolar gates. For example, the minimum direct voltage V_1 seen by the drain-bulk capacitance of M2 occurs when the input voltage v_i is high, hence the PMOS drain voltage is equal to (V_{DD}-$V_{SWING}/2$), thus resulting to V_1=-$V_{SWING}/2$, since the bulk is connected to the supply V_{DD}. Instead, when v_i is low, the PMOS drain voltage is V_{DD}, hence V_2=0.

From inspection of the MOS transistor layout in Fig. 6.2, junction perimeter and area result as $2(W_p+L_{d,p})$ and $W_pL_{d,p}$, respectively, being W_p the effective channel width and $L_{d,p}$ the other junction dimension (in Chapter 2, the symbol L_x was used instead of $L_{d,p}$).

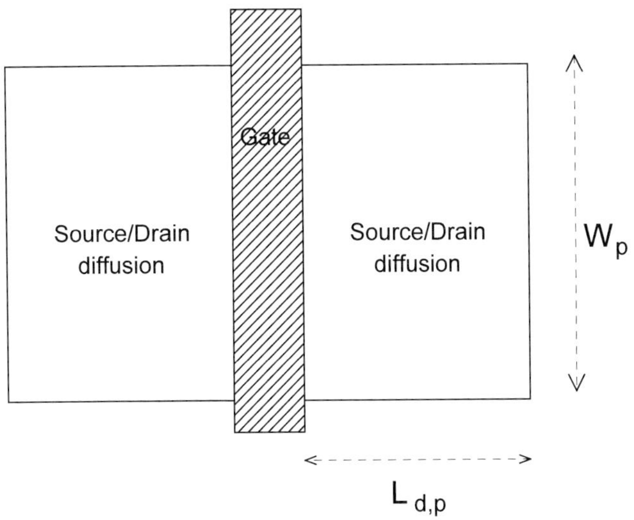

Fig. 6.2. Simplified layout of a MOS transistor.

From Fig. 6.2, the linearized drain-bulk capacitance of PMOS transistors results in

$$C_{db,p} = K_{j,p} C_{j,p} W_p L_{d,p} + 2K_{jsw,p} C_{jsw,p} \left(W_p + L_{d,p}\right) \tag{6.1}$$

In relationship (6.1), the bottom zero-bias capacitance $C_{j,p}$, as well as the built-in potential across the junction *pb* and the grading coefficient *mj* necessary to evaluate coefficient K_j, are standard BSIM3v3 model parameters. The same consideration holds for the parameters of the side-wall contribution (with subscript *sw*). Parameter $L_{d,p}$ is extrapolated from layout design rules.

The PMOS gate-drain capacitance $C_{gd,p}$ is equal to the sum of the overlap and the channel contributions. The former is equal to the product of the overlap capacitance C_{gd0}, that is a standard BSIM3v3 parameter, and the channel width W_p. The latter is the intrinsic contribution associated with the channel charge of the PMOS transistors working in the triode region, $C_{gd,p,int}$. Unfortunately, the effects of the channel charge in short channel transistors cannot be treated in the same manner as for long-channel devices, since the decomposition of channel capacitance into gate-source and gate drain capacitances no longer applies in submicron technologies [YEC83], [T98], [CH99]. In the BSIM3v3 capacitance model, the channel charge transfer is modeled by transcapacitances C_{ij}, defined as

$$C_{ij} = \frac{\partial Q_i}{\partial V_j} \tag{6.2}$$

where Q_i is the total charge associated with the MOSFET terminal *i*, and V_j is the voltage of the terminal *j*. Since transcapacitances C_{ij} are not reciprocal, i.e. $C_{ij} \neq C_{ji}$, in general the charge flow in submicron MOSFET transistors cannot be modeled by two-terminal capacitances [T98], [CH99]. However, for the specific case of the PMOS active load in SCL gates, an equivalent intrinsic capacitance $C_{gd,p,int}$ can still be introduced to model the channel charge flowing through the drain terminal, Q_d. The channel charge flow through the drain is only due to the variation of the drain-gate voltage V_d (the gate, source, and bulk voltages of transistors M3-M4 are fixed); therefore, it can be described by the capacitance C_{dd}, since by definition (6.2) it expresses the drain charge variation due to the variation of V_d. To evaluate the capacitance C_{dd} of PMOS transistors, let us consider the expression of Q_d in the strong inversion for a 40/60 charge partitioning (as indicated by parameter *Xpart* equal to 1 in the transistor model) [CH99]

$$Q_d = -W_p L_p C_{OX}\left[\frac{V_{SG}-|V_T|}{2}-\frac{3A_{bulk}V_{SD}}{4}+\frac{(A_{bulk}V_{SD})^2}{8\left(V_{SG}-|V_T|-\frac{A_{bulk}}{2}V_{SD}\right)}\right]$$

$$\approx -W_p L_p C_{OX}\left[\frac{V_{SG}-|V_T|}{2}-\frac{3A_{bulk}V_{SD}}{4}\right] \tag{6.3}$$

where $V_{SD} << \frac{V_{SG}-|V_T|}{A_{bulk}}$ was assumed, and the BSIM3v3 parameter A_{bulk} is defined as

$$A_{bulk} = \frac{1}{1+K_{ETA}|V_{SB}|}\left\{1+\frac{K_{1OX}}{2\sqrt{\phi_S-|V_{SB}|}}\left[\frac{A_0 L_p}{L_p+2\sqrt{X_J X_{dep}}}\right.\right.$$

$$\left.\left.\cdot\left(1-A_{GS}\left(|V_{GS}|-|V_T|\right)\left(\frac{L_p}{L_p+2\sqrt{X_J X_{dep}}}\right)^2\right)+\frac{B_0}{W_p+B_1}\right]\right\} \tag{6.4}$$

which depends on W_p, L_p and various other BSIM3v3 model parameters.

The dependence of (6.3) on W_p and L_p can be further simplified. Indeed, it can be evaluated in the worst case where A_{bulk} is equal to its maximum value $A_{bulk,max}$. From relationship (6.4), parameter A_{bulk} is maximized by setting W_p to the minimum value allowed by the process (i.e. W_p=$W_{p,min}$) and by maximizing function (6.4) for L_p with straightforward calculations. As an example, for the PMOS transistor in a 0.35-μm CMOS process and V_{DD}=3.3 V, the maximum value of A_{bulk} is $A_{bulk,max}$=1.34, that is slightly greater than unity as expected [CH99].

By differentiating relationship (6.3) for V_D and approximating A_{bulk} to its maximum value $A_{bulk,max}$, capacitance C_{dd} results to

$$C_{dd} = \frac{\partial Q_d}{\partial V_d} = \frac{3}{4}A_{bulk,\max}W_p L_p C_{OX} \tag{6.5}$$

From inspection of relationship (6.5), it is apparent that the expression derived for long-channel devices (i.e., $C_{gd,p,int}$=½$W_pL_pC_{OX}$) is inadequate.

From the above considerations, each of the two PMOS transistors implementing the active load of an SCL gate will be represented in the following by the RC circuit in Fig. 6.3.

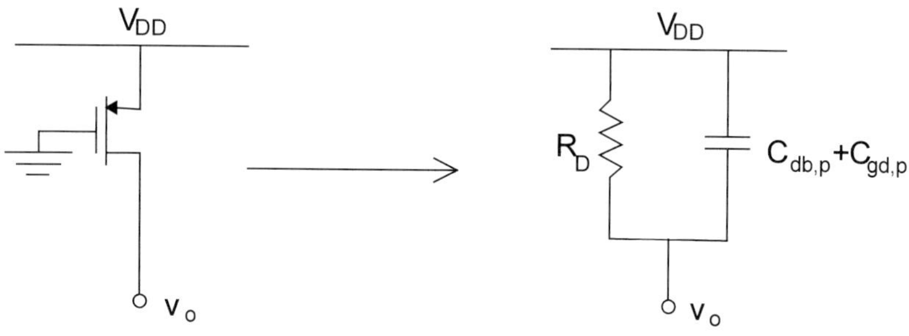

Fig. 6.3. Equivalent linear circuit of the PMOS active load.

6.2.2 Delay model of the SCL inverter for a step input

To model the propagation delay of the SCL inverter in Fig. 6.1, it is useful to observe that the NMOS transistors M1-M2 work in the saturation region most of the time, and their source voltage is the same for both input logic values, since it is fixed by the NMOS transistor in the ON state (i.e. drawing a current I_{SS}). Thus, as already discussed for Current-Mode bipolar gates, the circuit can be linearized around the bias point with v_i=0 and then simplified by applying the half-circuit concept, since the circuit is symmetrical and a differential input is applied.

The equivalent linear half circuit obtained is shown in Fig. 6.4, where transistor M1 (M2) is represented by its small-signal model (subscript n refers to NMOS transistors), while PMOS transistor is replaced by the equivalent circuit in Fig. 6.3. The capacitive effects associated with NMOS transistors consist of the drain-bulk junction capacitance $C_{db,n}$ and the gate-drain capacitance $C_{gd,n}$, that pertain to NMOS transistors working in the saturation region. The former is a junction capacitance and can be linearized in the same way as capacitance $C_{db,p}$, i.e. by multiplying its zero-bias value by a coefficient K_j evaluated with V_1=$-V_{DD}+V_{SWING}/2$ and V_2=$-V_{DD}$. Capacitance $C_{gd,n}$ is due to the overlap of the gate layer and the drain diffusion, thus it is equal to the product of the channel width W_n and the BSIM3v3 model parameter C_{gd0} representing the overlap gate-drain capacitance per unit channel width.

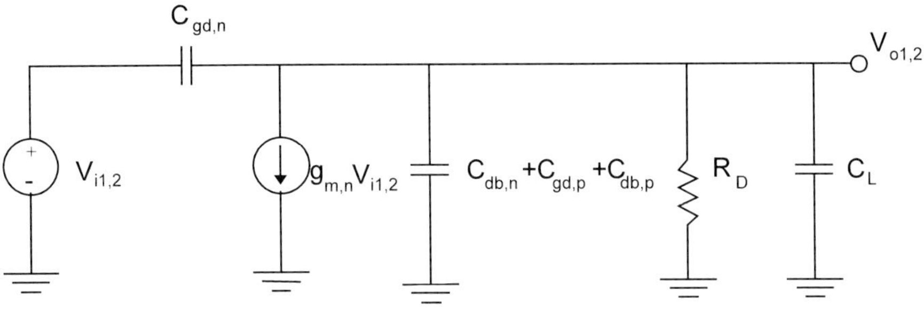

Fig. 6.4. Equivalent linear circuit of the SCL inverter gate.

The network in Fig. 6.4 is a first-order circuit with a time constant τ that can be evaluated by applying the open-circuit time constant method [CG73] and neglecting the high-frequency zero [LS94], whose effect on the transient behavior of the gate is a small negative initial overshoot during switching. By assuming the input waveform to be a unity step, the resulting delay is 0.69τ, hence the propagation delay $\tau_{PD,SCL}$ of the SCL gate results to be

$$\tau_{PD,SCL} = 0.69 R_D \left(C_{gd,n} + C_{db,n} + C_{gd,p} + C_{db,p} + C_L\right) \tag{6.6}$$

The expression (6.6) is simple and, hence, can be profitably used in pencil-and-paper calculations. Moreover, it shows how delay depends on design and process parameters and allows the designer to get the necessary intuitive insight into the circuit behavior.

To evaluate the accuracy of relationship (6.6), its results were compared to Spectre simulations by adopting a 0.35-μm CMOS process, whose main parameters are reported in Table 6.1, and BSIM3v3 parameters useful to evaluate capacitances in (6.6) are shown in Table 6.2.

TABLE 6.1

C_{OX}	4.6 fF/μm^2
$\mu_{no}C_{OX}$	175 μA/V^2
$\mu_{po}C_{OX}$	60 μA/V^2
$(W/L)_{min}$	0.6 μm/0.3 μm
V_{Tn0} (short channel)	0.54 V
V_{Tp0} (short channel)	-0.72 V
maximum V_{DD}	3.3 V

TABLE 6.2

NMOS		PMOS	
pb	6.9 E-1	*pb*	1.02
mj	3.1 E-1	*mj*	5.5 E-1
cj	9.3 E-4	*cj*	1.42 E-3
pbsw	6.9 E-1	*pbsw*	1.02
mjsw	1.9 E-1	*mjsw*	3.9 E-1
cjsw	2.8 E-4	*cjsw*	3.8 E-4
Ld	1.1 E-6	*Ld*	1.1 E-6
cgs0	2.1 E-10	*cgs0*	2.1 E-10
cgd0	2.1 E-10	*cgd0*	2.1 E-10

Simulations were carried out under a variety of design and load conditions. The bias current was varied from 5 μA to 100 μA, the transistors aspect ratios were sized to obtain the typical values V_{SWING}=700 mV and A_V=4, and the load capacitance C_L was set to 0 F, 50 fF, 200 fF and 1 pF.

In Figs. 6.5a, 6.5b, 6.5c and 6.5d the simulated delay and that predicted by relationship (6.6) are plotted versus the bias current I_{SS} for a load capacitance equal to 0 F, 50 fF, 200 fF and 1 pF, respectively. As expected, delay is decreased by increasing the bias current I_{SS} and asymptotically tends to a constant value.

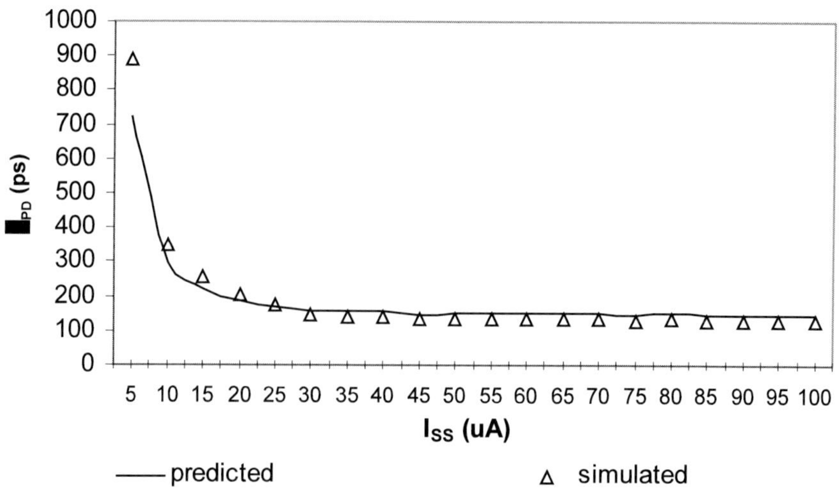

Fig. 6.5a. Simulated and theoretical delay vs. bias current I_{SS} with C_L=0 F.

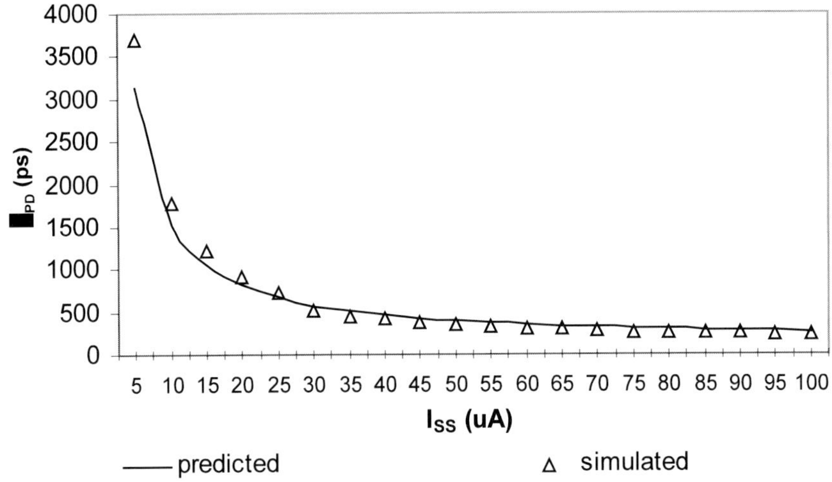

Fig. 6.5b. Simulated and theoretical delay vs. bias current I_{SS} with C_L=50 fF.

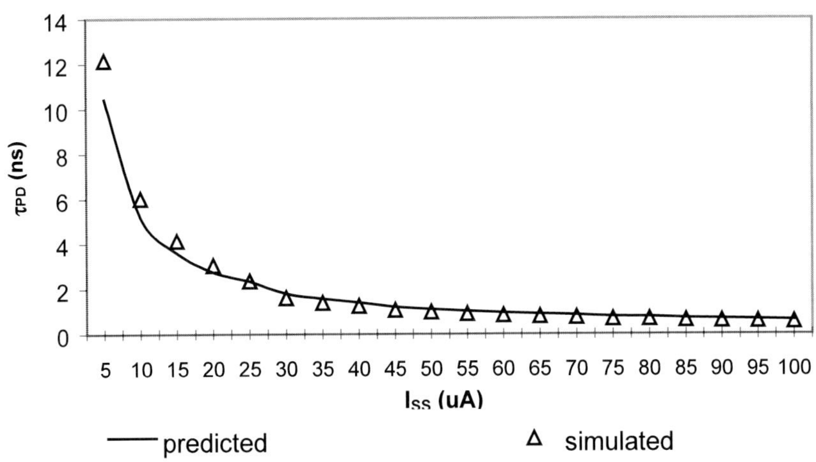

Fig. 6.5c. Simulated and theoretical delay vs. bias current I_{SS} with C_L=200 fF.

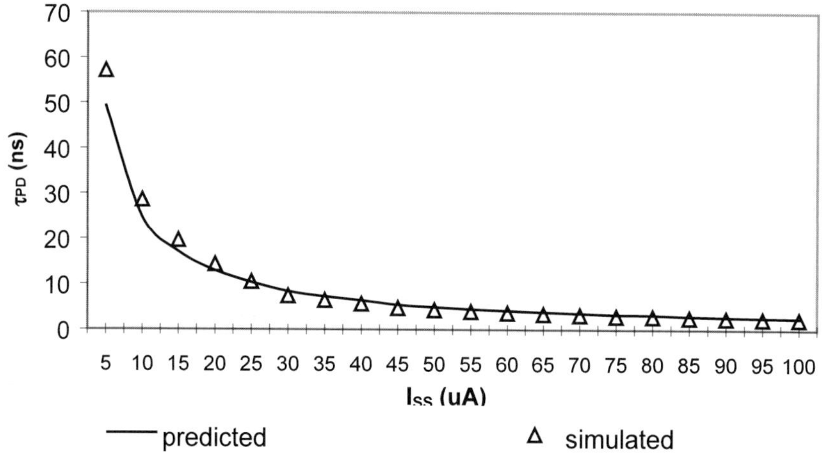

Fig. 6.5d. Simulated and theoretical delay vs. bias current I_{SS} with C_L=1 pF.

The model (6.6) is in agreement with simulated results, as is shown in Fig. 6.6, that reports the model error with respect to Spectre simulations versus I_{SS}, for the used values of the load capacitance. The maximum error in actual cases (i.e., with non-zero load capacitance) is equal to 15%, and it is as high as 19% in the nonrealistic case with C_L=0 F. To get an idea about reasonable values used in practical cases, consider an SCL gate with I_B=20 μA, C_L=50 fF, $(W/L)_n$=8/0.3, $(W/L)_p$=0.6/0.7 and V_{DD}=3.3 V. The predicted and simulated delay are respectively equal to 620 ps and 730 ps, and differ by 15%.

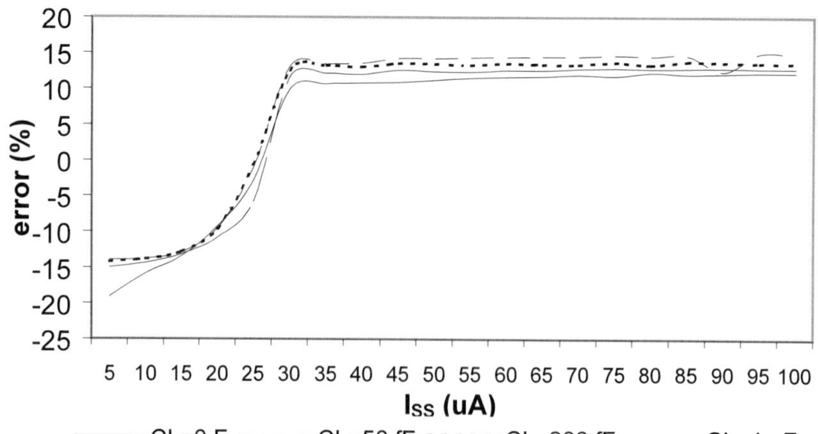

Fig. 6.6. Error of (6.5) with respect to simulated results vs. bias current I_{SS}.

As observed for CML gates, the analytical model presented so far could be further improved by introducing unknown coefficients that multiply the circuit capacitances and evaluating them through minimization of the error functional (4.6). However, the model accuracy obtained is better than that of CML circuits and it is sufficient for most purposes, thus the improved model is a less interesting option for SCL gates.

6.2.3 Extension of the delay model to arbitrary input waveforms

In general, the delay of a logic gate depends on the input signal waveform [R96]. For example, this dependence cannot be neglected in the case of static CMOS logic [ADR00], [BNK98]. Therefore, in the following some considerations are introduced to extend the validity of the model developed in Section 6.2.2 for a step input.

To understand the effect of a non-zero input rise time, let us first consider the case where a step input is applied to the SCL inverter. For the sake of simplicity, assume that the input voltage v_i abruptly switches from low to high. Before switching, the voltages v_{o1} and v_{o2} are equal to V_{DD} and (V_{DD}-V_{SWING}/2), while after switching the steady-state voltages are (V_{DD}-V_{SWING}/2) and V_{DD}, respectively. During the switching, the bias current is abruptly steered on M1-M3 due to the step input voltage, and the left-hand half circuit is equivalent to an RC circuit driven by a constant current I_{SS}, while the right-hand half one is an RC circuit with no input current[1].

When an input waveform with a non-zero rise time is applied, the current steering is more smoothed, thus slowing down the output voltage response and increasing the time $t_{out,50}$ required to reach its 50% value (i.e., delay tends to be increased with respect to the step input case). At the same time, the time $t_{in,50}$ required by v_i to reach its 50% value also increases as increasing the input rise time (i.e., delay tends to be reduced), therefore two opposite effects on delay (defined as ($t_{out,50}$-$t_{in,50}$)) are observed. In SCL gates, the two effects tend to compensate each other, thus leading to a weak delay dependence on the input rise time.

The weak dependence of SCL gates delay on the input rise time is confirmed by simulations. To be more specific, delay of a single SCL gate was evaluated by using the CMOS process described above, for a load capacitance equal to 0 F, 50 fF, 100 fF, 200 fF, 400 fF and 1pF, and a bias current ranging from 5 μA to 100 μA. The considered gate was driven by a

[1] This explains why the half circuit concept can be applied. Indeed, both half circuits are driven by a constant current and have the same time constant, thus their output voltages move toward each other in a symmetrical way with respect to the logic threshold.

realistic input waveform, obtained from the output voltage of either a single SCL gate or a chain made up of two to four SCL gates. A very wide range of conditions was considered, including the cases with an input waveform much faster or slower than the output waveform by up to two orders of magnitude. The results obtained are collected in the scattering plot of Fig. 6.7, where the step input delay of the considered SCL gate is reported in the x-axis and the actual delay for a realistic waveform on the y-axis. Analysis of Fig. 6.7 reveals that the actual delay is close to that with step input. More specifically, the difference between the two delay values is always lower than 24% even in unrealistically high input rise time, and it is typically much lower, since its average value is only 6.8%. This means that the delay of an SCL gate is not very sensitive to the input waveform, in contrast to an ideal first-order circuit behavior.

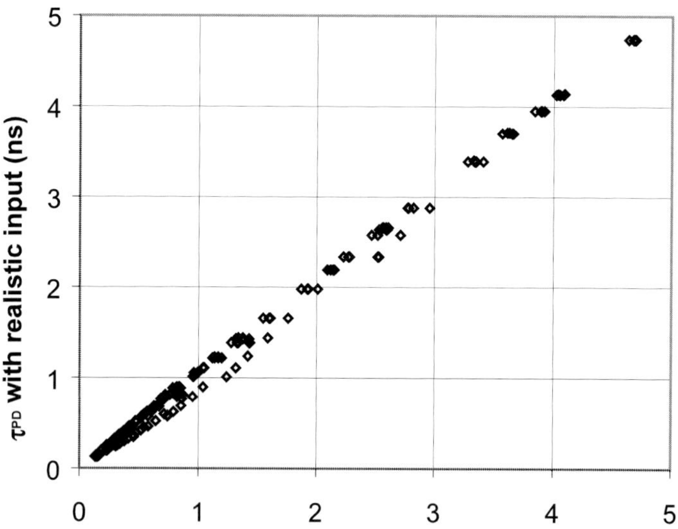

Fig. 6.7. Scattering plot of the SCL inverter delay with actual input waveform vs. delay with step input waveform.

6.3 MODELING OF THE SOURCE-COUPLED INVERTER WITH OUTPUT BUFFERS

To improve the driving capability of the gate and therefore its switching speed, or to shift the common-mode value of the output nodes voltage for reasons discussed in Chapter 2, an output buffer can be added to each output node, as shown in Fig. 6.8, where $v_o=v_{o1}-v_{o2}$ is the differential output voltage of the gate, and $v_{i,buf1}$ and $v_{i,buf2}$ are the input voltages of the two buffers, respectively. Each output buffer is a source-follower stage biased by the current source I_{SF}.

The propagation delay $\tau_{PD,SCLbuf}$ of an SCL gate with the output buffer in Fig. 6.8 can be evaluated by applying the methodology described in Section 6.2 and extending it as done for ECL gates in Section 4.5. To be more specific, the delay of the inverter gate in Fig. 6.8 can be decomposed into the contribution $\tau_{PD,SCL}$ of the internal SCL gate and that of the buffer $\tau_{PD,buf}$:

$$\tau_{PD,SCLbuf} = \tau_{PD,SCL} + \tau_{PD,buf} \tag{6.7}$$

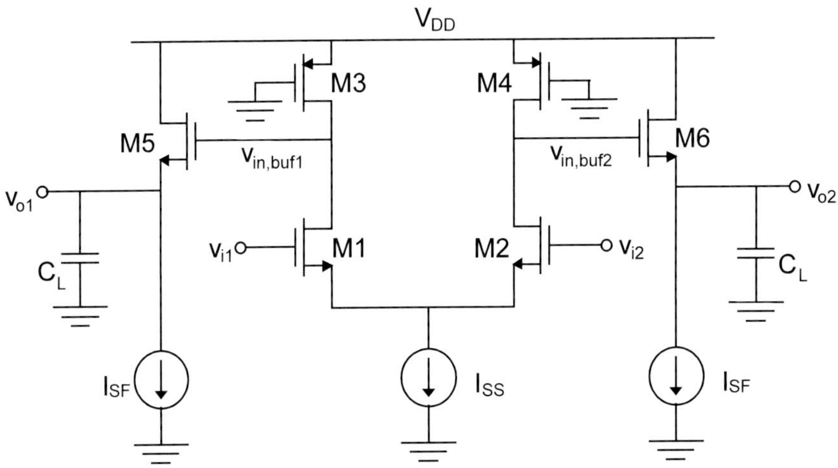

Fig. 6.8. Source-Coupled inverter gate with output buffers.

The delay contribution of the internal SCL gate, $\tau_{PD,SCL}$, is given by (6.6) in which C_L must be set to zero, while that of the buffer, $\tau_{PD,buf}$, is evaluated by driving the source-follower stage with the Thevenin equivalent circuit seen at the output of the internal SCL gate, modeled with a voltage source V_{th} and a resistance[2] R_{th}. Hence, $\tau_{PD,buf}$ can be evaluated by analyzing the

[2] In the SCL gate only the resistive contribution of the output impedance was considered, since the output capacitance does not contribute to the delay. To better understand this point, consider the overall transfer function of the circuit in Fig. 6.9, which can be approximated as $(1+a_1s)/(1+b_1s)$ in the specific case of MOS Current-Mode gates, as will be shown above. By applying the Elmore delay approximation,

circuit in Fig. 6.9, which depicts the linearized buffer circuit driven by the Thevenin equivalent circuit of the SCL gate.

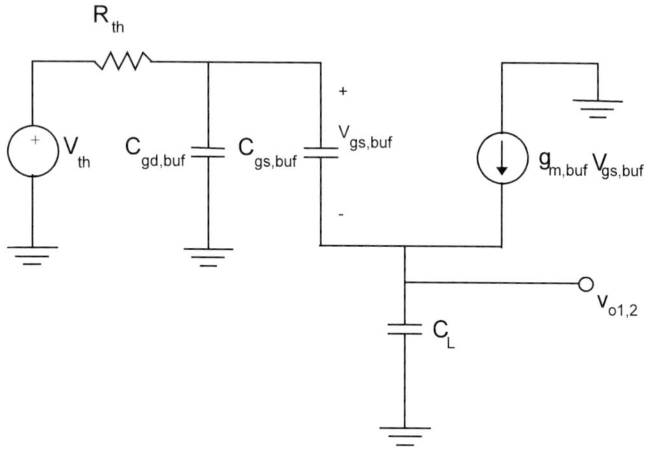

Fig. 6.9. Equivalent linear circuit of output buffers driven by the internal SCL inverter.

In Fig. 6.9, capacitors $C_{gd,buf}$ and $C_{gs,buf}$ represent the gate-drain and the gate-source contributions, and are evaluated as small-signal capacitances in the saturation region. To be more specific, $C_{gd,buf}$ is due to the overlap of the gate layer and the drain diffusion, thus it is equal to the product of the overlap capacitance per unit length, C_{gd0}, and the effective buffer transistors' channel width, W_{buf}. Regarding the capacitance $C_{gs,buf}$, noting that buffer transistors work in the saturation region most the time, it must be evaluated as $2/3 \cdot (W_{buf} L_{buf} C_{OX})$. Moreover, since it is between two nodes whose voltage gain in (2.50) is close to unity, we have to account for a Miller effect [MG87]. Hence, the equivalent Miller capacitance at the input of the buffer (i.e., in parallel to $C_{gd,buf}$) is given by $C_{gs,buf}\left[1-1\Big/\left(1+\frac{g_{mb,buf}}{g_{m,buf}}\right)\right]$, where $g_{mb,buf}$ and $g_{m,buf}$ are respectively the body effect transconductance and the transistor transconductance in (1.70) and (1.66), respectively, and their ratio is almost constant (typically, $g_{mb,buf}/g_{m,buf}$ ranges from 0.1 to 0.2, and for the 0.35-μm CMOS process used $g_{mb,buf}/g_{m,buf} \cong 0.13$).

---

delay results as $0.69(b_1-a_1)$ [E48]. Circuit analysis, for example using the extra element theorem [M89], shows that adding C_{out} to the internal SCL gate determines an equal increase in a_1 and b_1 by $R_D C_{out}$, therefore the delay is not affected by C_{out}.

The transconductance $g_{m,buf}$ of the transistor implementing the buffer can be evaluated from (1.66) as explained in Section 2.4.2, provided that in the effective electron mobility (2.33), V_{DS} and V_{SB} are approximated with their maximum values (i.e., $V_{GS}+V_{SWING}/2$ and $V_{DD}-V_{GS}$, respectively), and V_{GS} is underestimated by $V_{T,n}$.

The circuit in Fig. 6.9 has a transfer function with two poles and one high-frequency zero, that exhibits a dominant-pole behavior for practical values of I_{SS} and C_L. This is a fundamental difference with respect to ECL bipolar circuits, that have a second-order behavior, and is essentially due to the fact that MOSFET transistors usually have a much lower transconductance at a given bias current [LS94]. Detailed analysis shows that the second pole is greater than the first one at least by one order of magnitude.

From the previous considerations, the buffer circuit can be assimilated to a first-order system, whose delay is equal to 0.69 times its time constant. By applying the time constant method, the following expression of the buffer delay is obtained

$$\tau_{PD,buf} = 0.69\left[R_D\left(C_{gd,buf} + C_{gs,buf}\frac{\dfrac{g_{mb,buf}}{g_{m,buf}}}{1+\dfrac{g_{mb,buf}}{g_{m,buf}}}\right) + \frac{C_L + C_{gs,buf}}{g_{m,buf}}\right] \tag{6.8}$$

where the source-bulk capacitance was neglected with respect to C_L. From (6.8), the buffer delay is equal to the sum of two terms, whose circuital meaning is apparent: the first is proportional to R_D and models the loading effect of the buffer on the internal SCL gate, the second is inversely proportional to $g_{m,buf}$, and hence it depends on the buffer driving capability.

Accuracy of (6.7)-(6.8) was tested by extensive Spectre simulations using the 0.35 μm CMOS process described above, with bias current I_{SF} ranging from 5 μA to 100 μA, bias current I_{SS} set to 5 μA, 20 μA, 50 μA and 100 μA, and V_{DD} set to 3.3 V. Moreover, the internal SCL transistors' aspect ratios were sized to obtain the typical values V_{SWING}=700 mV and A_V=4, while the buffer transistors' aspect ratios were set to 0.6/0.3, 3/0.3 and 6/0.3, and the load capacitance C_L was set to 0 F, 50 fF, 200 fF and 1 pF. As an example, the resulting curves of delay predicted by (6.7)-(6.8) and that simulated for C_L=200 fF and $(W/L)_{buf}$=0.6/0.3 are plotted in Figs. 6.10a-6.10d, where I_{SS} is set to 5 μA, 20 μA, 50 μA and 100 μA, respectively.

From inspection of Figs. 6.10a-6.10d, the delay model agrees well with simulated results. Indeed, among the cases considered, the worst accuracy of 35% was found for C_L=200 fF, I_{SS}=5 μA and I_{SF}=100 μA. Note that such a strong current unbalance between the internal SCL and the output buffers is not encountered in practical cases, since it leads to a buffer delay much lower than the internal SCL one (i.e., a very inefficient distribution of the total bias current used between the internal SCL gate and output buffers).

For other load and bias conditions error was found to be always lower, and for realistic values of bias currents I_{SS} and I_{SF} that lead to similar delay values for the internal SCL gate and the output buffer, the error was always lower than 20%, and usually is much lower. Indeed, the average error for $(W/L)_{buf}$ equal to 0.6/0.3, 3/0.3 and 6/0.3 is 14.7%, 11.3% and 12.6%, respectively.

To have an idea about reasonable values used in practical cases, consider an SCL gate with I_{SS}=20 μA, I_{SF}=40 μA, C_L=50 fF, $(W/L)_n$=8/0.3, $(W/L)_p$=0.6/0.7, $(W/L)_{buf}$=3/0.3 and V_{DD}=3.3 V, whose predicted and simulated delay of 341 ps and 412 ps, respectively, differ by 17%.

As was observed for the SCL inverter without output buffers in Section 6.2.2, the accuracy of the analytical model (6.7)-(6.8) could be improved by following an approach analogous to that used in Section 4.5 for the ECL inverter, i.e. by introducing unknown coefficients to be evaluated through minimization of the error functional (4.9). However, as already highlighted for the SCL inverter, the model accuracy is better than that of ECL circuits and is usually adequate, thus the improved model is not developed for the SCL inverter with output buffers.

Regarding the effect of the input waveform, an even weaker than the SCL inverter dependence of delay on the input rise time was observed. Therefore, the same considerations as in Section 6.2.3 are valid, and the delay can be always be approximated to that for a step input.

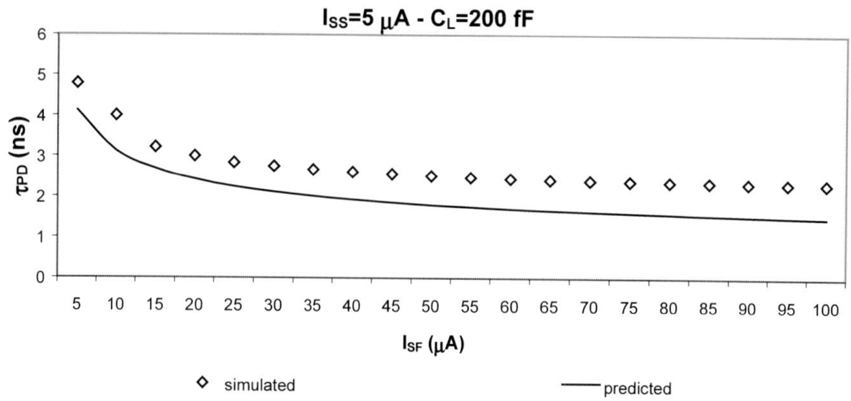

Fig. 6.10a. Predicted and simulated propagation delay vs. buffer bias current, I_{SF}, for an SCL inverter with output buffer for W/L_{buf}=0.6/0.3, C_L=200 fF and I_B=5 μA.

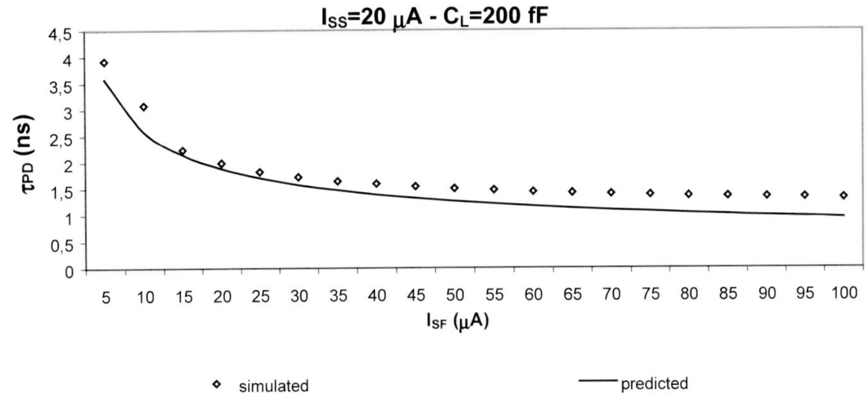

Fig. 6.10b. Predicted and simulated propagation delay vs. buffer bias current, I_{SF}, for an SCL inverter with output buffer for W/L_{buf}=0.6/0.3, C_L=200 fF and I_B=20 μA.

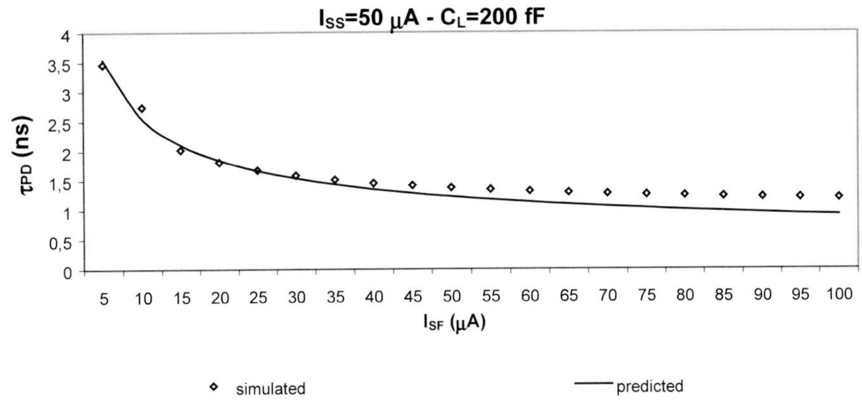

Fig. 6.10c. Predicted and simulated propagation delay vs. buffer bias current, I_{SF}, for an SCL inverter with output buffer for W/L_{buf}=0.6/0.3, C_L=200 fF and I_B=50 μA.

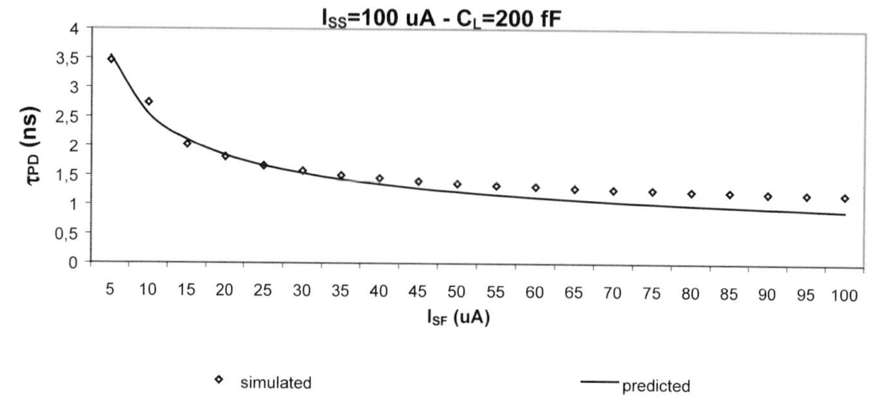

Fig. 6.10d. Predicted and simulated propagation delay vs. buffer bias current, I_{SF}, for an SCL inverter with output buffer for W/L_{buf}=0.6/0.3, C_L=200 fF and I_B=100 μA.

6.4 MODELING OF THE SOURCE-COUPLED MUX/XOR GATE

In this section, the approach introduced in Sections 6.2-6.3 for the SCL inverter is applied to evaluate the delay of the Source-Coupled MUX and XOR gate, both with and without output buffers. The approach is consistent to that used for CML/ECL gates in Chapter 4.

6.4.1. Delay model of the MUX/XOR gate without output buffer

Let us consider the SCL MUX and XOR gates shown in Figs. 6.11-6.12, respectively, for which two different delay values will be considered, depending on whether the switching input is applied to the lower or upper transistors. Between them, the worst-case delay occurs when the switching inputs v_i is applied to transistors that are at a lower level (i.e., M1-M2), for the same reasons discussed in Section 4.6 for CML gates.

Let us consider the case where input v_i is applied to transistors M1-M2, i.e. v_i represents signals ϕ and B in the MUX and XOR gate, respectively, while the others are kept constant. As clarified in Section 4.6, assume input A to be high (and input B of the MUX to be low to have an output transition) without loss of generality. Under this condition, biasing the gate at the logic threshold v_i=0, transistors M3 and M6 lie in the saturation region, while M4 and M5 are in cut-off. It is worth noting that the XOR gate has the same delay as the MUX, since its topology is obtained from the latter by setting $B = \overline{A}$, hence in the following only the MUX gate will be considered.

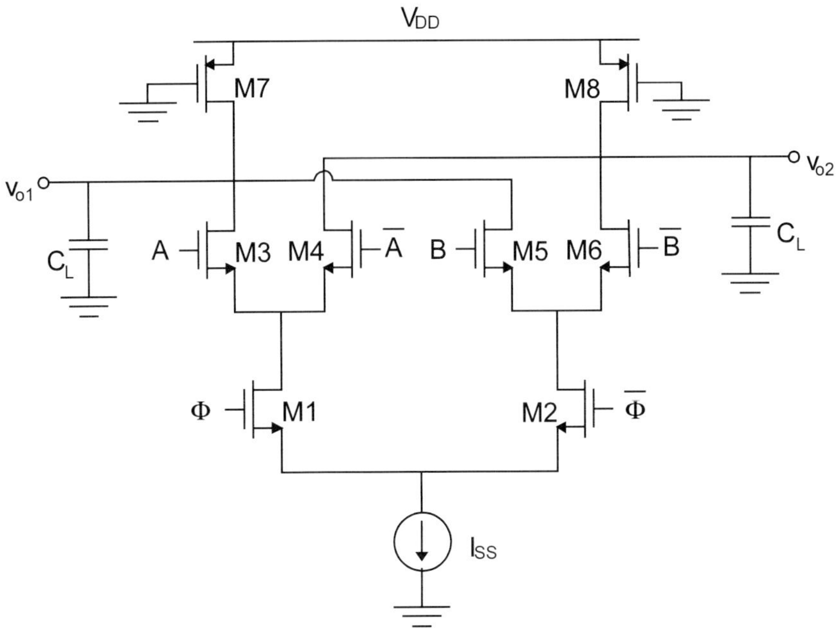

Fig. 6.11. SCL MUX gate.

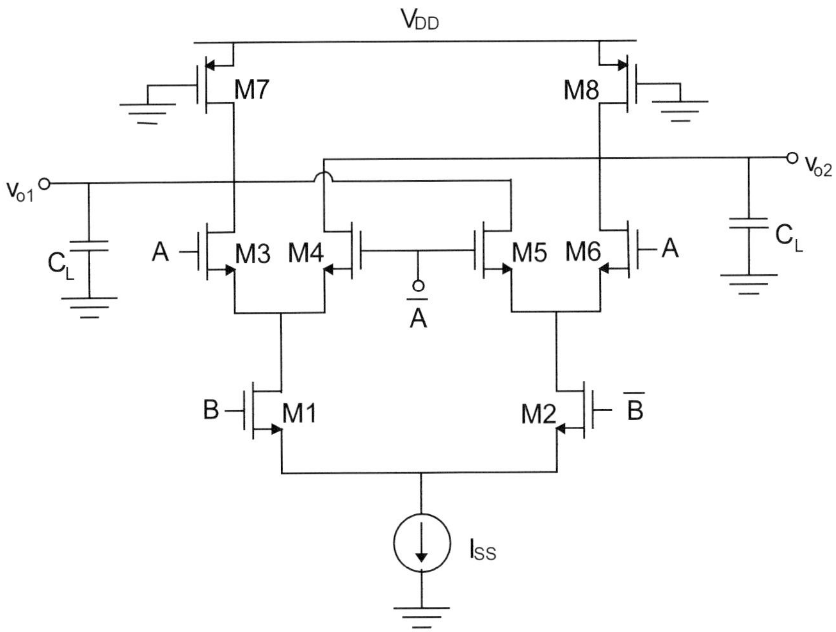

Fig. 6.12. SCL XOR gate.

As already observed in Section 6.2.2, transistors M1-M2 work in the saturation region most of the time, and their source voltage is the same for both input logic values, thus the circuit can be linearized around the logic threshold and then simplified by applying the half-circuit concept.

In the resulting linearized half-circuit, shown in Fig. 6.13, transistors M1-M3 (M2-M6) are represented by their small-signal model including an equivalent transconductance G_M, PMOS transistor M7 (M8) is replaced by its equivalent circuit in Fig. 6.3, and input A is supposed to be driven by an equal circuit, whose resistance R_D is connected to the gate of M3.

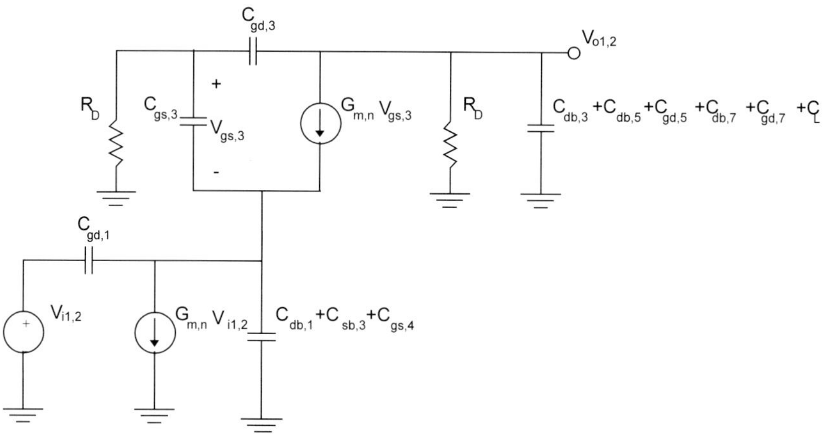

Fig. 6.13. Equivalent linear half circuit of the SCL MUX/XOR gate.

In Fig. 6.13, the capacitive contributions of the transistors are: C_{db} and C_{sb}, that represent the (equal) drain-bulk and source-bulk junction capacitances, C_{gd} that schematizes the channel and the overlap contribution between gate and drain, and C_{gs} that is the gate-source capacitance. An external load capacitance C_L accounts for the wiring capacitance and the input capacitance of the driven logic gates.

By approximating the circuit in Fig. 6.13 to a first-order network with time constant τ, that can be evaluated by resorting to the time constant method [CG73], its delay for a step input equals 0.69τ. Therefore, the propagation delay $\tau_{PD,MUX}$ of the SCL MUX gate results to

$$\tau_{PD,SCL} = 0.69\Big[R_D\left(2C_{gd,3} + C_{db,3} + C_{gd,5} + C_{db,5} + C_{gd,p} + C_{db,p} + C_L\right) \tag{6.9}$$

$$+\frac{1}{G_{M,n}}\left(C_{gd,1} + C_{db,1} + C_{gs,3} + C_{sb,3} + C_{gs,4} + C_{sb,4}\right)\Big]$$

where the equivalent transconductance $G_{M,n}$ of M3 cannot be approximated by its small-signal value $g_{m,n}$ in the saturation region, because when M3 switches its voltages vary greatly around the bias point considered. Rather, the equivalent NMOS transconductance $G_{M,n}$ of M3 can be evaluated as the ratio of the total variation of current $i_{D,3}$ and the total variation of gate-source voltage $V_{GS,3}$ during a complete switching. Considering that the current of M3 varies from I_{SS} to zero (or vice versa), and $V_{GS,3}$ varies from $V_{Tn,3} + \sqrt{I_{SS} \Big/ \left(\frac{\mu_{eff,n} C_{OX}}{2} \frac{W}{L}\Big|_3 \right)}$ to $V_{Tn,3}$, the resulting expression of $G_{M,n}$ is

$$G_{M,n} = \sqrt{\frac{\mu_{eff,n} C_{OX}}{2} \frac{W}{L}\Big|_3 I_{SS}} = \frac{g_{m,n}}{2} \tag{6.10}$$

where the expression (1.66) of the small-signal transconductance $g_{m,n}$ was used. As a result, in large-signal operation, the average transconductance of switching transistors is half of that in a small-signal condition around the logic threshold.

By substituting the expression of the small-signal voltage gain (2.44) and relationship (6.10), the MUX/XOR delay model (6.9) for an input applied to lower transistors can be rewritten as

$$\tau_{PD,SCL} = 0.69 R_D\Big[3C_{gd,1} + 2C_{db,3} + C_{gd,p} + C_{db,p} + C_L \tag{6.11}$$

$$+\frac{2}{A_V}\left(2C_{gd,1} + 3C_{db,1} + C_{gs,3}\right)\Big]$$

where, since all NMOS transistors have an equal aspect ratio, it was considered that $C_{db,5}=C_{db,3}$ and $C_{gd,5}=C_{gd,3}=C_{gd,1}$. Moreover, it was assumed that $C_{sb,3}=C_{sb,4}=C_{db,1}$ given that MOS transistor is symmetric, and $C_{gs,4}=C_{gd,1}$

because M4 is OFF and its gate-source capacitance is only due to the overlap contribution like $C_{gd,1}$.

A simple interpretation of relationship (6.11) can be given by comparison with the expression of the SCL inverter delay (6.6). Indeed, all the capacitive terms in brackets of (6.11) are multiplied by $0.69R_D$ and thus make the same contribution to the delay as if all of its correspondent capacitances were lumped at the output node. As a consequence, terms in brackets of (6.11) represent the equivalent capacitance at the output node due to all transistors. More specifically, terms $\left(3C_{gd,1}+2C_{db,3}+C_{gd,p}+C_{db,p}+C_L\right)$ are associated with capacitances already connected at the output nodes, while terms $\left(2C_{gd,1}+3C_{db,1}+C_{gs,3}\right)$ are lowered by a factor $2/A_V$ (that is usually lower than unity for typical values of A_V, as will be discussed in Chapter 7), since these capacitances see a small-signal resistance lower than R_D by the same factor ($1/G_{M,n}=2R_D/A_V$).

In relationship (6.11), drain-bulk and source-bulk junction capacitances are linearized according to relationship (6.1), where coefficient K_j are evaluated on the basis of the values of voltages V_1 and V_2 in Table 6.3. It is worth noting that capacitance $C_{db,1}$ experiences a small voltage change, since it is due to the variation of $V_{GS,3}$, therefore it can be evaluated in a small-signal condition around the direct voltage ($-V_{DD}+V_{GS,3}$), where $V_{GS,3}$ can be approximated to $V_{T,n}$ for the sake of simplicity.

TABLE 6.3

capacitance	V_1	V_2
$C_{db,3}$	$-V_{DD}$	$-V_{DD}+V_{SWING}/2$
$C_{db,p}$	$-V_{SWING}/2$	0
$C_{db,1}$	$-V_{DD}+V_{GS,3}\cong -V_{DD}+V_{T,n}$	*

* $C_{db,1}$ was evaluated in a small-signal condition

The NMOS capacitances C_{gd1}, C_{gd3} and C_{gd5} (C_{gs4}) make an overlap contribution equal to the product of the capacitance per unit channel width C_{gd0} (C_{gs0}) and their channel width, W_n. Moreover, capacitance C_{gs3} of transistor M3 is the one belonging to the saturation region, that is equal to $2/3\cdot W_nL_nC_{OX}$.

6.4.2. Validation of the model of MUX/XOR gate without output buffer

The delay model (6.11) of the SCL MUX/XOR gate was validated by extensive simulations with the 0.35-μm CMOS process previously used. To encompass a wide range of cases, the bias current was varied from 5 μA to 100 μA, while the transistors' aspect ratios were sized to obtain a logic swing ranging from 400 mV to the maximum value allowed for operation in the saturation region of M3-M6 (i.e., $2V_{T,n}\approx 1.4$ V, according to (2.43) and (2.45), and a voltage gain ranging from 2 to 7. In addition, the load capacitance C_L was set to 0 F, 50 fF, 200 fF and 1 pF.

Some of the results obtained are reported in Figs. 6.14a-6.14d, where transistors were sized to guarantee V_{SWING}=700 mV and A_V=4. In particular, Figs. 6.14a-6.14d report the plot of the MUX/XOR delay simulated and predicted by (6.11) with a load capacitance of 0 F, 50 fF, 200 fF and 1 pF, respectively.

The error of relationship (6.11) with respect to simulations under the same load conditions in Figs. 6.14a-6.14d is plotted in Fig. 6.15 versus I_{SS}. From inspection of Fig. 6.15, the error of the MUX/XOR gate models is always within 20% with C_L=0 F, and is typically lower. The same order of magnitude was found in the other simulations, thus the accuracy is usually adequate for modeling purposes, and the improvement of the model by introducing unknown coefficients as in Section 4.7 is rarely required.

In regard to the effect of the input waveform, the same results as in Section 6.2.3 were found, therefore the MUX/XOR SCL gate delay when the switching input is applied to upper transistors can be always be approximated to that for a step input in (6.11).

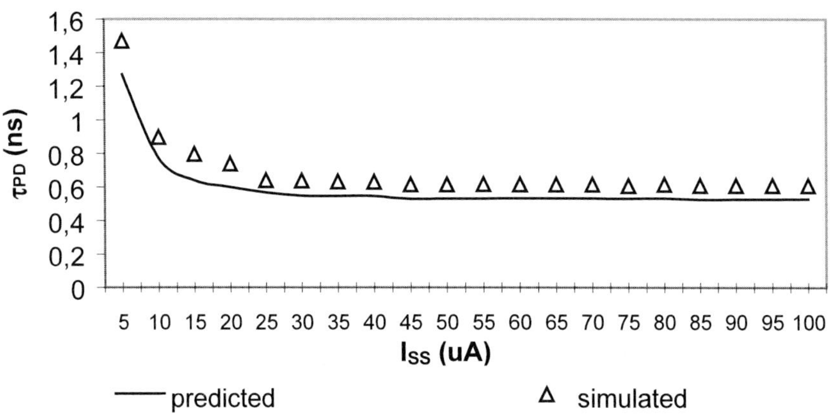

Fig. 6.14a. Simulated and predicted MUX/XOR delay vs. I_{SS} for C_L=0 F.

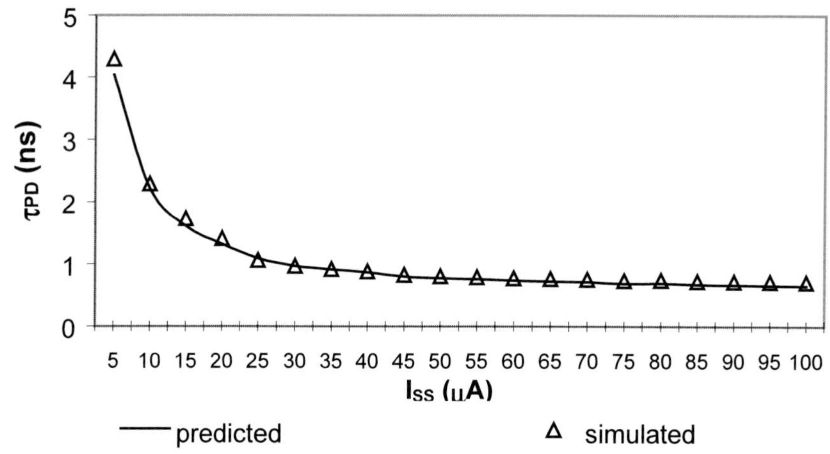

Fig. 6.14b. Simulated and predicted MUX/XOR delay vs. I_{SS} for C_L=50 fF.

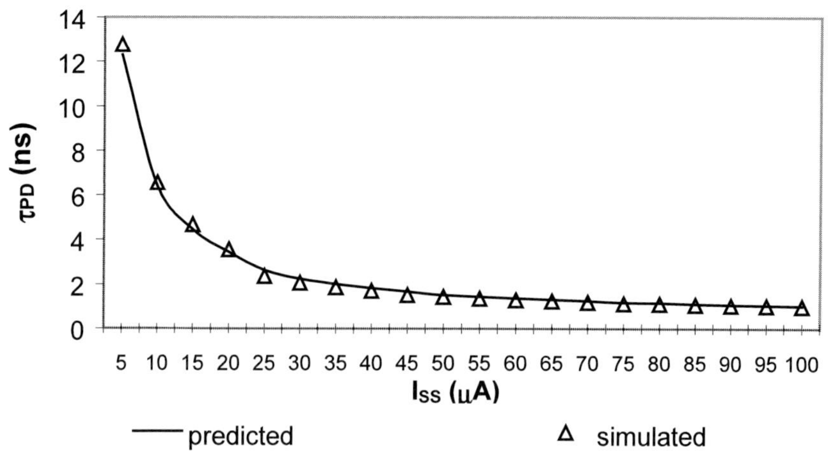

Fig. 6.14c. Simulated and predicted MUX/XOR delay vs. I_{SS} for C_L=200 fF.

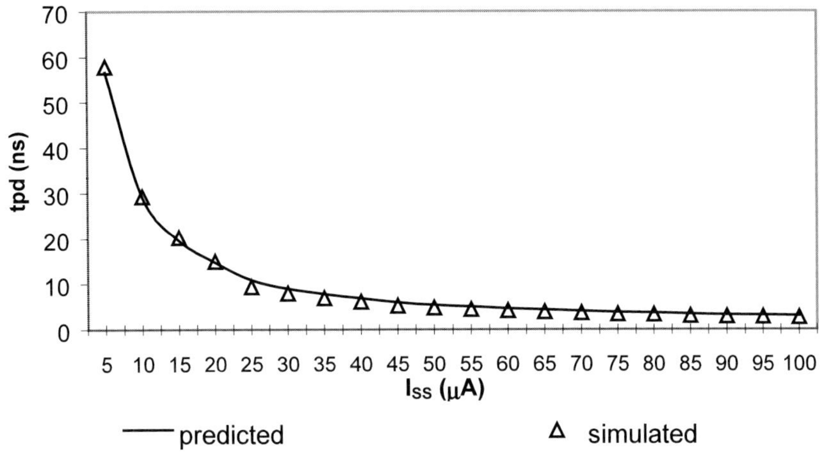

Fig. 6.14d. Simulated and predicted MUX/XOR delay vs. I_{SS} for C_L=1 pF.

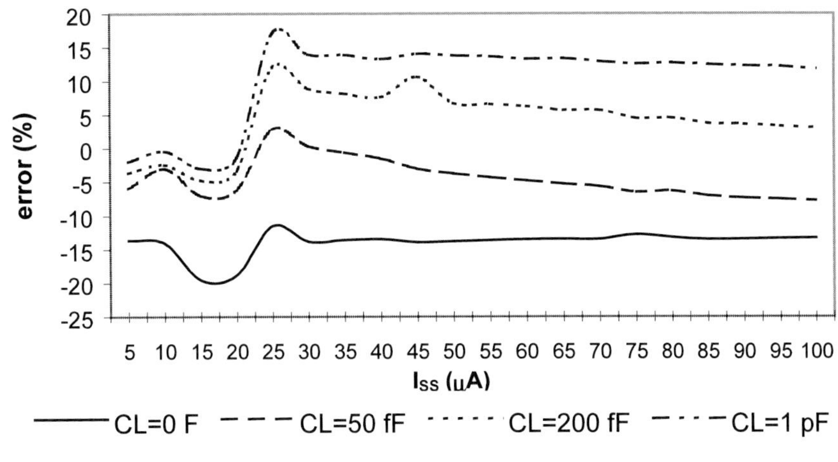

Fig. 6.15. Error of the MUX/XOR predicted delay with respect to simulations vs. I_{SS}.

6.4.3. MUX/XOR with the upper transistors switching

We have so far analyzed the case where the switching input v_i is applied to lower transistors M1-M2. Now, let us evaluate delay when v_i is applied to transistors at the upper level (M3-M6), which is expected to be lower than the former.

Let us evaluate the delay associated with signals driving the upper transistors by assuming that the bias current is already entirely steered to transistors M3-M4 through transistor M1. By applying the switching input v_i to M3-M4, this source-coupled pair can be analyzed like the inverter in Section 6.2 by linearizing the circuit and considering only a half circuit. As in Section 4.6, after defining C_{eq} as the parasitic capacitance contribution of M5-M6 (that are OFF) to the output nodes, the circuit can be schematized with the SCL inverter M3-M4 loaded by a capacitance C_{eq}+C_L. Therefore, the delay associated with the upper transistors is simply obtained from relationship (6.6) as

$$\tau_{PD,SCL} = 0.69 R_D \left(2C_{gd,n} + 2C_{db,n} + C_{gd,p} + C_{db,p} + C_L \right) \tag{6.12}$$

where it was considered that C_{eq}=$C_{gd,n}$+$C_{db,n}$, since M3-M6 have the same aspect ratio and thus the same capacitances $C_{gd,n}$ and $C_{db,n}$, regardless of whether they work in the saturation or in cut-off region. The error of (6.12) with respect to simulations was found to be equal to that of the inverter previously discussed, as expected. Regarding the effect of the input waveform, as for the case previously analyzed, the MUX/XOR SCL gate delay when the switching input is applied to upper transistors can be always be approximated to that for a step input (6.12).

6.4.4 Delay model of the MUX/XOR gate with output buffers

Output buffer can also be inserted in MUX/XOR gates to improve their driving capability or to shift the output voltage common-mode value.

To evaluate the delay of the MUX/XOR gate with output buffers, the same approach as in Section 6.3 can be used and generalized to all SCL gates with output buffers. More specifically, the delay of a generic SCL gate with output buffers can be decomposed into the contribution of the internal SCL gate, $\tau_{PD,SCL}$, and that of the buffer, $\tau_{PD,buf}$, as shown in relationship (6.7). The internal SCL gate delay depends on its topology (that in turn depends on the logic function implemented), and has to be evaluated by following the general methodology described in Section 6.2 and 6.4.1, i.e. as the delay for a zero load capacitance. The buffer contribution was already evaluated in Section 6.3 and expressed by relationship (6.8).

In the specific case of a MUX/XOR, the intrinsic SCL contribution is given by relationship (6.11)-(6.12) where C_L must be set to zero. The loading effect of the buffer on the internal SCL gate is accounted for in $\tau_{PD,buf}$

evaluation, since it is carried out by driving each buffer with the Thevenin equivalent circuit seen at the SCL output.

The delay model in (6.8) and (6.11) of SCL MUX/XOR gates with output buffers was tested by extensive Spectre simulations using the 0.35 μm CMOS process described above, under the same conditions as the inverter in Section 6.3. The internal SCL transistors' aspect ratios were sized to obtain the typical values V_{SWING}=700 mV and A_V=4, while the buffer transistors' aspect ratios were set to 0.6/0.3, 3/0.3 and 6/0.3. The delay curves obtained are not reported for the sake of brevity.

The worst-case error with respect to simulations occurred for C_L=200 fF, and is plotted in Fig. 6.16 versus the buffer bias current I_{SF} for different values of I_{SS}. Among these curves, the worst accuracy of 30% (even better than the inverter gate) was found for I_{SS}=5 μA and I_{SF}=100 μA, which is an unrealistic case, due to the strong current unbalance between the internal SCL and the output buffers. The typical error found is much lower than 30%, as can be deduced from the average error for C_L equal to 0 F, 50 fF, 200 F and 1pF, which is 14.7%, 18.8%, 16.8% and 12.1%, respectively.

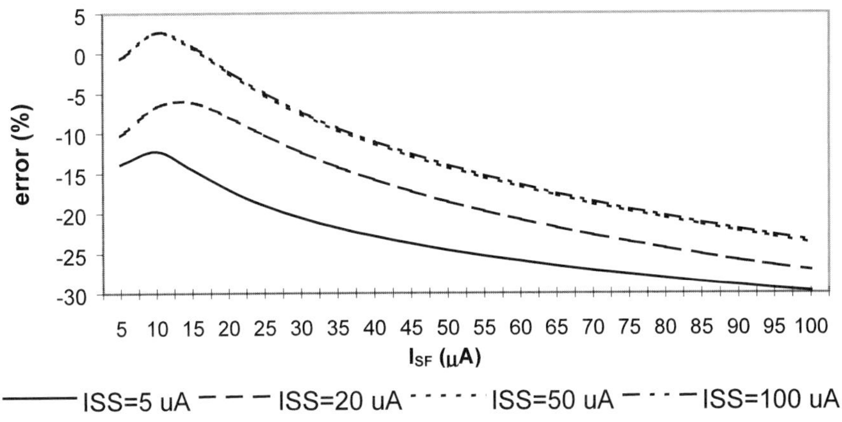

Fig. 6.16. Error of predicted delay of the MUX/XOR gate with output buffers with respect to simulations vs. I_{SF} for C_L=200 fF.

6.5 EVALUATION OF SCL GATES INPUT CAPACITANCE AND EXTENSION TO THE D LATCH

In the previous subsections, the delay of a single SCL gate was computed by assuming a generic load capacitance C_L. In real circuits consisting of cascaded logic gates, it is essential to evaluate the input capacitance of an SCL gate, since it loads the previous gates and affects their delay. For this reason, the input capacitance C_{input} seen by each gate terminal of an NMOS source-coupled pair is evaluated in the following.

The source voltage of a switching source-coupled pair is independent of the input logic value, thus the input capacitance C_{input} seen from the gate of each transistor can be assumed equal to its gate-source capacitance evaluated in the saturation region, given by

$$C_{input} = \frac{2}{3} W_n L_n C_{OX} \tag{6.13}$$

The accuracy of the input capacitance estimation (6.13) was tested through several simulations. In particular, the linear load capacitance that makes delay of a reference gate to be equal to the delay obtained when loaded by the considered gate was evaluated under many bias conditions. Results show that relationship (6.13) is always within 15% of the simulated value, and typically differs by less than 10%.

As already observed for the CML D latch gate, the input capacitance expression (6.13) permits to extend the delay model of the MUX/XOR gate developed in Section 6.4 to the D latch gate shown in Fig. 6.17.

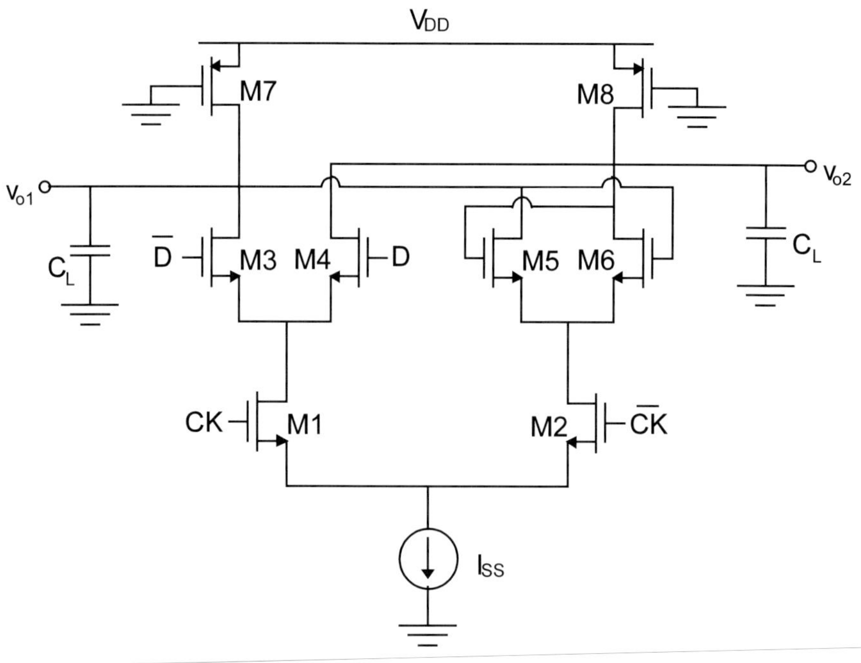

Fig. 6.17. SCL D latch gate.

The only difference between the SCL MUX/XOR gate and the D latch is due to the different load at the output nodes. Indeed, by neglecting the positive-feedback loop involving transistors M5-M6 as clarified in Section 4.9 for bipolar gates, these transistors load both output nodes through their capacitances $C_{gd,n}$ and $C_{db,n}$, as in the case of the MUX/XOR gate, and the capacitance seen from their gate terminal, that is given by relationship (6.13). Therefore, the D latch delay is equal to that of the MUX/XOR gate loaded by a capacitance equal to the sum of the external capacitance C_L and the input capacitance (6.13) of the source-coupled pair M5-M6.

From the previous considerations, the CK-Q latch delay is given by relationship (6.11) where the load capacitance C_L is replaced by an equivalent capacitance C_L'

$$C_L' = C_L + C_{input} \tag{6.14}$$

and, analogously, the D-Q latch delay is given by relationship (6.12) where (6.14) must be substituted. From (6.14), it is apparent that the delay of an SCL D latch gate is always greater than that of a MUX/XOR equally designed, and the speed penalty with respect to the latter is more pronounced for low values of C_L.

The delay model of the D latch discussed so far was validated by extensive simulations with the adopted 0.35-μm CMOS process, under the same conditions as for the MUX/XOR SCL gate. Predicted and simulated delay curves are reported in Fig. 6.18 only for the worst case with C_L=0 F for the sake of brevity, and for C_L equal to 50 fF, 200 fF and 1 pF the delay curves were found to be very close to those in Figs. 6.14b-6.14d, as expected. The error of the delay model with respect to simulations is plotted in Fig. 6.19 versus the bias current I_{SS} for a load capacitance of 0 F, 50 fF, 200 fF and 1 pF, respectively. From inspection of Fig. 6.19, the error of the D latch gate model is always within 23%, and is typically lower.

Regarding the effect of the input waveform, the same results as MUX/XOR SCL gates were found, thus the delay of an SCL D latch gate can be always be approximated to that for a step input in (6.11)-(6.12) with (6.14).

The D latch model can simply be extended to the case with output buffers by decomposing delay into the contribution of the internal SCL gate, $\tau_{PD,SCL}$, and that of the buffer, $\tau_{PD,buf}$. As explained in Section 6.3, the internal SCL gate delay of the D latch gate is evaluated with a zero load capacitance, while the buffer contribution must be evaluated through relationship (6.8).

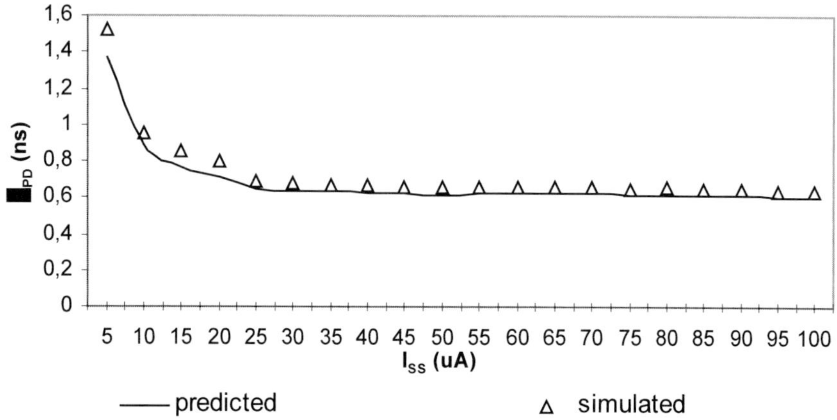

Fig. 6.18. Simulated and predicted D latch delay vs. I_{SS} for C_L=0 F.

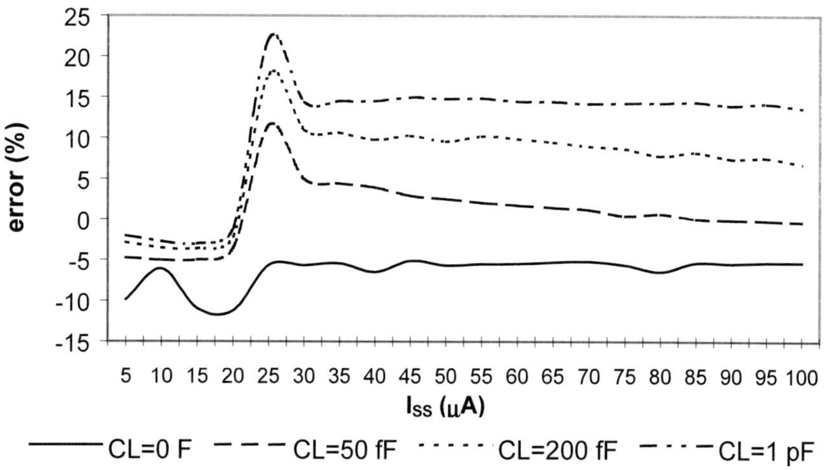

Fig. 6.19. Error of the D latch predicted delay with respect to simulations vs. I_{SS}.

The delay model of the SCL D latch with output buffers discussed above was validated by simulations with the adopted 0.35 μm CMOS process and under the same conditions used for the MUX/XOR gates in Section 6.4. The delay curves obtained are not reported for the sake of brevity, and the error is very close to that found for the MUX/XOR SCL gates in Section 6.4. Comparison with delay values obtained from realistic input waveforms confirms the weak dependence of delay on the input rise time, thus the model discussed so far is also valid for general input waveforms.

Chapter 7

OPTIMIZED DESIGN OF MOS CURRENT-MODE GATES

In this chapter, a general methodology to maximize the speed performance and to manage the power-delay trade-off in SCL gates is presented and applied to several fundamental gates.

7.1 INTRODUCTION TO OPTIMIZED DESIGN OF SCL GATES

In the previous chapter, a general delay model for SCL gates has been developed and applied to some widely used gates. This model may be used to support the design process, but it is not very effective to this purpose since delay is expressed as a function of the PMOS equivalent resistance R_D, as well as on transistors' capacitances, that in turn depend on their aspect ratio.

In practical design cases, transistors' aspect ratio and bias current must be set to meet given specifications on noise margin, as well as on delay and power dissipation, or a trade-off of the two. Therefore, to clearly express design trade-offs, it would be more useful to have an explicit delay dependence on noise margin *NM* and power dissipation *P*. To this aim, in this chapter the dependence of transistors' aspect ratio on *NM* and *P* is first evaluated and then substituted into the delay model described in the previous chapter. Thus, interdependence of design parameters is made explicit through a single equation, that is useful to explore the design space and to manage the possible trade-offs.

Regarding the noise margin performance of SCL gates, the results obtained in Chapter 2 will be used. As far as the power consumption is concerned, as already discussed for CML/ECL gates in Chapter 5, the power consumption P of an SCL gate is essentially static and is determined by its bias current I_{SS}

$$P \cong V_{DD} I_{SS} \tag{7.1}$$

which has to be kept as low as possible for a given speed performance and an acceptable noise immunity. Therefore, once an adequate noise margin is ensured, a strategy to consciously manage the power-delay trade-off is crucial in design of SCL gates.

In this chapter, a systematic design procedure to size the bias current and transistors' aspect ratio to satisfy assigned requirements on noise margin and delay is introduced. In particular, criteria are provided to design SCL gates in typical design cases, where a high speed, a low power consumption or an optimum trade-off between the two is required. Results are successively extended to the case with output buffers, by providing optimum size of their transistors and bias current to minimize delay, assuming practical design conditions.

The design strategy proposed gives simple closed-form expressions of design parameters and explicitly relates delay and bias current (i.e., power consumption from (7.1)), thereby providing the designer with the required understanding of the trade-offs involved in the design. The delay dependence on the logic swing is also investigated, and in contrast with the usual assumption, the results show that the delay is not necessarily reduced by reducing the logic swing. The strategies developed are valid for all SCL gates and are independent of the CMOS process used, thus the guidelines provided afford a deeper understanding of SCL gates from a design point of view.

7.2 OPTIMIZED DESIGN METHODOLOGY IN SCL GATES WITHOUT OUTPUT BUFFERS

Consider an SCL gate without output buffers having a bias current I_{SS} and a load capacitance C_L, with transistors' aspect ratios properly sized to obtain assigned values of logic swing V_{SWING} and voltage gain A_V that ensure a desired value of noise margin NM to be achieved, as will be explained more clearly in Section 7.3.1.

It will be demonstrated in the next sections that the SCL gate considered has a delay $\tau_{PD,SCL}$ given by

$$\tau_{PD,SCL} = 0.35 \cdot V_{SWING} \left(\frac{a}{V_{SWING}^2} + b \frac{V_{SWING}}{I_{SS}^2} + \frac{c + C_L}{I_{SS}} \right) \tag{7.2}$$

where coefficients a, b and c depend on the gate considered, the CMOS process adopted (through standard BSIM3v3 model parameters) and the power supply voltage V_{DD}. In addition, coefficient a is an increasing function of the voltage gain A_V previously assigned.

Relationship (7.2) simply relates delay to the bias current I_{SS}, that determines the power consumption (7.1), as well as to the logic swing, that in turn defines noise margin for a given A_V. Therefore, eq. (7.2) can be suitably used to design SCL gates, as it expresses the trade-offs among delay, power consumption and noise margin. In particular, it shows that the delay reduces as increasing bias current (or, equivalently, power consumption), whereas for $I_{SS}\rightarrow\infty$ it asymptotically tends to an ideal minimum value given by

$$\tau_{PD,SCL,\min} = 0.35\frac{a}{V_{SWING}} \tag{7.3}$$

In order to derive simple design criteria, three practical cases will be considered in the following subsections: power-efficient (7.2.1), high-speed (7.2.2) and low-power design (7.2.3). Results are summarized and commented in Section 7.2.4.

7.2.1 Power-efficient design

According to relationship (7.2), the delay of an SCL gate can be reduced by increasing bias current (or, equivalently, the power consumption). The power-delay product *PDP* [R96] can suitably measure the trade-off efficiency between delay and power dissipation. For an SCL gate, *PDP* is equal to the product of expressions (7.1) and (7.2), and results to

$$PDP_{SCL} \approx 0.35 V_{DD} V_{SWING}\left(\frac{a}{V_{SWING}^2} I_{SS} + b\frac{V_{SWING}}{I_{SS}} + c + C_L\right) \tag{7.4}$$

An optimum balance between delay and power consumption is accomplished when (7.4) is minimum, i.e. for the bias current I_{SS,opt_PDP} given by

$$I_{SS,opt\_PDP} = \sqrt{\frac{b}{a}}\left(V_{SWING}\right)^{\frac{3}{2}} \tag{7.5}$$

that was obtained by setting the derivative of (7.4) equal to zero and solving for I_{SS}. It is worth noting that the bias current (7.5) makes the first two terms in bracket into (7.2) equal. From relationship (7.5), the bias current that optimizes the power-delay trade-off is independent of the load. Moreover, by substituting relationship (7.5) into (7.4), the minimum power-delay product results to be

$$PDP_{opt,SCL} \approx 0.35 V_{DD} V_{SWING} \left(2 \frac{\sqrt{ab}}{\sqrt{V_{SWING}}} + c + C_L \right) \quad (7.6)$$

that increases when V_{SWING} increases. As a consequence, from (7.6) it is apparent that the logic swing has to be chosen as low as possible for a power-efficient design (more details on the acceptable range of V_{SWING} will be provided in Section 7.3.1).

7.2.2 High-speed design

When a high-speed performance is the principal concern in designing an SCL gate, two situations may occur. In the first one, a delay constraint $\tau_{PD,SCL}$ derived from considerations at the gate level has to be met by properly setting the bias current. Analytically, the bias current is found by inverting (7.2) for I_{SS}, leading to

$$I_{SS} = 0.17 \cdot V_{SWING} \frac{c + C_L}{\tau_{PD,SCL} - \tau_{PD,SCL,\min}} \left[1 + \sqrt{1 + 11.4 \cdot b \frac{\tau_{PD,SCL} - \tau_{PD,SCL,\min}}{(c + C_L)^2}} \right] \quad (7.7)$$

where relationship (7.3) was substituted to make expression more compact. Obviously, no solution exists if the required delay is lower than $\tau_{PD,SCL,min}$.

In the second case, no delay constraint is given and the delay has to be kept as close to the asymptotic value $\tau_{PD,SCL,min}$ as possible, in order to exploit the speed potential of the circuit and the process used, while keeping the bias current within a reasonable range. However, it is not evident what is an acceptable range of I_{SS} leading to a delay almost equal to the asymptotic minimum, as relationship (7.2) always decreases as increasing I_{SS}. In practical cases, where a high speed design is targeted, $\tau_{PD,SCL}$ must be close to $\tau_{PD,SCL,min}$, and bias current I_{SS} should only be increased as long as a significant speed improvement is achieved, which no longer holds when delay approaches $\tau_{PD,SCL,min}$. A good compromise between the two opposite requirements can be found from analysis of relationship (7.2), as will be explained in the following.

The expression in brackets into relationship (7.2) consists of a constant term a/V_{SWING}^2 (which determines the asymptotic minimum (7.3)) and one that decreases as increasing bias current I_{SS} (i.e., $b\frac{V_{SWING}}{I_{SS}^2}+\frac{c+C_L}{I_{SS}}$). For values of I_{SS} sufficiently high such that the constant term dominates

$$\frac{a}{V_{SWING}^2} \geq b\frac{V_{SWING}}{I_{SS}^2}+\frac{c+C_L}{I_{SS}} \tag{7.8}$$

a further increase in the bias current does not lead to a significant speed advantage, and a high speed is achieved since delay is close to its minimum value. Instead, for lower values of I_{SS} such that (7.8) does not hold (i.e., the left-hand side is lower than the right-hand side), the terms depending on I_{SS} dominate over the constant one, and delay is highly sensitive to a bias current increase. However, in this case a high speed is not achieved, since delay is far from being minimum. A good compromise between the two opposite cases is achieved when the equality is considered in (7.8), i.e. when the terms depending on I_{SS} equate the constant one. Indeed, in this case delay is easily found to be only twice the minimum achievable

$$\tau_{PD,SCL} = 0.69\frac{a}{V_{SWING}} \tag{7.9}$$

and a further bias current increase is not so beneficial. As a consequence, the strict equality in (7.8) represents a reasonable condition for keeping the delay close to its minimum value and avoiding wasting a uselessly high bias current. This condition is achieved for I_{SS} equal to

$$I_{SS,opt\_delay} = \frac{c+C_L}{2a}V_{SWING}^2\left(1+\sqrt{1+4\frac{ab}{c+C_L}\frac{1}{V_{SWING}}}\right) \tag{7.10}$$

which is always greater than I_{SS,opt_PDP} in (7.5), especially for a high load capacitance C_L. This can be seen by observing that relationship (7.10) equates (7.5) if C_L=c=0 is assumed, while for practical values of C_L and c the other two terms associated with them tend to increase it, especially for high values of C_L. As a consequence, design criteria (7.10) allows for a high speed performance at the cost of a worse *PDP* than the minimum achievable (7.6).

Relationship (7.9) clearly expresses the effect of logic swing on delay in high-speed design. To be more specific, in high-speed design[1] the logic swing has to be set as high as possible to improve the speed performance (more details on the acceptable range of V_{SWING} will be provided in Section 7.3.1). Surprisingly, this in contrast with the usual belief that the high-speed feature of SCL gates is due to the small logic swing [T01], that probably is due to a superficial extension of well-known properties of CML bipolar gates (see Section 5.2 for the effect of logic swing on their power-delay trade-off). As a result, regarding the delay dependence on the logic swing, MOS Current-Mode gates have a completely different behavior compared to bipolar ones, and an in-depth explanation of this difference is provided in Section 7.4.5. This substantial difference between CMOS and bipolar Current-Mode gates in high-speed applications must therefore be taken into account, in order to consciously exploit the speed potential of each technology.

7.2.3 Low-power design

In low-power design, e.g. when the gate speed is not the main target, or even in high performance applications for gates that do not lie in the critical path, the power consumption per gate allowed is usually an assigned parameter that is derived from requirements at the system level. Therefore, bias current is set to a very low value chosen from system considerations, and the only design parameter is the logic swing. More specifically, usual values of bias current are surely lower than the value (7.5) which balances power consumption and delay. From simple inspection of (7.2), condition $I_{SS} << I_{SS,opt\_PDP}$ leads the term bV_{SWING}/I_{SS}^2 to dominate over $(c+C_L)/I_{SS}$. Moreover, under the same condition, considerations in Section 7.2.2 also lead the former term to dominate over a/V_{SWING}^2, since inequality $I_{SS} << I_{SS,opt\_delay}$ surely holds (by remembering that $I_{SS,opt\_PDP} < I_{SS,opt\_delay}$). Thus, under low-power design, the delay expression can be approximated as

$$\tau_{PD,SCL} \cong 0.35 \cdot b\left(\frac{V_{SWING}}{I_{SS}}\right)^2 \qquad (7.11)$$

[1] From relationship (7.9), this holds for both the delay obtained with high-speed criteria discussed, as well as for the minimum delay achievable (7.3) that only differs from (7.9) by a factor of 2.

which shows that in low-power design the logic swing has to be set as low as possible, as in the case of power-efficient design.

7.2.4 Remarks on the delay dependence on bias current and logic swing

To better understand the delay dependence on bias current and logic swing in SCL gates, it is useful to summarize results presented in previous sections for specific design cases, by collecting them in a unified treatment.

Following the considerations carried out above, the delay dependence on bias current can be approximated in three regions, according to

$$\tau_{PD,SCL} = \begin{cases} 0.35 \cdot b \left(\dfrac{V_{SWING}}{I_{SS}} \right)^2 & \text{if } I_{SS} << I_{SS,opt\_PDP} \\ 0.35 \cdot V_{SWING} \dfrac{c + C_L}{I_{SS}} & \text{if } I_{SS,opt\_PDP} << I_{SS} << I_{SS,opt\_delay} \\ 0.35 \cdot \dfrac{a}{V_{SWING}} & \text{if } I_{SS} >> I_{SS,opt\_delay} \end{cases} \tag{7.12}$$

which is represented in Fig. 7.1.

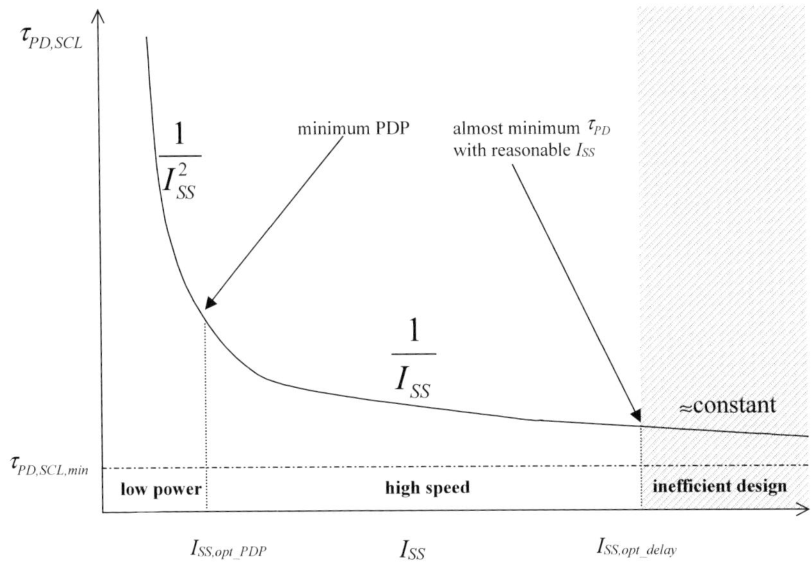

Fig. 7.1. General delay dependence on bias current in SCL gates.

It is observed that, in low-power design (i.e. for $I_{SS} << I_{SS,opt\_PDP}$) the delay can be rapidly decreased by increasing I_{SS}, since it is proportional to $1/I_{SS}^2$. In high-speed design, a less substantial delay reduction is achieved by increasing I_{SS}, since delay is proportional to $1/I_{SS}$. Accordingly, the best power-delay compromise is found for the intermediate value of bias current I_{SS,opt_PDP}. A high speed (i.e. half the maximum achievable for $I_{SS} \rightarrow \infty$) is obtained for $I_{SS} = I_{SS,opt\_delay}$. For greater values of I_{SS}, delay tends to be almost constant, thus small speed improvements are very expensive in terms of bias current, and hence of power consumption.

Regarding the delay dependence on logic swing, summarizing the three cases dealt with in the previous subsections, logic swing must be set as high as possible when a high speed is required, while it has to be reduced as much as possible when a low power consumption or a power-delay trade-off is targeted.

7.3 TRANSISTOR SIZING TO MEET NOISE MARGIN SPECIFICATION

In this section, general delay expression (7.2) of SCL gates will be derived by applying the modeling strategy described in Chapter 6. To be more specific, the delay of SCL gates was there expressed as a function of the PMOS equivalent resistance R_D and transistors' capacitances, that in turn depend on their aspect ratios. In practical cases, aspect ratios have to be set to meet the noise margin requirements, for a given bias current. Therefore, to make delay dependence on bias current and logic swing explicit, it is necessary to find the expression of aspect ratios. To this aim, design criteria to find aspect ratios according to the noise margin required for an assigned bias current are discussed in the following.

7.3.1 Design criteria for V_{SWING} and A_V to meet a noise margin specification

Let us consider an SCL gate with an assigned bias current I_{SS} and a noise margin requirement *NM*. From expression (2.47) of noise margin, this constraint can be met by suitably setting logic swing V_{SWING} and voltage gain A_V.

To understand design aspects related to the choice of A_V, it is useful to observe that from (2.47) we have to choose A_V sufficiently greater than $\sqrt{2}$, in order to avoid an excessive noise margin degradation with respect to the maximum value $V_{SWING}/2$. However, from inspection of (2.44), high values of A_V are achieved by increasing NMOS transistors' aspect ratio, for a given value of I_{SS} and V_{SWING}, thereby increasing parasitic capacitances and slowing down the circuit switching. In addition, (6.13) indicates that this also leads to an increase in the input capacitance, thus slowing down the driving gate as well. A good compromise to achieve an adequate noise margin without excessively degrading speed performance is obtained by setting A_V equal to 4 or slightly greater (for instance, by choosing A_V=4, relationship (2.47) gives NM=$0.65 \cdot V_{SWING}/2$). Therefore, in the following, A_V is assumed to have been chosen in advance according to this criteria.

Regarding the logic swing, it must belong to a well-defined range $[V_{SWING,min}, V_{SWING,max}]$ to allow the circuit to operate correctly. Indeed, V_{SWING} must be greater than its minimum allowed value $V_{SWING,min}$ that strictly satisfies the noise margin requirement NM and that is found by inverting relationship (2.47)

$$V_{SWING,\min} \approx \frac{2NM}{1-\frac{\sqrt{2}}{A_V}} \tag{7.13}$$

Moreover, as discussed in Section 2.4.2, a high logic swing may lead the NMOS transistors to work in the linear region, thus degrading both speed and noise margin performance[2]. More specifically, the maximum value of V_{SWING} allowed, $V_{SWING,max}$, is

$$V_{SWING,\max} \approx 2V_{T,n} \tag{7.14}$$

Summarizing, the voltage gain A_V has to be chosen equal to 4 or slightly greater, while V_{SWING} must be set according to design strategies developed in Section 7.2, provided that it belongs to $[V_{SWING,min}, V_{SWING,max}]$. More specifically, in design cases where power consumption is of concern, logic swing has to be set equal to $V_{SWING,min}$, while for high-speed design it has to be chosen equal to $V_{SWING,max}$. Accordingly, in the following both V_{SWING} and A_V are supposed to be assigned on the basis of the noise margin specification.

---

[2] This is because the transconductance in the linear region is lower than in the saturation region. In other terms, the transistor driving capability is reduced, thus determining a decrease in both speed and voltage gain from (2.44) (and hence in the noise margin (2.47)).

Once desired values of V_{SWING} and A_V are chosen, they are obtained by properly setting transistors' aspect ratios, as explained in the following sections.

7.3.2 Transistor sizing versus I_{SS}

Let us consider an SCL gate with a given bias current I_{SS} and assume that V_{SWING} and A_V have been chosen from the noise margin specification, as clarified in the previous subsection. Now, let us find design equations to size transistors' aspect ratio yielding the desired values of V_{SWING} and A_V.

For a given I_{SS} value, an assigned value of the logic swing can be achieved by appropriately setting the PMOS equivalent resistance R_D. More specifically, by inverting the logic swing expression (2.43), the suitable value of R_D is

$$R_D = \frac{V_{SWING}}{2I_{SS}} \tag{7.15}$$

from which the proper value W_p/L_p of the PMOS aspect ratio must be evaluated. This value of R_D may be higher or lower than that obtained for the minimum transistor size, $W_{p,min}/L_{p,min}$, depending on the assigned value of I_{SS}. To understand this point, we first define R_{D,min_size} as the PMOS resistance obtained for minimum transistor size, and I_{HIGH} as the corresponding current that gives the desired logic swing

$$I_{HIGH} = \frac{V_{SWING}}{2R_{D,\min\_size}} \tag{7.16}$$

that, for example, is equal to 24.6 μA for the 0.35-μm technology adopted and setting typical value V_{SWING}=700 mV.

If $I_{SS}<I_{HIGH}$, or equivalently $R_D>R_{D,min\_size}$, the PMOS aspect ratio must be lower than the minimum value $W_{p,min}/L_{p,min}$. Therefore, we have to set W_p to its minimum value $W_{p,min}$, while setting L_p to a value greater than $L_{p,min}$ that can be evaluated by inverting the expression of R_D in relationship (2.38). More specifically, to simplify the inversion of (2.38) for L_p, this relationship can be expanded in Taylor series truncated to the first order around zero

$$R_D \approx R_{\text{int}}\left(1+\frac{R_{DS}}{R_{\text{int}}}\right)$$
$$= \frac{L_p\left[1+\mu_{eff,p}C_{OX}\frac{1}{L_p}\left(V_{DD}-\left|V_{T,p}\right|\right)R_{DSW}10^{-6}\right]}{\mu_{eff,p}C_{OX}W_{p,\min}\left(V_{DD}-\left|V_{T,p}\right|\right)} \tag{7.17}$$

whose parameters have been already defined in Chapter 2. It is worth noting that series truncation to first order is justified by the observation that the resistance in the linear region of MOS transistors is usually dominated by the contribution of intrinsic resistance R_{int} (i.e., R_D/R_{int}<<1). Relationship (7.17) can be easily solved for L_p, leading to

$$L_p \approx W_{p,\min}\mu_{eff,p}C_{OX}\left(V_{DD}-\left|V_{T,p}\right|\right)\left(\frac{V_{SWING}}{2I_{SS}}-\frac{R_{DSW}10^{-6}}{W_{p,\min}}\right) \tag{7.18}$$

If $I_{SS}>I_{HIGH}$, or equivalently $R_D<R_{D,min\_size}$, the PMOS aspect ratio must be greater than the minimum value $W_{p,min}/L_{p,min}$. Therefore, we have to set L_p to its minimum value $L_{p,min}$, while W_p is found by inverting (2.38)

$$W_p = \frac{2I_{SS}}{V_{SWING}} \cdot \frac{L_{p,\min}}{\mu_{eff,p}C_{OX}\left(V_{DD}-\left|V_{T,p}\right|\right)\left\{1-\frac{R_{DSW}10^{-6}}{L_{p,\min}}\left[\mu_{eff,p}C_{OX}\left(V_{DD}-\left|V_{T,p}\right|\right)\right]\right\}} \tag{7.19}$$

Now, let us discuss design criteria to guarantee a given voltage gain A_V. From relationship (2.44), once logic swing has been set by the PMOS transistor sizing according to the assigned bias current, voltage gain A_V only depends on NMOS transistors' aspect ratio, that will be assumed to be the same for all NMOS transistors. Indeed, this is the usual case where voltage gain is equal for all source-coupled pairs, i.e. voltage gain and noise margin are the same for all inputs. The very uncommon case with different NMOS aspect ratios can be easily analyzed with slight modifications.

From the expression of the voltage gain (2.44), the NMOS aspect ratio W_n/L_n must be increased as increasing I_{SS} to guarantee an assigned gain A_V, as can be seen by inverting (2.44) for W_n/L_n

$$\frac{W_n}{L_n} = \frac{4}{\mu_{eff,n} C_{OX}} \left(\frac{A_V}{V_{SWING}} \right)^2 I_{SS} \tag{7.20}$$

showing direct proportionality to the bias current used, for an assigned value of A_V and V_{SWING}. Moreover, (7.20) confirms theoretical considerations in Section 7.3.1, for which a high value of A_V leads to an increase of the NMOS aspect ratio and thus of their parasitic capacitances.

Substituting (2.43) into relationship (7.20), the resulting W_n/L_n could be lower than $W_{n,min}/L_{n,min}$. This occurs when the bias current is lower than the value I_{LOW} that inverting (7.20) results to be

$$I_{LOW} = \frac{1}{4} \frac{W_{n,\min}}{L_{n,\min}} \mu_{eff,n} C_{OX} \left(\frac{V_{SWING}}{A_V} \right)^2 \tag{7.21}$$

For example, for the 0.35-μm technology adopted and setting typical values V_{SWING}=700 mV and A_V=4, I_{LOW} is equal to 1.45 μA (i.e., an impractically low value). However, actually it does not make sense to set the NMOS aspect ratio W_n/L_n lower than its minimum (i.e. with $W_n=W_{n,min}$ and $L_n>L_{n,min}$), since defining the NMOS to be minimum sized (i.e. with $W_n=W_{n,min}$ and $L_n=L_{n,min}$) leads to a lower input capacitance, thus reducing the delay of the driving gate, and even keeps voltage gain higher than the desired value A_V, which is somewhat beneficial in terms of noise margin.

Summarizing the above considerations, the NMOS channel length L_n always has to be set equal to the minimum value $L_{n,min}$ allowed by the CMOS process used. The NMOS channel width W_n has to be sized according to relationship (7.20) for $I_{SS}>I_{LOW}$, while it must be set at its minimum for $I_{SS} \leq I_{LOW}$.

7.3.3 Summary and remarks on the transistor sizing versus I_{SS}

In the previous subsection, design criteria were provided to set transistors' aspect ratios in order to achieve desired values of V_{SWING} and A_V evaluated in Section 7.3.1. Expressions found are evaluated for a given (but unknown) bias current, therefore their dependence on I_{SS} is made explicit.

For the sake of simplicity, design criteria to size aspect ratios of PMOS transistors to achieve an assigned value of V_{SWING} are summarized below as a function of I_{SS}

$$\begin{cases} W_p = W_{p,\min} \\ L_p = W_{p,\min}\mu_{eff,p}C_{OX}\left(V_{DD} - \left|V_{T,p}\right|\right)\left(\dfrac{V_{SWING}}{2I_{SS}} - \dfrac{R_{DSW}10^{-6}}{W_{p,\min}}\right) \end{cases} \tag{7.22a}$$

if $I_{SS}<I_{HIGH}$ and

$$\begin{cases} W_p = \dfrac{2I_{SS}}{V_{SWING}} \cdot \\ \qquad \dfrac{L_{p,\min}}{\mu_{eff,p}C_{OX}\left(V_{DD} - \left|V_{T,p}\right|\right)\left\{1 - \dfrac{R_{DSW}10^{-6}}{L_{p,\min}}\left[\mu_{eff,p}C_{OX}\left(V_{DD} - \left|V_{T,p}\right|\right)\right]\right\}} \\ L_p = L_{p,\min} \end{cases} \tag{7.22b}$$

if $I_{SS}>I_{HIGH}$, where I_{HIGH} is given by relationship (7.16). The NMOS sizing found for assigned values of A_V and V_{SWING} as a function of the bias current is

$$W_n = W_{n,\min} \tag{7.23a}$$

if $I_{SS}\leq I_{LOW}$ and

$$W_n = 4\frac{L_{n,\min}}{\mu_{eff,n}C_{OX}}\left(\frac{A_V}{V_{SWING}}\right)^2 I_{SS} \tag{7.23b}$$

if $I_{SS}>I_{LOW}$, where I_{LOW} is given by relationship (7.21), and the channel length L_n is always set to its minimum value $L_{n,min}$. Intuitively, an increase of I_{SS} determines a decrease of R_D to maintain the same V_{SWING}, thus W_p/L_p must be increased. Accordingly, to keep A_V to the desired value, the decrease of R_D must be compensated by increasing the NMOS transconductance through their aspect ratio W_n/L_n.

It is worth noting that the condition $I_{LOW}<I_{HIGH}$ is always satisfied. This can be shown by approximating R_D to its intrinsic contribution R_{int} in (2.36), which dominates over the contribution due to parasitic source/drain resistance. Under this approximation, the ratio I_{HIGH}/I_{LOW} results to

$$\frac{I_{HIGH}}{I_{LOW}} = 2(A_V)^2 \frac{\mu_{eff,p}}{\mu_{eff,n}} \frac{\frac{W_{p,\min}}{L_{p,\min}}}{\frac{W_{n,\min}}{L_{n,\min}}} \frac{V_{DD} - |V_{T,p}|}{V_{SWING}}$$

$$\approx (A_V)^2 \frac{V_{DD} - |V_{T,p}|}{V_{SWING}} > 1 \qquad (7.24)$$

where the mobility ratio between PMOS and NMOS transistor is in the order of ½ for current technologies, the ratio of minimum aspect ratio of PMOS and NMOS transistors is about unity, A_V is surely greater than unity and $V_{DD} - |V_{T,p}| >> V_{SWING}$. Therefore, (7.24) is largely greater than unity, confirming that $I_{LOW}<I_{HIGH}$.

Design criteria to size transistors' aspect ratios versus the bias current are summarized in Table 7.1. In particular, design equations of transistor sizes change according to the three possible bias current ranges: low current (L) for $I_{SS}<I_{LOW}$, medium current (M) for $I_{LOW}\leq I_{SS}\leq I_{HIGH}$, and high current (H) for $I_{SS}>I_{HIGH}$. In practical cases, logic gates are typically biased in the M or H range.

TABLE 7.1

	L ($I_S<I_{LOW}$)	**M** ($I_{LOW}\leq I_B\leq I_{HIGH}$)	**H** ($I_B>I_{HIGH}$)
W_n	$W_{n,min}$	eq. (7.23b)	eq. (7.23b)
L_n	$L_{n,min}$	$L_{n,min}$	$L_{n,min}$
W_p	$W_{p,min}$	$W_{p,min}$	eq. (7.22b)
L_p	eq. (7.22a)	eq. (7.22a)	$L_{p,min}$

7.4 OPTIMIZED DESIGN OF THE SOURCE-COUPLED INVERTER

In this section, the design of the SCL inverter gate is addressed. To this aim, it is demonstrated that its delay expression can be expressed in the form of relationship (7.2) as a function of bias current, as well as parameters V_{SWING} and A_V previously assigned according to the strategy discussed in Section 7.3.

Delay expression (6.6) derived in Chapter 6 is a function of PMOS equivalent resistance R_D and transistors' capacitances. Therefore, to make delay dependence on I_{SS}, V_{SWING} and A_V explicit, transistor aspect ratio expressions evaluated in the previous section must be substituted in parasitic capacitances expressions. In particular, parasitic capacitances dependence on I_{SS} changes according to the range (L, M or H) which the bias current belongs to.

7.4.1 Delay expression versus bias current and logic swing in region M

Consider an SCL inverter gate biased in the region M, where, according to Table 7.1, W_n and L_p are given by (7.23b) and (7.22a), respectively, while L_n and W_p are minimum. By substituting these transistor sizes into capacitances described in Section 6.2, expressions of capacitances $C_{gd,n}$, $C_{db,n}$, $C_{gd,p}$ and $C_{db,p}$ can be shown to have the same dependence on the bias current, that can be expressed in the following compact way

$$C_{xy}^{M} = \frac{a_{xy}^{M}}{V_{SWING}^{2}} I_{SS} + b_{xy}^{M} \frac{V_{SWING}}{I_{SS}} + c_{xy}^{M} \tag{7.25}$$

where x and y are the terminals that capacitance considered refers to, and the superscript refers to biasing region M. For instance, $C_{gd,n}^{M}$ represents the NMOS gate-to-drain capacitance expression for I_{SS} ranging in the interval M, and its dependence on I_{SS} is described by the associated coefficients $a_{gd,n}^{M}$, $b_{gd,n}^{M}$ and $c_{gd,n}^{M}$. As an example, coefficients of capacitance $C_{gd,n}^{M}$ in the region M are evaluated by substituting expression (7.23b) of W_n into the general expression of $C_{gd,n}$ discussed in Section 6.2.2

$$C_{gd,n} = C_{gd0,n} W_n = 4C_{gd0,n} \frac{L_{n,\min}}{\mu_{eff,n} C_{OX}} \left(\frac{A_V}{V_{SWING}} \right)^2 I_{SS} \tag{7.26}$$

which, by comparison with relationship (7.25), leads to $b_{gd,n}^{M}$ and $c_{gd,n}^{M}$ equal to zero, and $a_{gd,n}^{M}$ equal to expression reported in Table 7.2. To analyze another example, consider expression of capacitance $C_{gd,p}^{M}$ evaluated in Section 6.2.1 and made it more explicit by substituting PMOS transistors' sizes (7.22a) pertaining to region M, thus obtaining

$$C_{gd,p} = \frac{3}{8} A_{bulk,\max} \mu_{eff,p} C_{OX}^2 W_{p,\min}^2 \left(V_{DD} - \left|V_{T,p}\right|\right) \frac{1}{I_{SS}^2} + C_{gd0,p} W_{p,\min} +$$

$$- \frac{3}{4} A_{bulk,\max} \mu_{eff,p} C_{OX}^2 W_{p,\min} \left(V_{DD} - \left|V_{T,p}\right|\right) R_{DSW} 10^{-6} \tag{7.27}$$

from which, by comparison to (7.25), coefficient $a_{gd,p}^M$ results equal to zero, while $b_{gd,p}^M$ and $c_{gd,p}^M$ result equal to the relationships reported in Table 7.2.

By following the same procedure, analytical expressions of the three coefficients for all the transistor capacitances are explicitly reported in Table 7.2 (those not included are equal to zero).

TABLE 7.2

$a_{gd,n}^M$	$4 A_V^2 C_{gd0,n} \dfrac{L_{n,\min}}{\mu_{eff,n} C_{OX}}$
$a_{db,n}^M$	$4 A_V^2 \left(K_{j,n} C_{j,n} L_{d,n} + 2K_{jsw,n} C_{jsw,n}\right) \dfrac{L_{n,\min}}{\mu_{eff,n} C_{OX}}$
$c_{db,n}^M$	$2K_{jsw,n} C_{jsw,n} L_{d,n}$
$b_{gd,p}^M$	$\dfrac{3}{8} A_{bulk,\max} \mu_{eff,p} C_{OX}^2 W_{p,\min}^2 \left(V_{DD} - \left|V_{T,p}\right|\right)$
$c_{gd,p}^M$	$C_{gd0,p} W_{p,\min} - \dfrac{3}{4} A_{bulk,\max} \mu_{eff,p} C_{OX}^2 W_{p,\min} \left(V_{DD} - \left|V_{T,p}\right|\right) R_{DSW} 10^{-6}$
$c_{db,p}^M$	$K_{j,p} C_{j,p} L_{d,p} W_{p,\min} + 2K_{jsw,p} C_{jsw,p} \left(L_{d,p} + W_{p,\min}\right)$

(zero-valued coefficients are not reported)

The general capacitance expression (7.25) could have been expected before detailed calculations by considering that design criteria for transistors' aspect ratios in the previous section lead to one dimension to be minimum and the other which varies with I_{SS}. In particular, for I_{SS} belonging to the region M, from Table 7.1 and relationship (7.23b) it is apparent that in NMOS transistors the channel width proportionally varies with I_{SS}, while from (7.22a) the PMOS channel length inversely proportionally depends on I_{SS}. As a consequence, each capacitance of NMOS transistors consists of a constant term and one proportional to I_{SS}, while the PMOS capacitances contain a constant term and one inversely proportional to I_{SS}. These considerations provide an intuitive understanding of relationship (7.25).

From inspection of Table 7.2, coefficients a_{xy}^{M}, b_{xy}^{M} and c_{xy}^{M} only depend on the previously assigned value of A_V, as well as the process used and supply voltage that affects the evaluation of junction capacitances, hence in the design they are constant.

Relationship (7.25) can be suitably used to express the sum of capacitances in the delay model (6.6) of the SCL inverter as an explicit function of bias current and logic swing, assuming the gate to be biased in region M. Indeed, by using capacitance expression (7.25) with coefficient evaluated in Table 7.2, the sum of capacitances $C_{gd,n}+C_{db,n}+C_{gd,p}+C_{db,p}$ can be written as

$$C_{gd,n}+C_{db,n}+C_{gd,p}+C_{db,p}=\frac{a^{M}}{\left(V_{SWING}\right)^{2}}I_{SS}+b^{M}\frac{V_{SWING}}{I_{SS}}+c^{M} \tag{7.28}$$

where coefficients a^M, b^M and c^M are defined as

$$a^{M}=a_{gd,n}^{M}+a_{db,n}^{M}+a_{gd,p}^{M}+a_{db,p}^{M}=a_{gd,n}^{M}+a_{db,n}^{M} \tag{7.29a}$$

$$b^{M}=b_{gd,n}^{M}+b_{db,n}^{M}+b_{gd,p}^{M}+b_{db,p}^{M}=b_{gd,p}^{M} \tag{7.29b}$$

$$c^{M}=c_{gd,n}^{M}+c_{db,n}^{M}+c_{gd,p}^{M}+c_{db,p}^{M}=c_{gd,n}^{M}+c_{gd,p}^{M}+c_{db,p}^{M} \tag{7.29c}$$

and those coefficients equal to zero were omitted. By substituting relationships (7.29) and (7.15) into eq. (6.6), delay of an SCL inverter biased in region M can be expressed as

$$\tau_{PD,SCL}=0.35\cdot V_{SWING}\left[\frac{a^{M}}{\left(V_{SWING}\right)^{2}}+b^{M}\frac{V_{SWING}}{I_{SS}^{2}}+\frac{c^{M}+C_{L}}{I_{SS}}\right] \tag{7.30}$$

that has the same form as relationship (7.2), as anticipated in Section 7.2.

7.4.2 Delay expression versus bias current and logic swing in region L and H

Now let us consider an SCL gate biased in the region L, whose NMOS transistors have minimum size and PMOS transistors have $W_p = W_{p,min}$ and L_p given by relationship (7.22a), from Table 7.1. By substituting these sizes into capacitances' expressions discussed in Section 6.2 and following the same procedure as that used in Section 7.4.1, expressions of capacitances $C_{gd,n}$, $C_{db,n}$, $C_{gd,p}$ and $C_{db,p}$ are shown to have the same dependence on the bias current as in relationship (7.25), where superscript M must be replaced by L. Analytical expressions of the non-zero coefficients for all the transistor capacitances are reported in Table 7.3, whose inspection shows that coefficients in Table 7.3 again depend only on the process used and supply voltage.

TABLE 7.3

$c^L_{gd,n}$	$C_{gd0,n} W_{n,\min}$
$c^L_{db,n}$	$K_{j,n} C_{j,n} L_{d,n} W_{n,\min} + 2K_{jsw,n} C_{jsw,n} \left(L_{d,n} + W_{n,\min}\right)$
$b^L_{gd,p}$	$b^M_{gd,p}$
$c^L_{gd,p}$	$c^M_{gd,p}$
$c^L_{db,p}$	$c^M_{db,p}$

(zero-valued coefficients are not reported)

By reiterating the same procedure developed in region M, the sum of capacitances $C_{gd,n} + C_{db,n} + C_{gd,p} + C_{db,p}$ in region L results to

$$C_{gd,n} + C_{db,n} + C_{gd,p} + C_{db,p} = \frac{a^L}{V^2_{SWING}} I_{SS} + b^L \frac{V_{SWING}}{I_{SS}} + c^L \tag{7.31}$$

where coefficients a^L, b^L and c^L are

$$a^L = a^L_{gd,n} + a^L_{db,n} + a^L_{gd,p} + a^L_{db,p} = 0 \tag{7.32a}$$

$$b^L = b^L_{gd,n} + b^L_{db,n} + b^L_{gd,p} + b^L_{db,p} = b^L_{gd,p} \tag{7.32b}$$

$$c^L = c^L_{gd,n} + c^L_{db,n} + c^L_{gd,p} + c^L_{db,p} \tag{7.32c}$$

By substituting relationships (7.31)-(7.32) and (7.15) into eq. (6.6), the delay of an SCL inverter biased in region L can be written as

$$\begin{aligned} \tau_{PD,SCL} &= 0.35 \cdot V_{SWING}\left(\frac{a^L}{V^2_{SWING}} + b^L \frac{V_{SWING}}{I^2_{SS}} + \frac{c^L + C_L}{I_{SS}}\right) \\ &= 0.35 \cdot V_{SWING}\left(b^L \frac{V_{SWING}}{I^2_{SS}} + \frac{c^L + C_L}{I_{SS}}\right) \end{aligned} \tag{7.33}$$

whose dependence on the bias current is easily understandable by considering that L_p in relationship (7.22a) consists of a constant term and one inversely proportional to the bias current, while other transistors' dimensions are minimum.

By reiterating the same procedure and substituting transistor sizes reported in Table 7.1, the delay of an SCL inverter biased in region H can be expressed as

$$\tau_{PD,SCL} = 0.35 \cdot V_{SWING}\left(\frac{a^H}{V^2_{SWING}} + b^H \frac{V_{SWING}}{I^2_{SS}} + \frac{c^H + C_L}{I_{SS}}\right) \tag{7.34}$$

where coefficients a^H, b^H and c^H are

$$a^H = a^H_{gd,n} + a^H_{db,n} + a^H_{gd,p} + a^H_{db,p} \tag{7.35a}$$

$$b^H = b^H_{gd,n} + b^H_{db,n} + b^H_{gd,p} + b^H_{db,p} = 0 \tag{7.35b}$$

$$c^H = c^H_{gd,n} + c^H_{db,n} + c^H_{gd,p} + c^H_{db,p} = c^H_{db,n} + c^H_{db,p} \tag{7.35c}$$

whose terms are explicitly reported in Table 7.4.

TABLE 7.4

$a_{gd,n}^{H}$	$a_{gd,n}^{M}$
$a_{db,n}^{H}$	$a_{db,n}^{M}$
$c_{db,n}^{H}$	$c_{db,n}^{M}$
$a_{gd,p}^{H}$	$\dfrac{2\left(C_{gd0,p}+\frac{3}{4}A_{bulk,\max}C_{OX}L_{p,\min}\right)V_{SWING}}{L_{p,\min}} \cdot \dfrac{1}{\mu_{eff,p}C_{OX}\left(V_{DD}-\lvert V_{T,p}\rvert\right)\left[1-\frac{\mu_{eff,p}C_{OX}}{L_{p,\min}}R_{DSW}10^{-6}\left(V_{DD}-\lvert V_{T,p}\rvert\right)\right]}$
$a_{db,p}^{H}$	$\dfrac{2\left(K_{j,p}C_{j,p}L_{d,p}+2K_{jsw,p}C_{jsw,p}\right)V_{SWING}}{L_{p,\min}} \cdot \dfrac{1}{\mu_{eff,p}C_{OX}\left(V_{DD}-\lvert V_{T,p}\rvert\right)\left[1-\frac{\mu_{eff,p}C_{OX}}{L_{p,\min}}R_{DSW}10^{-6}\left(V_{DD}-\lvert V_{T,p}\rvert\right)\right]}$
$c_{db,p}^{H}$	$2K_{jsw,p}C_{jsw,p}L_{d,p}$

(zero-valued coefficients are not reported)

7.4.3 Extension of the delay model in the region M to region L and H: a unified expression of delay and remarks

From relationships (7.30), (7,33) and (7.34), the delay of an SCL inverter in each biasing region can be expressed in the same form. However, a compact delay model (i.e., defined by a single equation) would be more useful for design, and to this aim it will be shown that delay of an SCL inverter can be expressed in the compact form of relationship (7.2) regardless of the biasing region.

To obtain a compact delay model, it is useful to observe that the delay expression of the SCL inverter biased in region M can be approximately extended to other regions. Indeed, according to Table 7.1, for low bias currents (both in the range L and at the beginning of region M), the dominant contribution to delay is that of $C_{gd,p}$ due to the high value of L_p, whose expression (7.22a) is the same in both ranges L and M. Therefore, delay equation in the range M can be extended to region L without significant error. Analogously, for high bias currents (both in the range H and at the end of range M), capacitances $C_{db,n}$ and $C_{gd,n}$ are dominant due to the high value of W_n, whose expression (7.23b) is the same in both ranges M and H, allowing to extend the model valid in the range M to the interval H. As a result, the delay model (7.30), valid in region M, well approximates the delay expression regardless of the biasing region. Thus the validity of relationship (7.2) for the SCL inverter, using coefficients a, b and c of the region M is confirmed

$$a = a^M = a^M_{gd,n} + a^M_{db,n} \tag{7.36a}$$

$$b = b^M = b^M_{gd,p} \tag{7.36b}$$

$$c = c^M = c^M_{db,n} + c^M_{gd,p} + c^M_{db,p} \tag{7.36c}$$

As an example, for the 0.35-μm CMOS process used and assuming V_{SWING}=700 mV, A_V=4 and V_{DD}=3.3 V, the model developed in region M used for the other regions leads to the error in Fig. 7.2 (which is plotted versus I_{SS} in logarithmic scale and under the worst case C_L=0).

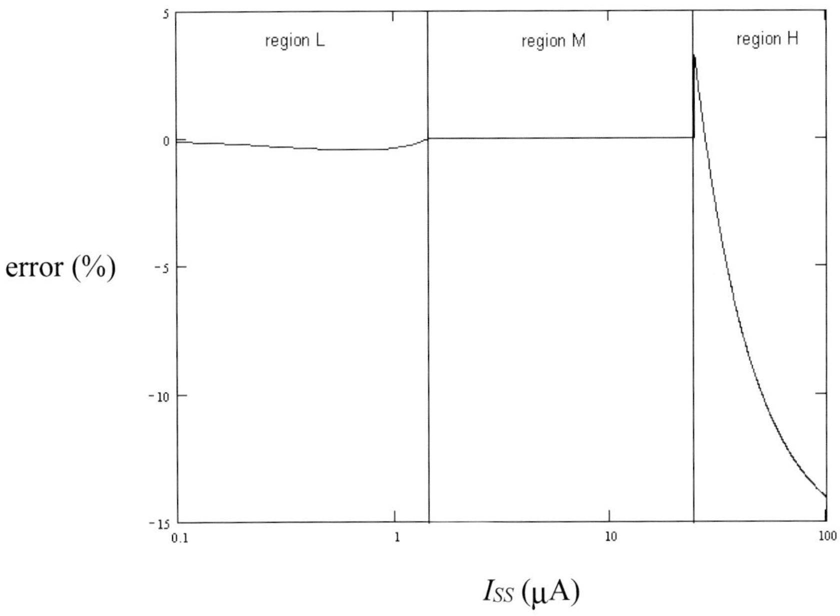

Fig. 7.2. Error of the delay model along the three region using that in region M for the worst case C_L=0 F.

From Fig. 7.2, it is apparent that the error is always lower than 14%, and it can be shown that for more realistic load conditions it is in the order of a few percentage points.

From an analytical point of view, extension of relationship (7.30) to region H is easily understood by comparing coefficients a, b and c in the three regions. To be more specific, extension from region M to H is justified by observing that $a^H \approx a^M$ (since at the boundary the term proportional to a dominates over that proportional to b), while extension from region M to L is understood by observing that $b^L = b^M$ (at the boundary the term proportional to b dominates over that proportional to a). As an example, these results are confirmed by numerical values of coefficients in the three regions for the 0.35-μm CMOS process used and assuming V_{SWING}=700 mV, A_V=4 and V_{DD}=3.3 V, reported in Table 7.5.

TABLE 7.5

L	M	H
a^L=0	a^M=2.26 E-10	a^H=2.7 E-10
b^L=7.47 E-20	b^M=7.47 E-20	b^H=0
c^L=2.48 E-15	c^M=1.82 E-15	c^H=1.23 E-15

Now, let us introduce a circuit interpretation of terms included in the general delay model (7.30) of an SCL inverter, since their meaning is less evident than the model (6.6) developed in the previous chapter.

From relationship (7.29), term a^M is due only to NMOS transistors, while b^M only to PMOS transistors. Moreover, in relationship (7.30) term c^M which is due to both devices and can in general be neglected[3]. As a consequence, the term $a^M I_{SS}/V_{SWING}^2 \approx C_{gd,n} + C_{db,n}$ in (7.30) models the NMOS transistors' capacitance at the output node. Analogously, the term $b^M V_{SWING}/I_{SS} \approx C_{gd,p} + C_{db,p}$ represents the equivalent capacitance at the output node associated with PMOS transistors.

From the circuit interpretation of addends in delay model (7.30), some interesting observations can be derived. First, capacitances at the output

[3] For example, in the CMOS process considered, c^M is in the order of a gate capacitance of a minimum NMOS transistor, which is usually much lower than typical load capacitances in (7.30).

node associated with NMOS and PMOS transistors are equal when $a^M I_{SS}/(V_{SWING})^2 = b^M V_{SWING}/I_{SS}$. Solving for I_{SS}, this means that NMOS and PMOS capacitances are equal for I_{SS} given by relationship (7.5) which minimizes the power-delay product. Hence, we can interpret the power-efficient design criteria as that leading to an equal contribution to the delay of NMOS and PMOS transistors.

In the high-performance design (i.e., $I_{SS}>>I_{SS,opt\_PDP}$), the term $a^M I_{SS}/V_{SWING}^2$ dominates over $b^M V_{SWING}/I_{SS}$, and delay is essentially due to NMOS capacitances and the design criteria (7.8) with strict equality brings the equivalent NMOS capacitance to be equal to the load capacitance C_L (thus, no significant advantage would be achieved for higher values of I_{SS}, since the NMOS capacitances would excessively self-load the gate). In contrast, in the low-power design (i.e., $I_{SS}<<I_{SS,opt\_PDP}$) the dominant contribution is that from the PMOS transistors.

7.4.4 Design criteria and examples

As was demonstrated in the previous section, delay of an SCL gate can be modeled by (7.2) (or, equivalently, (7.30)). As a consequence, design criteria introduced in Section 7.2 can be applied to size design parameters of an SCL inverter gate.

Regarding the power-efficient design in Section 7.2.1, the bias current (7.5) that leads to an optimal power-delay trade-off becomes

$$I_{SS,opt\_PDP} = 0.31\sqrt{\mu_{eff,n}\mu_{eff,p}}\, C_{OX}^{\frac{3}{2}} \frac{1}{A_V} \cdot \sqrt{\frac{W_{p,\min}^2}{L_{n,\min}} \frac{A_{bulk,\max}\left(V_{DD}-\left|V_{T,p}\right|\right)}{C_{gd0,n}+K_{j,n}C_{j,n}L_{d,n}+2K_{jsw,n}C_{jsw,n}}} \left(V_{SWING}\right)^{\frac{3}{2}} \tag{7.37}$$

where relationship (7.29) together with relationships in Table 7.2 were substituted. Regarding process parameters, bias current (7.37) for a power-efficient design is proportional to the geometric average of the NMOS and PMOS mobility, and depends on the oxide capacitance as $C_{OX}^{3/2}$. Moreover, regarding design parameters, (7.37) is proportional to $V_{SWING}^{3/2}$ and inversely proportional to the voltage gain A_V, as qualitatively predicted in Section 7.3.1. It is worth noting that capacitance at the denominator within the square root can be interpreted as the capacitance per unit channel width of NMOS transistors (i.e. it includes both $C_{gd,n}$ and $C_{db,n}$).

By substituting (7.29)-(7.30) into (7.6), the resulting (minimum) power-delay product under bias current (7.37) is

$$PDP_{opt,SCL} = 0.35 \cdot V_{DD} I_{SS,opt\_PDP} V_{SWING} \cdot \left[\frac{a^M}{\left(V_{SWING}\right)^2} + b^M \frac{V_{SWING}}{I^2_{SS,opt\_PDP}} + \frac{c^M + C_L}{I_{SS,opt\_PDP}} \right]$$

$$\approx 0.35 V_{DD} V_{SWING} \left(C_L + 2\frac{\sqrt{ab}}{\sqrt{V_{SWING}}} \right)$$

$$= 0.35 V_{DD} V_{SWING} \left[C_L + 2.45 \frac{A_V}{\sqrt{V_{SWING}}} \sqrt{\frac{\mu_{eff,p}}{\mu_{eff,n}} C_{OX} W^2_{p,\min} L_{n,\min}} \right.$$

$$\left. \cdot \sqrt{\left(C_{gd0,n} + K_{j,n} C_{j,n} L_{d,n} + 2K_{jsw,n} C_{jsw,n}\right) A_{bulk,\max}\left(V_{DD} - \left|V_{T,p}\right|\right)} \right] \quad (7.38)$$

where coefficient c^M was neglected. From (7.38), for a high load capacitance C_L that dominates over the second addend (i.e. C_L is much higher than the parasitic capacitance at the output node, consisting of NMOS and PMOS contributions both equal to $\sqrt{a^M b^M / V_{SWING}}$, from (7.28) and Table 7.2), power-delay product is roughly equal to $0.35 V_{DD} V_{SWING} C_L$, hence it can be proportionally reduced by reducing logic swing, while satisfying noise margin requirement. Analogously, the optimum *PDP* can be reduced by reducing supply voltage, while keeping it sufficiently high to allow a correct operation of an assigned number of series gating levels, as discussed in Section 2.5.4. When the load capacitance is negligible as compared to parasitic capacitances, (7.38) is proportional to voltage gain, as well as to square root of the oxide capacitance and the logic swing.

As an example, using the 0.35-µm technology used and setting A_V=4, V_{SWING}=700 mV, I_{SS,opt_PDP} results in 10.7 µA, with aspect ratio of NMOS and PMOS transistors equal to 4.2/0.3 and 0.6/1.8, respectively. The sum of equal PMOS and NMOS parasitic capacitances $2\sqrt{a^M b^M / V_{SWING}}$ is equal to 9.8 fF. Assuming a load capacitance of 50 fF, delay predicted by (7.30) with data in Table 7.5 is 1.42 ns, that differs by 20% with respect to simulated value of 1.77 ns.

Regarding the high-speed design discussed in Section 7.2.2, the bias current (7.10) that leads to an almost minimum delay with reasonable power consumption can be simplified in two cases which differ for the value of the load capacitance, C_L.

When a high value of C_L loads the gate, such that $4ab/[(c+C_L)^2 V_{SWING}] \ll 1$ (for the CMOS process considered, this occurs when C_L is greater than the minimum gate capacitance by more than one order of magnitude), the optimum bias current results to

$$I_{SS,opt\_delay} \approx \frac{C_L}{4\frac{L_{n,\min}}{\mu_{eff,n}C_{OX}}\left(C_{gd0,n}+K_{j,n}C_{j,n}L_{d,n}+2K_{jsw,n}C_{jsw,n}\right)}\left(\frac{V_{SWING}}{A_V}\right)^2 \quad (7.39)$$

Relationship (7.39) is found by substituting (7.29) into (7.10) and neglecting the second addend under the square root. From inspection of (7.39), bias current for high speed increases proportionally to C_L, V_{SWING}^2 and $1/A_V^2$.

In the other case, when a very low load capacitance is assumed, I_{SS,opt_delay} equates I_{SS,opt_PDP} given by relationship (7.37), as already discussed in Section 7.2.2. From (7.9) and (7.29), the delay achieved for high-speed design in an SCL inverter results to

$$\tau_{PD,SCL} = 2.76\frac{A_V^2\frac{L_{n,\min}}{\mu_{eff,n}C_{OX}}\left(C_{gd0,n}+K_{j,n}C_{j,n}L_{d,n}+2K_{jsw,n}C_{jsw,n}\right)}{V_{SWING}} \quad (7.40)$$

that is proportional to A_V^2, thereby confirming that reduction of the voltage gain is decisive in achieving a high speed performance.

As an example, under the same conditions of the power-efficient design, I_{SS,opt_delay} results in the high value of 110 μA, that leads to PDP=82 fJ (greater than value of 50 fJ given by (7.38) in the power-efficient case). The aspect ratios of NMOS and PMOS transistors are 41/0.3 and 2.4/0.3, respectively, while the predicted and simulated delay is 226 ps and 196.5 ps.

Finally, regarding the low-power design case discussed in Section 7.2.3, the delay of an SCL inverter results to

$$\tau_{PD,SCL} \cong 0.13 \cdot A_{bulk,\max}\mu_{eff,p}C_{OX}^2 W_{p,\min}^2\left(V_{DD}-\left|V_{T,p}\right|\right)\left(\frac{V_{SWING}}{I_{SS}}\right)^2 \quad (7.41)$$

7.4.5 Intuitive understanding of the delay dependence on logic swing and voltage gain in practical design cases

In the previous section, the delay dependence on parameters V_{SWING} and A_V was analytically discussed, and design criteria for such parameters were derived. In this section, an intuitive explanation of such results is provided to gain insight into design aspects of SCL gates.

In low-power design, as observed in Section 7.4.3, the capacitive contribution of PMOS transistors to the output node is much greater than that of NMOS transistors. This is because the PMOS channel length L_p is much greater than its minimum value to ensure the desired logic swing. By reducing the logic swing for an assigned bias current, from (7.15) a smaller equivalent PMOS resistance R_D must be implemented, which means that a lower values of L_p must be used. As a consequence, the logic swing must be kept as low as possible to improve the speed performance, as analytically pointed out in Section 7.4.4. Analogous considerations hold for the power-efficient design.

In a high-speed design, NMOS capacitances are the main contribution to the capacitance at the output node, since from Table 7.1 NMOS channel width W_n is much greater than its minimum value, to guarantee the desired A_V in (2.44). From this relationship, it becomes apparent that the NMOS channel width (and hence NMOS parasitic capacitances) is reduced by increasing logic swing for a given bias current. This delay dependence on logic swing in a high-speed design makes the fundamental difference between CMOS Current-Mode gates and bipolar ones. Indeed, bipolar gates display a better speed performance when logic swing is reduced at a given bias current, since this leads to a reduction of load resistance R_C and thus of time constants associated with capacitances at the output nodes.

In infrequent design cases where bias current is assigned regardless of the criteria discussed so far, an optimum value of logic swing that minimizes delay exists. For the sake of completeness, a brief discussion on this subject will be introduced in Section 7.6.3.

Regarding the delay dependence on the voltage gain, in a power-efficient design the voltage gain should be kept low according to (7.38), since for an assigned value of I_{SS} and V_{SWING} this allows reducing NMOS transistors' aspect ratio and thus their parasitic capacitances.

It is worth noting that the choice of A_V does not heavily affect *PDP*, as can be noticed from the linear dependence in (7.38). Instead, in a high-speed design it is essential to set A_V as low as possible since delay in (7.40) increases proportionally to A_V^2. This is easily explained by observing that in such design cases NMOS capacitances are the dominant contribution, hence from (2.44) a reduction of A_V by a given factor x entails a reduction of

NMOS aspect ratio by x^2. Besides, the decrease in the noise margin (2.47) due to the reduction in A_V is not of concern, since it can be compensated by the increase in V_{SWING} required by the high-speed design criteria previously discussed. In low-power design cases, delay (7.41) is not affected by A_V since the PMOS dominant capacitive contribution only depends on the logic swing, as already discussed in Section 7.3.2.

For the sake of clarity, all guidelines presented until now are summarized in Fig. 7.3 to make design of SCL gates easier.

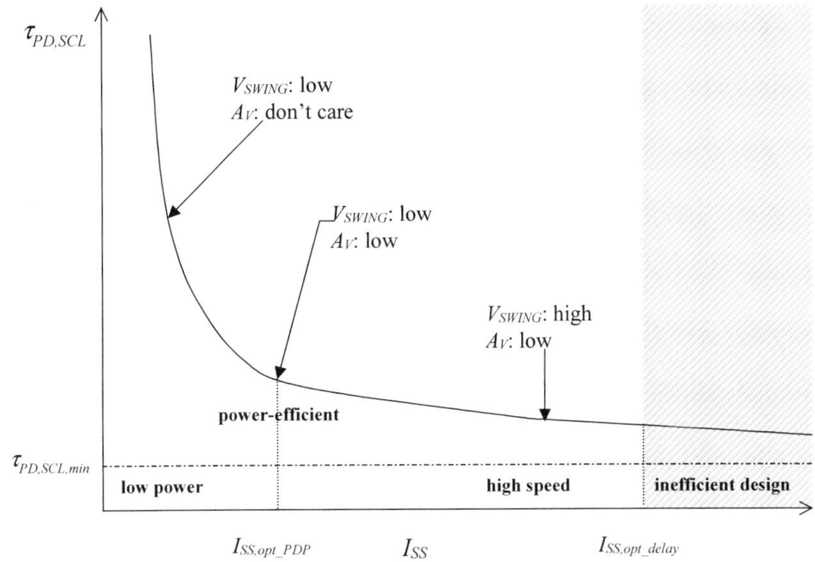

Fig. 7.3. Summary of delay dependence on logic swing and voltage gain in SCL gates.

7.5 OPTIMIZED DESIGN OF THE SOURCE-COUPLED INVERTER WITH OUTPUT BUFFERS

As discussed in Chapter 2, the noise margins of SCL gates with and without output buffer are very close, therefore the expressions of aspect ratios of transistors in the internal SCL inverter introduced in Section 7.3 are still valid. As a consequence, the delay of an SCL inverter is given by relationship (6.7), where (7.2) with C_L=0 and (6.8) must respectively be substituted to the delay of internal SCL and output buffers

$$\tau_{PD,SCLbuf} = 0.35 \cdot V_{SWING}\left(\frac{a^M}{V_{SWING}^2} + b^M \frac{V_{SWING}}{I_{SS}^2} + \frac{c^M}{I_{SS}}\right)$$

$$+0.69\left[R_D\left(C_{gd,buf} + hC_{gs,buf}\right) + \frac{C_L + C_{gs,buf}}{g_{m,buf}}\right]$$

$$= 0.35 \cdot V_{SWING}\left(\frac{a^M}{V_{SWING}^2} + b^M \frac{V_{SWING}}{I_{SS}^2}\right.$$

$$\left. + \frac{c^M + C_{gd,buf} + hC_{gs,buf}}{I_{SS}}\right) + \frac{C_L + C_{gs,buf}}{g_{m,buf}} \tag{7.42}$$

In the following, expressions of buffer transistors' aspect ratio is derived and then applied to develop a design strategy of the SCL inverter with output buffers. In particular, there are two other design parameters compared to the SCL inverter without buffers: the bias current I_{SF} of the buffer and the aspect ratio W_{buf}/L_{buf} of its transistors, whose appropriate sizing depends on whether buffers are added to intentionally introduce a level shift (Section 7.5.1) or to improve the speed performance (Section 7.5.2).

7.5.1 Buffer used as a level shifter

When the buffer is used to implement a level shifter for reasons discussed in Section 2.5, it reduces the common-mode output voltage of an SCL gate by a V_{GS} voltage equal to

$$V_{SHIFT} = V_{T,n} + \sqrt{\frac{2I_{SF}}{\mu_{eff,buf} C_{OX} \frac{W_{buf}}{L_{buf}}}} \tag{7.43}$$

By inverting this expression, we determine the value of the ratio W_{buf}/I_{SF} needed to achieve the assigned shift voltage V_{SHIFT}, resulting in

$$\frac{W_{buf}}{I_{SF}} = \frac{2L_{buf,\min}}{\mu_{eff,buf} C_{OX}\left(V_{SHIFT} - V_{T,n}\right)^2} \tag{7.44}$$

where L_{buf} has been assumed minimum sized, as occurs for practical values of V_{SHIFT}.

Now, as done for the SCL inverter without buffer, let us assume the current per gate I_{gate} to be assigned. This current must be split into I_{SS} and I_{SF} according to a factor $\gamma=I_{SS}/I_{gate}$ that defines the amount of the total bias current used in the internal SCL (accordingly, $I_{SF}=0.5(1-\gamma)I_{gate}$). For a given I_{gate}, factor γ must be optimally chosen to minimize the delay, and extensive numerical analysis shows that in practical cases γ is significantly lower than unity (typically 0.1÷0.3) and hence I_{SS} is so small ($I_{SS} \ll I_{HIGH}$) that the term proportional to $1/I_{SS}^2$ dominates over the others in the internal SCL gate delay model (7.42). Therefore, the delay can be approximated to

$$\tau_{PD,SCLbuf} = 0.35 b^M \frac{V_{SWING}^2}{I_{SS}^2} + 0.69 \frac{C_L}{g_{m,buf}} \tag{7.45}$$

where $C_{gs,buf}$ was neglected with respect to load capacitance C_L. By using (7.44), substituting $I_{SS}=\gamma I_{gate}$ and $I_{SF}=0.5(1-\gamma)I_{gate}$, relationship (7.45) can be minimized for γ by evaluating its derivative $\partial\tau_{PD,SCLbuf}/\partial\gamma$

$$\frac{\partial \tau_{PD,SCLbuf}}{\partial \gamma} \approx -0.35 \frac{b^M V_{SWING}^2}{\gamma^3 I_{gate}^2} + 0.69 \frac{C_L (V_{SHIFT} - V_{T,n})}{I_{gate}} \tag{7.46}$$

where $\gamma \ll 1$ has been assumed, as previously discussed. Equating (7.46) to zero and solving for γ, the value that minimizes $\tau_{PD,SCLbuf}$ is approximately given by

$$\gamma_{opt} \approx \sqrt[3]{\frac{b^M V_{SWING}}{C_L I_{gate} (V_{SHIFT} - V_{T,n})}} \tag{7.47}$$

which shows that increasing the load capacitance or the gate current, the fraction of I_{gate} used for I_{SS} decreases as $(I_{gate})^{-1/3}$ and $(C_L)^{-1/3}$. Extensive verification of (7.47) shows that it agrees well with optimum γ evaluated numerically from exact expression (7.42), and their difference is always lower than 25%, and typically lower than 10%. The effect of this difference on delay is even lower and typically about a few percentage points since the minimum of $\tau_{PD,SCLbuf}$ with respect to γ is rather flat, as it can be observed in Fig. 7.4. This figure shows delay versus γ assuming C_L=1 pF, I_{gate}=30 μA, V_{SHIFT}=1.2 V, under conditions explained in the previous design examples. In this case, (7.47) provides γ=0.16, which differs from the exact minimum

0.148 by 7%, while the optimum delay (9 ns) is overestimated only by 0.15%.

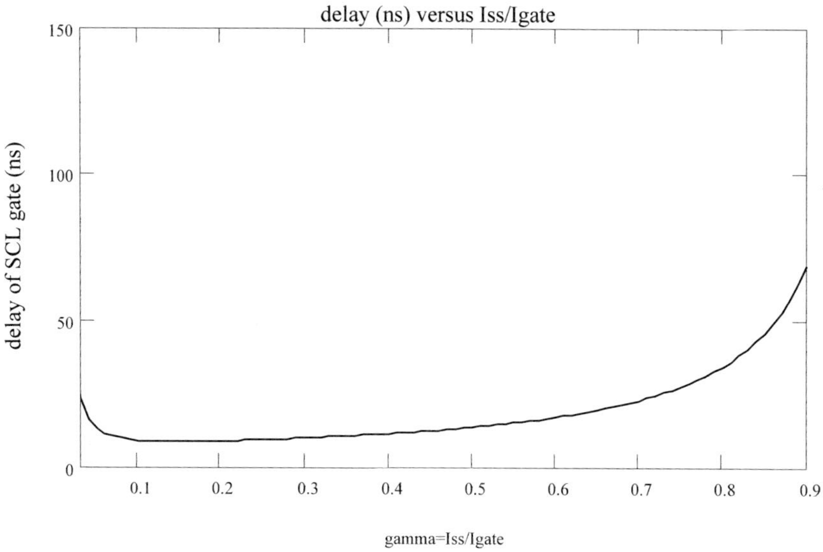

Fig. 7.4. Delay of an SCL inverter with buffer versus γ for C_L=1 pF, I_{gate}=30 μA, V_{SHIFT}=1.2 V.

The resulting minimum delay (7.45), with optimum γ in (7.47) assumed to be much lower than unity, results to

$$\tau_{PD,SCLbuf,op} \approx 0.35\sqrt[3]{b^M C_L^2 \left(V_{SHIFT} - V_{T,n}\right)^2}\left(\frac{V_{SWING}}{I_{gate}}\right)^{\frac{4}{3}} + 0.69\frac{C_L}{I_{gate}}\left(V_{SHIFT} - V_{T,n}\right) \tag{7.48}$$

which expresses the delay of the SCL inverter versus its total bias current I_{gate} after optimally distributing it between the internal SCL gate and output buffers. From (7.48), it is apparent that it is always necessary to keep logic swing as low as possible, while A_V does not significantly affect delay. Moreover, delay for very low values of I_{gate} is proportional to $I_{gate}^{-4/3}$, while for high bias current values it is inversely proportional to I_{gate}. For impractically high values of I_{gate}, the delay tends to a non-zero value that is not accounted for (7.48). It is worth noting that from (7.48) it is possible to derive a power-efficient design criteria by minimizing $PDP=V_{DD}\cdot I_{SS}\cdot \tau_{PD,SCLbuf\_op}$ as already done for the SCL inverter without output buffers.

7.5.2 Buffer used to improve speed

In some cases, for a given total gate current I_{gate}, the choice of an SCL gate with buffers (after properly splitting I_{gate} into I_{SS} and I_{SF}) leads to a better speed performance than an SCL gate without buffer (biased with $I_{SS}=I_{gate}$). This occurs when the available gate current I_{gate} is low and the load capacitance C_L is large. Indeed, for very high values of C_L, delay (7.30) of an SCL gate is proportional to $1/I_{SS}$, while the delay with output buffers (7.42) is proportional to $1/\sqrt{I_{gate}}$ (since $g_{m,buf}$ evaluated in Section 6.3 is proportional to $\sqrt{I_{gate}}$).

When buffers are used to improve the speed performance, the design parameters I_{SS}, I_{SF} and W_{buf} must be evaluated. More specifically, design criteria are needed to split the assigned current per gate I_{gate} into I_{SS} and I_{SF} as well as to size buffer transistor channel width W_{buf} (differently from previous subsection, no constraint between W_{buf} and I_{SF} exists). To this aim, let us rewrite expression of $\tau_{PD,SCLbuf}$ to make its dependence on W_{buf} more explicit

$$\tau_{PD,SCLbuf} \cong 0.35V_{SWING}\left[\frac{a^M}{V_{SWING}^2}+b^M\frac{V_{SWING}}{I_{SS}^2}+\frac{c^M+\left(C_{gd,buf,\min}+hC_{gs,buf,\min}\right)w}{I_{SS}}\right]+0.69\frac{C_L+wC_{gs,buf,\min}}{\sqrt{w}g_{m,buf,\min}} \tag{7.49}$$

that has been obtained by expressing into (7.42) capacitances $C_{gd,buf}$, $C_{gs,buf}$ as the product of their value at minimum W_{buf} (named $C_{gd,buf,min}$, $C_{gs,buf,min}$) and the factor

$$w=\frac{W_{buf}}{W_{buf,\min}} \tag{7.50}$$

that represents the channel width normalized to the minimum allowed. Moreover, from (1.66) and (2.44), $g_{m,buf}$ results in its value $g_{m,buf,min}$ at minimum W_{buf} multiplied by $\sqrt{w}$ (minimum channel length has been assumed, as before). Since (7.50) has to be optimized both for γ and w, for the sake of simplicity first assume w=1 (i.e., minimum W_{buf}) and evaluate the optimum ratio $\gamma=I_{SS}/I_{gate}$ as in the previous subsection, and then optimize w itself.

In the following, let us assume term $b^M V_{SWING}/I_{SS}^2$ in (7.49) being dominant over the others in brackets (as in the previous subsection) and C_L being much greater than $C_{gs,buf}$ (generally satisfied in practical cases since addition of buffers makes sense only for high C_L). Accordingly, the derivative of (7.49) with respect to γ with w=1 results to

$$\frac{\partial \tau_{PD,SCLbuf}}{\partial \gamma} \approx \frac{\partial}{\partial \gamma}\left[0.35V_{SWING}\frac{b^M}{\gamma^2 I_{gate}^2}+\frac{0.69C_L}{\sqrt{\mu_{eff,buf}C_{OX}\frac{W_{buf,\min}}{L_{buf,\min}}(1-\gamma)I_{gate}}}\right]$$

$$\approx -0.35V_{SWING}\frac{2b^M}{\gamma^3 I_{gate}^2}+\frac{0.69C_L}{2\sqrt{\mu_{eff,buf}C_{OX}\frac{W_{buf,\min}}{L_{buf,\min}}I_{gate}}} \tag{7.51}$$

that, equated to zero, gives the following value of γ that minimizes $\tau_{PD,SCLbuf}$

$$\gamma_{opt} \approx \sqrt[3]{\frac{2b^M V_{SWING}}{C_L}\sqrt{\mu_{eff,buf}C_{OX}\frac{W_{buf}}{L_{buf}}}\frac{1}{\sqrt{I_{gate}}}} \tag{7.52}$$

From (7.52), the fraction γ of I_{gate} used in I_{SS} (i.e., in the internal SCL gate) decreases as $(C_L)^{-1/3}$ as increasing load capacitance, while it decreases as $(I_{gate})^{-1/2}$ as increasing the gate current. Verification of (7.52) shows that, in the cases of interest (i.e., when the SCL gate with buffers is faster than that without buffer) it agrees well with optimum γ, being the difference always lower than 30% of the latter (typically lower than 15%), and the difference of associated delay of a few percentage points. As an example, Fig. 7.5 shows delay of SCL gate versus γ assuming C_L=1 pF, I_{gate}=30 µA, under conditions explained in the previous design examples. In this case, relationship (7.52) provides γ=0.18, while the exact values is 0.17, leading to about the same delay of 11.4 ns.

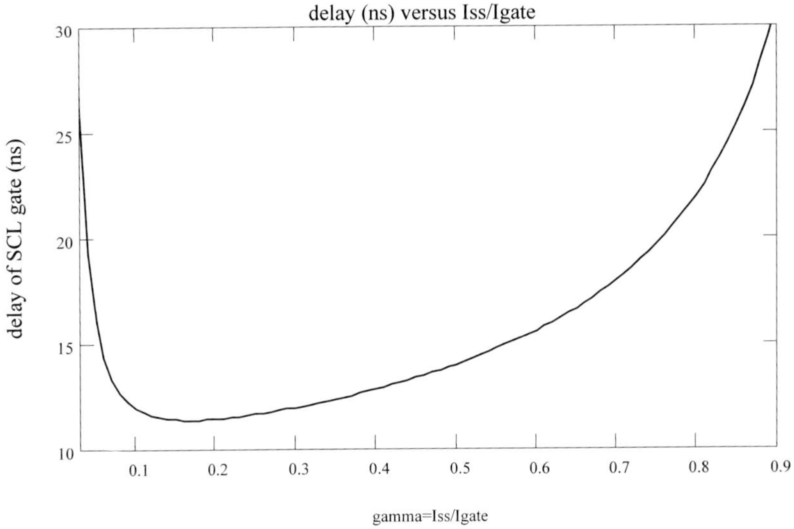

Fig. 7.5. Delay of an SCL gate with buffer versus γ for C_L=1 pF and I_{gate}=30 μA.

Once factor γ is optimized, a further optimization is possible by properly setting normalized transistor channel width, w, after substituting $I_{SS}=\gamma_{opt}I_{gate}$ and $I_{SF}=0.5(1-\gamma_{opt})I_{gate}$. Intuitively, the contribution of load capacitance to delay (7.49) can be reduced by increasing w, that in turn determines an increase of the terms associated with $C_{gd,buf}$ and $C_{gs,buf}$. Hence, an optimum value of w exists, and can be found by differentiating (7.49) for w

$$\frac{\partial}{\partial w}\tau_{PD,SCLbuf} \approx \frac{\partial}{\partial w}\left[0.35\frac{V_{SWING}}{\gamma_{opt}I_{gate}}\left(C_{gd,buf,\min}+hC_{gs,buf,\min}\right)w+\frac{0.69C_L}{\sqrt{wg_{m,buf,\min}}}\right]$$

$$\approx 0.35\frac{V_{SWING}}{\gamma_{opt}I_{gate}}\left(C_{gd,buf,\min}+hC_{gs,buf,\min}\right) \qquad (7.53)$$

$$-\frac{0.69C_L}{2w^{\frac{3}{2}}\sqrt{\mu_{eff,buf}C_{OX}\frac{W_{buf,\min}}{L_{buf,\min}}(1-\gamma_{opt})I_{gate}}}$$

Setting (7.53) to zero and solving for w leads to its optimum value, w_{opt}, that minimizes delay

$$w_{opt} \approx \sqrt[3]{\frac{\gamma_{opt}^2}{1-\gamma_{opt}} \frac{C_L^2 I_{gate}}{\mu_{eff,buf} C_{OX} \frac{W_{buf,\min}}{L_{buf,\min}} V_{SWING}^2 \left(C_{gd,buf,\min} + hC_{gs,buf,\min}\right)^2}} \tag{7.54}$$

In the cases of interest, (7.54) exhibits an error lower than 25% with respect to numerical evaluation, and delay is within 9% of numerically minimized results due to the flat minimum, as can be noticed in Fig. 7.6, that shows delay of SCL gate versus w assuming C_L=1 pF, I_{gate}=30 µA, under conditions explained in the previous design examples. In this case, equation (7.54) gives α=68, while the exact values is 63, and the corresponding delay values are 2.65 ns and 2.74 ns, respectively, differing by 3.3%.

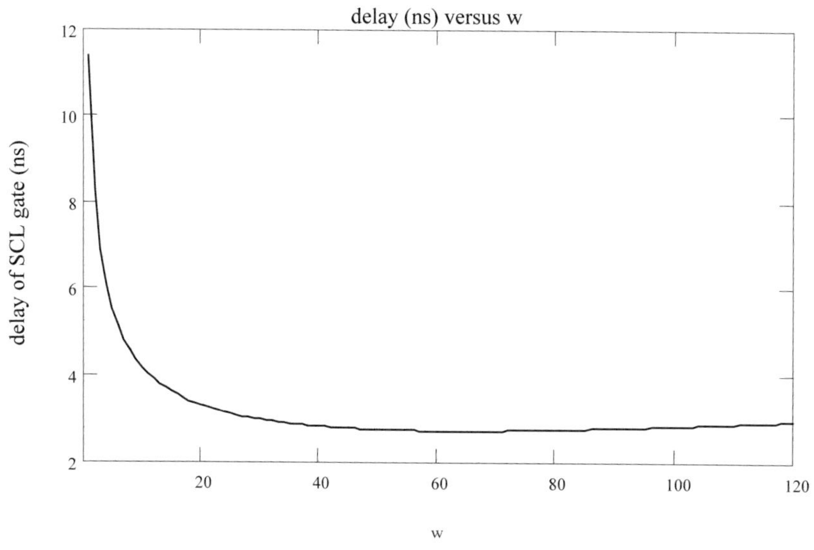

Fig. 7.6. Example of delay of an SCL gate with buffer versus α for C_L=1 pF and I_{gate}=30 µA.

It is worth noting that the optimization of buffer transistors' size has led to a delay reduction by a factor of four, as compared to the case with minimum devices treated before in this subsection.

In general, (7.54) provides values in the order of several tens, which are not always acceptable due to the area increase. However, the minimum of $\tau_{PD,SCLbuf}$ with respect to w is very flat, as shown in Fig. 7.6, therefore w can be reduced without a significant delay increase. Typically, reducing w by a factor of two leads to a delay increase of 10%, while with a factor of four the delay increase is about 30%. To accurately estimate the delay increase for a given reduction of w with respect to (7.54), it is preferable to resort to (7.49).

7.6 OPTIMIZED DESIGN OF THE SOURCE-COUPLED MUX/XOR AND D LATCH

As anticipated in Section 7.2, delay of SCL gates can be expressed in the form reported in relationship (7.2). In particular, in the following this will also be demonstrated for MUX, XOR and D latch gates.

TABLE 7.6

$a^M_{gd,1}$	$4A_V^2 C_{gd0,n} \dfrac{L_{n,\min}}{\mu_{eff,n} C_{OX}}$
$a^M_{db,1}$ †	$4A_V^2 \left(K_{j,n} C_{j,n} L_{d,n} + 2K_{jsw,n} C_{jsw,n}\right) \dfrac{L_{n,\min}}{\mu_{eff,n} C_{OX}}$
$c^M_{db,1}$ †	$2K_{jsw,n} C_{jsw,n} L_{d,n}$
$a^M_{gs,3}$	$\dfrac{8}{3} A_V^2 \dfrac{L^2_{n,\min}}{\mu_{eff,n}}$
$b^M_{gd,p}$	$\dfrac{3}{8} A_{bulk,\max} \mu_{eff,p} C^2_{OX} W^2_{p,\min} \left(V_{DD} - \lvert V_{T,p} \rvert\right)$
$c^M_{gd,p}$	$C_{gd0,p} W_{p,\min} - \dfrac{3}{4} A_{bulk,\max} \mu_{eff,p} C^2_{OX} W_{p,\min} \left(V_{DD} - \lvert V_{T,p} \rvert\right) R_{DSW} 10^{-6}$
$c^M_{db,p}$	$K_{j,p} C_{j,p} L_{d,p} W_{p,\min} + 2K_{jsw,p} C_{jsw,p} \left(L_{d,p} + W_{p,\min}\right)$

(zero-valued coefficients are not reported)

† Expressions of $a^M_{db,3}$, $b^M_{db,3}$ and $c^M_{db,3}$ differ from $a^M_{db,1}$, $b^M_{db,1}$ and $c^M_{db,1}$ only for the value of coefficients K, due to different bias conditions

7.6.1 MUX/XOR delay expression versus bias current and logic swing with the lower transistors switching

As discussed in Chapter 2, static parameters V_{SWING}, A_V and NM of a generic SCL gate are the same as those found in an inverter with equal bias current and transistors' aspect ratios. Thus, criteria to size the latter ones to meet assigned values of such parameters in Section 7.3 are still valid. As a result, the design equations for transistors' aspect ratios reported in Table 7.1 are still valid, according to the three biasing regions L, M and H (obviously, their boundaries (7.16) and (7.21) are equal to those of the inverter). Accordingly, capacitances in the delay model in (6.11) of the MUX/XOR SCL gate can easily be shown to be expressed in the same form as relationship (7.25). To be more specific, non-zero coefficients a_{xy}^M, b_{xy}^M and c_{xy}^M of capacitances in (7.25) in region M of the MUX/XOR gates with input applied to the lower transistors are reported in Table 7.6.

As already observed for the SCL inverter, Table 7.6 shows that coefficients a_{xy}^M, b_{xy}^M and c_{xy}^M of the MUX/XOR gate only depend on A_V, the process used and supply voltage, hence in the design they are constant. Relationship (7.25) with coefficients reported in Table 7.6 can be profitably used to express the sum of capacitances in (6.11), that becomes

$$3C_{gd,1}+2C_{db,3}+C_{gd,p}+C_{db,p}+\frac{2}{A_V}\left(2C_{gd,1}+3C_{db,1}+C_{gs,3}\right)$$
$$=\frac{a^M}{V_{SWING}^2}I_{SS}+b^M\frac{V_{SWING}}{I_{SS}}+c^M \tag{7.55}$$

where coefficients a^M, b^M and c^M are defined as

$$a^M=3a_{gd,1}^M+2a_{db,3}^M+a_{gd,p}^M+a_{db,p}^M+\frac{2}{A_V}\left(2a_{gd,1}^M+3a_{db,1}^M+a_{gs,3}^M\right)$$
$$=3a_{gd,1}^M+2a_{db,3}^M+\frac{2}{A_V}\left(2a_{gd,1}^M+3a_{db,1}^M+a_{gs,3}^M\right) \tag{7.56a}$$

$$b^M = 3b^M_{gd,1} + 2b^M_{db,3} + b^M_{gd,p} + b^M_{db,p} + \frac{2}{A_V}\left(2b^M_{gd,1} + 3b^M_{db,1} + b^M_{gs,3}\right)$$

(7.56b)

$$= b^M_{gd,p}$$

$$c^M = 3c^M_{gd,1} + 2c^M_{db,3} + c^M_{gd,p} + c^M_{db,p} + \frac{2}{A_V}\left(2c^M_{gd,1} + 3c^M_{db,1} + c^M_{gs,3}\right)$$

(7.56c)

$$= 2c^M_{db,3} + c^M_{gd,p} + c^M_{db,p} + \frac{6}{A_V}c^M_{db,1}$$

and where coefficients equal to zero were omitted. By substituting relationships (7.55)-(7.56) and (2.43) into (6.11), the delay of a MUX/XOR biased in region M can be expressed as in relationship (7.30) found for the inverter gate. For the 0.35-μm CMOS process considered and assuming V_{DD}=3.3 V and A_V=4, evaluation of relationships (7.56) leads to a^M=8.96 E-10, b^M=7.47 E-20, c^M=2.93 E-15, as reported in Table 7.9.

Again, when the gate is biased in region L, capacitances can be expressed as in (7.25), thus delay can be written in the same form (7.2), in which coefficients *a*, *b* and *c* are given by relationship (7.57) and data in Table 7.7.

$$a^L = 3a^L_{gd,1} + 2a^L_{db,3} + a^L_{gd,p} + a^L_{db,p} + \frac{2}{A_V}\left(2a^L_{gd,1} + 3a^L_{db,1} + a^L_{gs,3}\right)$$

(7.57a)

$$= 0$$

$$b^L = 3b^L_{gd,1} + 2b^L_{db,3} + b^L_{gd,p} + b^L_{db,p} + \frac{2}{A_V}\left(2b^L_{gd,1} + 3b^L_{db,1} + b^L_{gs,3}\right)$$

(7.57b)

$$= b^L_{gd,p}$$

$$c^L = 3c^L_{gd,1} + 2c^L_{db,3} + c^L_{gd,p} + c^L_{db,p} + \frac{2}{A_V}\left(2c^L_{gd,1} + 3c^L_{db,1} + c^L_{gs,3}\right) \quad (7.57c)$$

TABLE 7.7

$c^{L}_{gd,1}$	$C_{gd0,n}W_{n,\min}$
$c^{L}_{db,1}$ †	$K_{j,n}C_{j,n}L_{d,n}W_{n,\min}+2K_{jsw,n}C_{jsw,n}\left(L_{d,n}+W_{n,\min}\right)$
$c^{L}_{gs,3}$	$\frac{2}{3}W_{n,\min}L_{n,\min}C_{OX}$
$b^{L}_{gd,p}$	$b^{M}_{gd,p}$
$c^{L}_{gd,p}$	$c^{M}_{gd,p}$
$c^{L}_{db,p}$	$c^{M}_{db,p}$

(zero-valued coefficients are not reported)

† Expression of $c^{L}_{db,3}$ differ from $c^{L}_{db,1}$ only for the value of coefficients K, due to different bias conditions.

Analogously, when the gate is biased in region H, delay can be written in the form (7.2), where coefficients a, b and c are given by relationship (7.58) and data in Table 7.8.

$$a^{H}=3a^{H}_{gd,1}+2a^{H}_{db,3}+a^{H}_{gd,p}+a^{H}_{db,p}+\frac{2}{A_V}\left(2a^{H}_{gd,1}+3a^{H}_{db,1}+a^{H}_{gs,3}\right) \quad (7.58a)$$

$$\begin{aligned} b^{H}&=3b^{H}_{gd,1}+2b^{H}_{db,3}+b^{H}_{gd,p}+b^{H}_{db,p}+\frac{2}{A_V}\left(2b^{H}_{gd,1}+3b^{H}_{db,1}+b^{H}_{gs,3}\right) \\ &=0 \end{aligned} \quad (7.58b)$$

$$\begin{aligned} c^{H}&=3c^{H}_{gd,1}+2c^{H}_{db,1}+c^{H}_{gd,p}+c^{H}_{db,p}+\frac{2}{A_V}\left(2c^{H}_{gd,1}+3c^{H}_{db,1}+c^{H}_{gs,3}\right) \\ &=2c^{H}_{db,1}+c^{H}_{db,p}+\frac{6}{A_V}c^{H}_{db,1} \end{aligned} \quad (7.58c)$$

Numerical values of coefficients in the three biasing regions for the 0.35-μm CMOS process considered and assuming V_{DD}=3.3 V and A_V=4 are reported in Table 7.9. By comparing data in Table 7.5 and 7.9, it is apparent that coefficient a^M of the MUX/XOR gate is about four times that of the inverter gate, while b^M is equal.

TABLE 7.8

$a^H_{gd,1}$	$a^M_{gd,1}$
$a^H_{db,1}$	$a^M_{db,1}$
$c^H_{db,1}$	$c^M_{db,1}$
$a^H_{gs,3}$	$a^M_{gs,3}$
$a^H_{gd,p}$	$2\left(C_{gd0,p}+\frac{3}{4}A_{bulk,\max}C_{OX}L_{p,\min}\right) \cdot \frac{L_{p,\min}}{\mu_{eff,p}C_{OX}\left(V_{DD}-\lvert V_{T,p}\rvert\right)\left[1-\frac{\mu_{eff,p}C_{OX}}{L_{p,\min}}R_{DSW}10^{-6}\left(V_{DD}-\lvert V_{T,p}\rvert\right)\right]} \cdot V_{SWING}$
$a^H_{db,p}$	$2\left(K_{j,p}C_{j,p}L_{d,p}+2K_{jsw,p}C_{jsw,p}\right) \cdot \frac{L_{p,\min}}{\mu_{eff,p}C_{OX}\left(V_{DD}-\lvert V_{T,p}\rvert\right)\left[1-\frac{\mu_{eff,p}C_{OX}}{L_{p,\min}}R_{DSW}10^{-6}\left(V_{DD}-\lvert V_{T,p}\rvert\right)\right]} \cdot V_{SWING}$
$c^H_{db,p}$	$2K_{jsw,p}C_{jsw,p}L_{d,p}$

(zero-valued coefficients are not reported)

TABLE 7.9

L	M	H
a^L=0	a^M=8.96 E-10	a^H=9.4 E-10
b^L=7.47 E-20	b^M=7.47 E-20	b^H=0
c^L=5.61 E-15	c^M=2.93 E-15	c^H=2.35 E-15

For the same reasons discussed for the inverter, the delay expression derived in region M can be extended to the other regions. As an example, for the 0.35-μm CMOS process used and for A_V=4, V_{SWING}=700 mV, the error of delay model derived in region M with respect to the expressions rigorously derived in each region is plotted versus I_{SS} (reported in logarithmic scale) in Fig. 7.7 assuming the worst case C_L=0. Even for impractically high values of I_{SS}, the error is always lower than 8% and rapidly decreases to a few percentage points for more realistic load capacitance values.

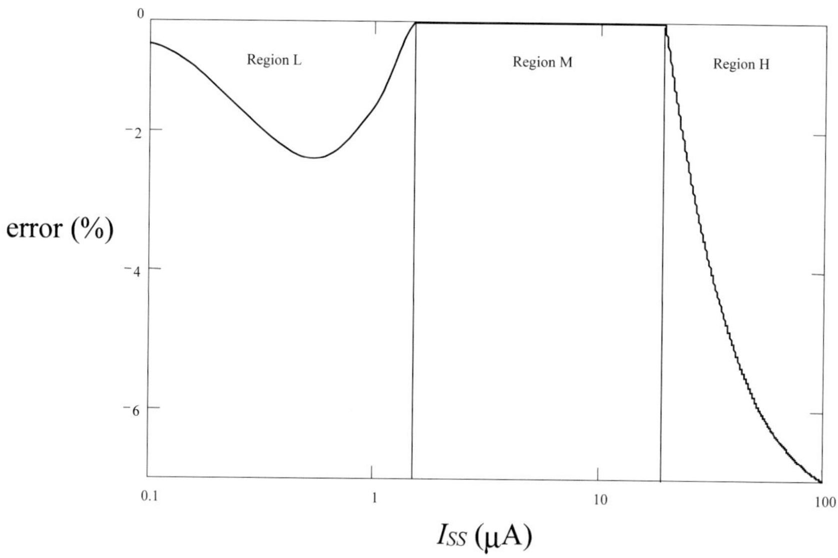

Fig. 7.7. Error between the delay derived in region M with respect to rigorously derived expressions in the region L and H versus I_{SS} for the worst case C_L=0 F.

Now, let us give a circuit interpretation of terms in the delay model of an SCL MUX/XOR gate, as done for the inverter in Section 7.4.3. Inspection of Table 7.6 shows that the term $a=a^M$ in (7.2) is due only to NMOS transistors, $b=b^M$ only to PMOS transistors and $c=c^M$ to both devices. Since the last one can be neglected with respect to the other two addends[4], by using eq. (7.55) the terms $a^M I_{SS}/V_{SWING}^2$ and $b^M V_{SWING}/I_{SS}$ model the NMOS and PMOS

[4] For example, for the MUX/XOR gate in the CMOS process considered and under the conditions discussed above, neglecting c^M leads to an error which is always lower than 10% even for the worst case C_L=0 F. For practical load capacitance values, this error is even lower.

transistors' capacitances, respectively. To be more specific, each of these capacitive terms in (7.2) is multiplied by $0.35V_{SWING}/I_{SS}$=$0.69R_D$, which is the equivalent resistance seen by the output node to ground, thus they make the same contribution to the delay as if all of its correspondent capacitances were lumped at the output node. As a consequence, terms $a^M I_{SS}/V_{SWING}^2$ and $b^M V_{SWING}/I_{SS}$ represent the equivalent capacitance at the output node associated with NMOS and PMOS transistors, respectively. The same interpretation in terms of dominant capacitances in power-efficient, high-speed or low-power design criterion as the inverter in Section 7.4.3 still hold (for example, the power-efficient leads to equal contributions to delay of NMOS and PMOS capacitances).

7.6.2 MUX/XOR delay expression versus bias current and logic swing for input applied to upper transistors

Let us evaluate delay when the input signal is applied to transistors at the upper level (M3-M6 in Figs. 6.11 and 6.12), that is equal to that of an inverter with modified capacitive contributions in (6.12), as observed in Section 6.4.3. Moreover, as discussed in the previous section, delay can be approximated to that in region M regardless of the biasing region. Thus, delay is again written in the form (7.2) with coefficients a, b and c equal to

$$a^M = 2a_{gd,n}^M + 2a_{db,n}^M + a_{gd,p}^M + a_{db,p}^M = 2a_{gd,n}^M + 2a_{db,n}^M \tag{7.59a}$$

$$b^M = 2b_{gd,n}^M + 2b_{db,n}^M + b_{gd,p}^M + b_{db,p}^M = b_{gd,p}^M \tag{7.59b}$$

$$c^M = 2c_{gd,n}^M + 2c_{db,n}^M + c_{gd,p}^M + c_{db,p}^M = 2c_{db,n}^M + c_{gd,p}^M + c_{db,p}^M \tag{7.59c}$$

where (7.25) was used to express capacitances. By comparing (7.59) and coefficients (7.29) of the inverter, it is apparent that coefficient a^M of the MUX/XOR gate is twice as that of the inverter, while b^M has the same value. This means that the optimum bias current for a power-efficient or an high-speed design is lower than the inverter gate. The error of (7.2) using (7.59) compared to simulations was found to be equal to that of the inverter previously discussed.

7.6.3 Delay dependence on logic swing

Until now, the delay dependence on bias current has been evaluated for an assigned value of logic swing. For the sake of completeness, let us consider the delay dependence on the logic swing for a given bias current I_{SS}. From general SCL delay expression (7.2), it is apparent that an optimum value of logic swing can be found by differentiating for V_{SWING} and setting the result to zero. The resulting optimum value of V_{SWING} that minimizes delay for a bias current I_{SS} is the solution of the following equation

$$\frac{2b^M}{I_{SS}^2}V_{SWING}^3 + \frac{c^M + C_L}{I_{SS}}V_{SWING}^2 - a^M = 0 \qquad (7.60)$$

which in general cannot be solved in a closed-form, and thus requires numerical analysis. For example, for I_{SS}=50 μA and C_L=200 fF, solving (7.60) for the MUX/XOR (with coefficients in Table 7.9) assuming an input applied to lower transistors results in 221 mV. Simulations confirm this result, as can be deduced from Fig. 7.8 that reports the plot of the delay versus the logic swing, where the minimum delay is achieved for V_{SWING} equal to about 250 mV.

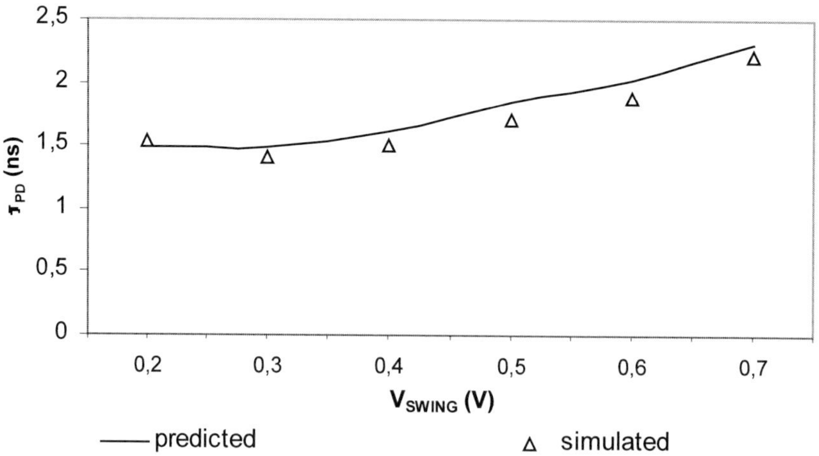

Fig. 7.8. MUX/XOR delay versus logic swing for I_{SS}=50 μA and C_L=200 fF.

To validate the dependence of the delay in (7.2) on the logic swing, simulations were also performed for C_L equal to 0 F, 50 fF, 200 fF and 1 pF, with bias current equal to 5 μA, 20 μA, 50 μA and 100 μA, and the resulting curves are not shown for the sake of brevity. Maximum error of (7.2) using coefficients in (7.56) is 17%, and typically much lower, thus confirming the validity of the model discussed.

Even though in general relationship (7.60) cannot be solved in a closed form, in practical design cases optimum value of V_{SWING} can be easily found. Indeed, it can be seen that optimum V_{SWING} tends to zero for $I_{SS}\rightarrow 0$, thus according to low-power design considerations in Section 7.2.3, the logic swing is given by relationship (7.13). Moreover, optimum V_{SWING} tends to infinity for $I_{SS}\rightarrow\infty$, thus according to high-speed design considerations in Section 7.2.2, the logic swing is given by relationship (7.14). For intermediate bias current values, the optimum value of V_{SWING} belongs to the acceptable range $[V_{SWING,min}, V_{SWING,max}]$.

7.6.4 Extension to D latch

The D latch delay can be found as for the MUX/XOR gate firstly by assuming the gate to be biased in the region M, and then extending it to the other intervals in a similar manner. As discussed in Section 6.5, the delay of a D latch gate is equal to that of MUX/XOR with an equivalent load capacitance (6.14) that accounts for the input capacitance of a source-coupled NMOS pair (6.13). By substituting relationship (7.25) and data in Table (7.6), input capacitance (6.13) can be expressed[5] as the gate-source capacitance in an NMOS transistor working in saturation region, e.g. $C_{gs,3}$

$$C_{input} = C_{gs,3} = \frac{a_{gs,3}^{M}}{V_{SWING}^{2}} I_{SS} \tag{7.61}$$

where from Table 7.6 it was observed that $b_{gs,3}^{M} = c_{gs,3}^{M} = 0$. By substituting (7.61) into (6.14), and both into the delay expression (7.2), it is apparent that the D latch delay has the same expression as the MUX/XOR with coefficient a^{M} being replaced by $a^{M'}$

$$a^{M'} = a^{M} + a_{gs,3}^{M} \tag{7.62}$$

[5] Actually, this expression holds in regions M and H which gates are usually biased in.

that results to 1.06 E-9 under the conditions discussed above, in the case with switching input applied to lower transistors. Analogously, coefficient (7.62) results to 4.52 E-10 when input is applied to upper transistors.

Simulation results validate the extension. Indeed, the error was found to be in the same order of magnitude of the inverter and MUX/XOR. In conclusion, unless for the small change discussed, all the design strategies discussed for the MUX/XOR gates still hold.

7.7 OPTIMIZED DESIGN OF THE SOURCE-COUPLED MUX/XOR AND D LATCH WITH OUTPUT BUFFERS

Regarding the case with output buffers, design techniques introduced in Section 7.5 for the inverter can be extended to other SCL gates in a straightforward way. Indeed, delay of SCL gates without output buffers is always expressed by (7.42) regardless of the specific gate considered, being the sum of the internal SCL circuit and buffer delay contributions. The only difference between such gates is in the expression of coefficients a^M, b^M and c^M in the internal SCL delay. As a consequence, the same design criteria for the two practical design cases dealt with in Section 7.5 apply. Obviously, different gates have in general a different value of optimum bias current distribution γ, as well as a different optimum buffer transistors' aspect ratio factor w.

When buffers are used to implement a level shifter in a MUX/XOR or D latch gate, the optimum value of γ in relationship (7.47) is the same as the inverter, since coefficient b^M in (7.47) results in the same value for the three gates. In other words, the amount of the bias current used in the internal SCL gate (and buffers) for an assigned gate current in a MUX/XOR or D latch is the same as the inverter. Moreover, the resulting minimum delay in (7.48) for an assigned I_{gate} is equal to that of the MUX/XOR, D latch and inverter gates. This is because the internal SCL gate has the same bias current (7.47) and is usually biased in its low-power region (see Fig. 7.1), thus its delay is mainly due to PMOS capacitances that are independent of the gate for assigned values of A_V, V_{SWING} and I_{SS}. From (7.47) it is apparent that the buffer delay does not depend on the gate considered, since its bias current (equal to γI_{gate}) is the same for all gates for a given I_{gate}.

The same observations hold when buffers are used to improve speed performance. Indeed, optimum current distribution (7.52) depends only on coefficient b^M, thus γ does not depend on the gate considered. In addition, optimum transistor aspect ratio (7.54) of the buffer and resulting minimum delay is the same, regardless of the specific gate analyzed.

7.8 COMPARISON OF GATES ANALYZED AND EXTENSION TO ARBITRARY SCL LOGIC GATES

In the previous section, it has been shown that the delay of various SCL gates can be written in the form (7.2), which simply shows the design tradeoffs among speed, power consumption and noise immunity. In - particular, this has been demonstrated for the inverter, MUX, XOR and D latch by observing that parasitic capacitances associated with transistors can always be expressed as in relationship (7.25), as well as evaluating the equivalent resistance seen by each capacitance. By reiterating the considerations reported in Section 6.4.1, in general equivalent resistances are equal to R_D (i.e. when a capacitance is connected to the output nodes) or $1/G_M$ (when a capacitance is not connected to the output nodes). As a consequence, for an arbitrary SCL gate, time constants have the same dependence on logic swing and bias current as the gates considered. Since the delay is proportional to the sum of time constants, it follows that the delay dependence in an SCL gate on V_{SWING} and I_{SS} is independent of the specific SCL gate considered, and therefore (7.2) is always valid. The same observations hold for gates with output buffers, since their delay simply adds to that of the internal SCL gate, as shown in relationship (7.42).

Now, it is useful to compare performance achievable by the gates considered, in order to understand how performance depends on their complexity. In particular, assuming inputs to be applied at lower transistors (i.e., we are considering the worst-case delay), coefficients of gates considered in relationship (7.2) are summarized in Table 7.10. Since coefficient a^M depends on NMOS transistors, it obviously increases when a more complex logic gate (i.e., including a greater number of NMOS transistors) is considered, as confirmed by Table 7.10. Therefore, the minimum delay achievable for a high-speed design, modeled with relationship (7.9), increases for more complex gates, while the bias current needed decreases, according to (7.10). Regarding coefficient b^M, it only depends on PMOS transistors, therefore it is independent of the specific gate considered for assigned values of V_{SWING} and I_{SS}. As a result, the speed performance (7.11) in a low-power design is equal for all SCL gates, for given values of V_{SWING} and I_{SS}. Finally, in a power-efficient design the required bias current I_{SS,opt_PDP} in (7.5) decreases when a more complex gate is considered, while power-delay product gets worse due to the increase in coefficient a^M (this means that the delay increase is greater than the power saving).

TABLE 7.10

Inverter	MUX/XOR (upper transistors)	D latch
a^M=2.26 E-10	a^M=8.96 E-10	a^M=1.06 E-9
b^M=7.47 E-20	b^M=7.47 E-20	b^M=7.47 E-20
c^M=1.82 E-15	c^M=2.93 E-15	c^M=2.93 E-15

Finally, it is useful to note that delay model (7.2) suggests an alternative approach to characterize SCL logic cells in terms of design trade-offs without analytically evaluate coefficients a, b and c. In other words, the approach introduced for the accurate model of bipolar gates can also be applied to SCL gates. Indeed, coefficients a, b and c only depend on process parameters, supply voltage and voltage gain A_V. Therefore, if such parameters are preliminarily set before designing logic gates, it is possible to evaluate coefficients by performing only three simulation runs for different bias current values widely distributed to cover the low-power, power-efficient and high-speed regions in Fig. 7.1 for a given logic swing. To be more specific, the procedure based on the minimization of functional S in Section 4.4 can be used. Once coefficients a, b and c are found, delay dependence on bias current and logic swing is expressed by (7.2), and thus all design considerations discussed so far can be applied with no modification. This approach can be very useful when a library of standard cells implemented with a new process must be characterized in terms of the delay dependence on bias current and logic swing.

Chapter 8

APPLICATIONS AND REMARKS ON CURRENT-MODE DIGITAL CIRCUITS

In the previous chapters, design and modeling strategies of Current-Mode gates have been developed for single logic gates. Actual circuits are made up of cascaded logic gates, therefore their overall delay is equal to the sum of the delay contributions of gates belonging to the path considered, each of which is evaluated by representing the following gate as an equivalent linear load capacitor. In this chapter some circuits currently used in typical applications, such as ring oscillators and frequency dividers, are explicitly analyzed in Sections 8.1 and 8.2, respectively, by applying results presented in previous chapters. To be more specific, their speed performance is evaluated as a function of process and design parameters. Besides, an alternative approach to implement CML gates with a low supply voltage is dealt with in Section 8.3.

Regarding the design of general logic circuits consisting of cascaded gates, criteria developed for single logic gates are generalized in Section 8.4 to relate design variables to the overall performance.

8.1 RING OSCILLATORS

Ring oscillators have recently attracted the interest of circuit designers [WKG94], [M971], [SL98], [HR99], [HLL99]. Indeed, they are widely used in PLLs that are key elements of RF circuits and microprocessors to provide synchronization, data recovery and perform frequency synthesis [R96]-[R961].

A ring oscillator is a closed-loop chain of equal inverter gates (or inverting amplifiers), as shown in Fig. 8.1, with a negative feedback to provide oscillation. To guarantee oscillation, the negative feedback is necessary but not sufficient. To allow the oscillation to start, the unique stationary state (in which all gates are biased around their logic threshold) has to be unstable, and this happens when the circuit linearized around the bias point exhibits a negative phase margin, m_ϕ.

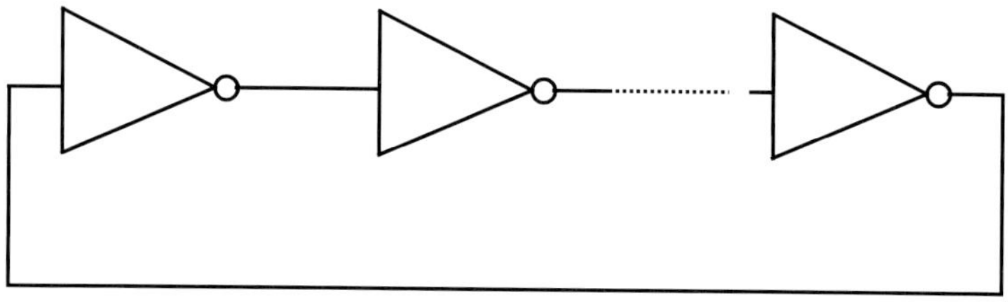

Fig. 1. Ring oscillator.

As done in the previous chapter, by assuming each (inverting) stage to be modeled by a single-pole transfer function with time constant τ, (i.e., a transfer function $H(s)$=-$A_V/(1+s\tau)$) the loop gain $L(s)$ (evaluated as the opposite of the transfer function obtained by breaking the loop at the input of any stage) results to

$$L(s) = (-1)^{n+1}\left[\frac{A_V}{1+s\tau}\right]^n = \left[\frac{A_V}{1+s\tau}\right]^n \qquad (8.1)$$

where term $(-1)^{n+1}$ must necessarily be positive to ensure a negative feedback, hence an odd number of inverting stages must be used. However, since Current-Mode gates have a differential output, an even number of stages can also be used, provided that the output nodes of one stage are inverted. Utilizing an even number of stages is particularly useful when both a reference periodic signal and its quadrature component must be generated.

By definition, the phase margin expressed in degrees is 180 plus the phase of the loop gain at the transition frequency (i.e., the frequency where the magnitude of $L(j\omega)$ is equal to unity). Hence, from relationship (8.1) we get

$$m_\phi = 180 + \angle L(j\omega_t) = 180 - n \cdot \angle L(1 + j\omega_t\tau) \qquad (8.2)$$

where the phase is denoted by symbol $\angle$, and the transition frequency ω_t is equal to

$$\omega_t = \frac{\sqrt{A_V^2 - 1}}{\tau} \approx \frac{A_V}{\tau} \tag{8.3}$$

The simplification holds for A_V greater than unity, such that A_V^2>>1. Therefore, by substituting relationship (8.1) and (8.3) into (8.2), the phase margin becomes

$$m_\phi \cong\approx 180 - n \cdot \angle \arctan(A_V) \tag{8.4}$$

From relationship (8.4), to guarantee a negative phase margin, which causes oscillation, the number of stages n must be sufficiently high, according to the following condition

$$n > \frac{180}{\arctan(A_V)} \tag{8.5}$$

that only depends on the voltage gain A_V of the inverter gate, and is always greater than 2 (since arctan(A_V)<90°), thus in practical cases it is necessary to use at least three stages. It is worth noting that relationship (8.5) slightly overestimates the minimum number of stages because higher poles and positive zeros tend to reduce the phase margin. The right-hand side of relationship (8.5) is plotted versus A_V in Fig. 8.2, whose inspection shows that three stages ensure oscillation to occur for a voltage gain greater than 1.7. Extensive simulations show that oscillation does not generally take place for n=2, except for a few bias current value for which the oscillation amplitude is very low (in the order of 10 mV), due to the contribution of higher-order poles and zeroes.

The frequency oscillation of the ring oscillator is given by [R96]

$$f = \frac{1}{2n\tau_{PD}} \tag{8.6}$$

where τ_{PD} is the propagation delay of the fundamental gate. Therefore, the frequency estimation is reduced to the evaluation of the gate delay, that has to be carried out by remembering that each gate has an input rise time equal to that at the output and is loaded by an equal stage. In the following, the delay model developed in the previous chapters is used to evaluate the oscillations frequency model in ring oscillators based on bipolar (Section

8.1.1) and CMOS (Section 8.1.4) Current-Mode gates. From a design point of view, this model allows to predict performance achievable before the design, to correctly design the ring oscillator and to predict the oscillation frequency inaccuracy due to tolerances in model parameters.

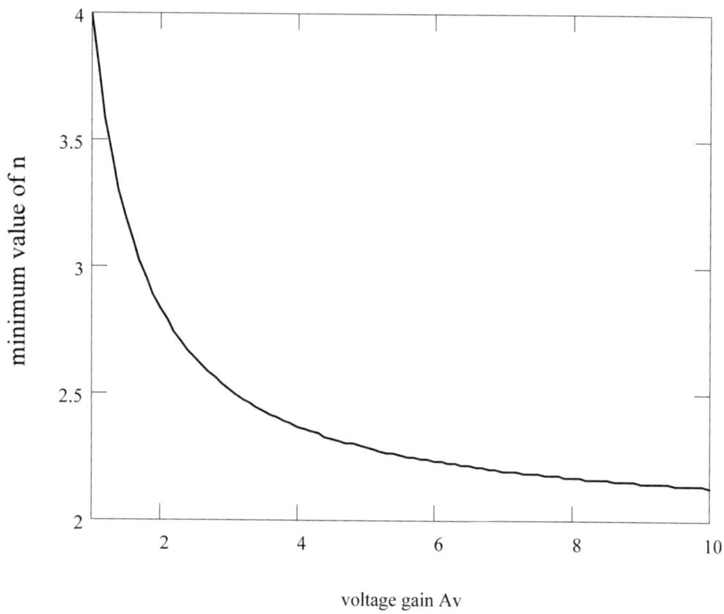

Fig. 8.2. Minimum value of n given by relationship (8.5) versus A_V.

8.1.1 Bipolar CML ring oscillators

Let us consider a ring oscillator based on the CML inverter in Fig. 4.2, whose small-signal gain A_V is given by relationship (2.10), and typically ranges from 4 to 10. Choosing the typical value V_{SWING}=500 mV for high-speed applications, the value of A_V results to be 5 from (2.10), and thus, from Fig. 8.2, three stages are sufficient to ensure oscillation (relationship (8.5) gives n>2.3).

Now, let us evaluate the oscillation frequency (8.6), or equivalently the gate delay under the constraint that the input rise time is equal to that of the output waveform. Since the CML inverter gate has a first-order behavior with a time constant given by relationship (4.7) (actually, from (4.4) it must be divided by 0.69), the delay dependence on the input rise time of a one-pole system is analyzed in the following.

Consider a one-pole system with a linear ramp input $v_{in}(t)$ having a rise time T and a maximum value equal to unity, as represented by the following relationship

$$v_{in}(t)=\begin{cases}\dfrac{t}{T} & t\le T\\[2ex] 1 & t>T\end{cases} \tag{8.7}$$

The output response, $y(t)$, of the system is given by [AP99]

$$y(t_n)=\begin{cases}t_n-\tau_n\left(1-e^{-\frac{t_n}{\tau_n}}\right) & t_n\le 1\\[2ex] 1-\tau_n\left(e^{\frac{1}{\tau_n}}-1\right)e^{-\frac{t_n}{\tau_n}} & t_n>1\end{cases} \tag{8.8}$$

where t_n and τ_n are the time and the time constant normalized to T, respectively (i.e., $t_n=t/T$ and $\tau_n=\tau/T$).

The normalized propagation delay $\tau_{PDn}=\tau_{PD}/T$ is the difference between $\tau_{PDn,out}$ and $\tau_{PDn,in}$, which are the time normalized to T when the output and the input reach half of their final value, respectively. The output propagation delay $\tau_{PDn,out}$ can be evaluated by setting relationship (8.8) equal to 0.5. The input propagation delay $\tau_{PDn,in}$ is equal to 0.5. It is worth noting that, although $\tau_{PDn,out}$ depends only on τ_n, it is difficult to solve $y(\tau_{PDn,out})=0.5$ for $\tau_{PDn,out}$ because (8.8) is a nonlinear function. Moreover, since we do not know whether $\tau_{PDn,out}$ is greater or lower than unity, we have to solve both the two parts of relationship (8.8), and then discard the one without practically meaningful solution.

In order to evaluate the relation between the delay and the time constant, the ratio τ_{PDn}/τ_n versus τ_n is evaluated numerically and plotted in Fig. 8.3 (with dashed line). As expected, if $\tau_n\to\infty$ (i.e., we have an ideal step input) the ratio τ_{PDn}/τ_n tends to the well-know value 0.69. In the opposite case, when $\tau_n\to 0$, which is the case of a very slow input ramp, the ratio τ_{PDn}/τ_n tends to 1. The latter result is reached by solving the equation in the case $t_n\le 1$, and approximating $y_n(t_n)\cong t_n-\tau_n$ in (8.8), since the exponential term is very low. From inspection of Fig. 8.3, if $T>\tau$, approximating the propagation delay of a ramp input with the value obtained assuming a step input can lead to an error greater than 30% (the error is lower than 5% for $T<\tau$).

To achieve a simple and useful closed-form expression of the ratio τ_{PDn}/τ_n, one must rewrite this ratio as $0.69[1+g(\tau_n)]$. Hence, we have to approximate $g(\tau_n)$ with an equivalent function which tends to zero and to 0.45 for $\tau_n\rightarrow+\infty$ and $\tau_n\rightarrow 0$, respectively. A suitable function $g(\tau_n)$ which satisfies the asymptotic behavior is

$$g(\tau_n) = 0.45\frac{1+A\tau_n}{1+B\tau_n^2} \tag{8.9}$$

where parameters A and B, found by a numerical fitting of the curve, are set equal to 0.66 and 7.8, respectively. Thus, function $f(\tau_n)$ which approximates the ratio τ_{PDn}/τ_n is given by

$$f(\tau_n) = 0.69\left[1+0.45\frac{1+0.66\tau_n}{1+7.8\tau_n^2}\right] \tag{8.10}$$

Relationship (8.10) is plotted with a solid line in Fig. 8.3. The proposed approximation fits ratio τ_{PDn}/τ_n well. Indeed, the error found, plotted in Fig. 8.4, is always lower than 2%.

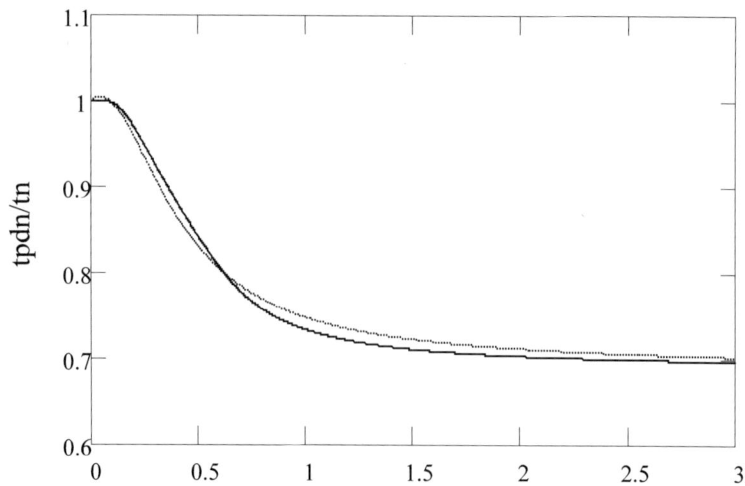

Fig. 8.3. Ratio of τ_{PDn}/τ_n versus τ_n: exact (dashed line); expression (8.10) (solid line).

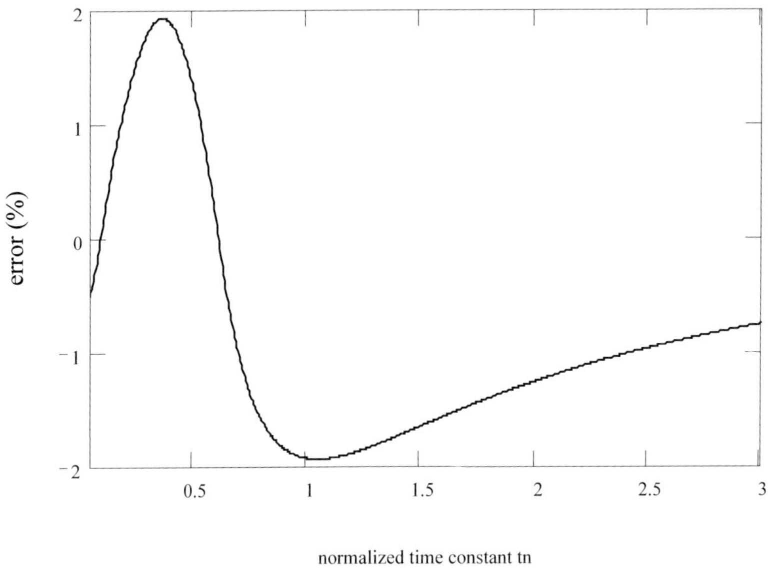

Fig. 8.4. Percentage error of expression (8.10) versus τ_n.

Denormalizing (8.10), the general expression for the propagation delay of a first-order system versus the input rise time T and the time constant τ is equal to

$$\tau_{PD} = \tau f\left(\frac{\tau}{T}\right) \cong 0.69\tau\left[1+0.45\frac{1+0.66\dfrac{\tau}{T}}{1+7.8\left(\dfrac{\tau}{T}\right)^2}\right] \tag{8.11}$$

To estimate the oscillation frequency of a ring oscillator, we can further simplify relationship (8.11) by introducing the condition $T=2\tau_{PD}$ that ensures equal output and input rise time values. Thus solving (8.11) for τ/τ_{PD} we get

$$\tau_{PD} = 0.8\tau \tag{8.12}$$

where time constant is given by relationship (4.7) (more precisely, it is equal to relationship (4.7) divided by 0.69). Of course, relationship (8.12) must be evaluated by properly modeling the loading effect of the subsequent gate with its input capacitance (4.23), i.e. by setting C_L equal to it in (4.7). The resulting expression of relationship (8.12) is

$$\tau_{PD} = 0.8\left[\frac{r_e + r_b}{1+g_m r_e} C_{be} + r_b C_{bci}\left(1+\frac{g_m (r_c + R_C)}{1+g_m r_e}\right) + \right. \tag{8.13}$$

$$\left. + (r_c + R_C)(C_{bci} + C_{bcx} + C_{cs}) + R_C C_{bcx}\left(1+\frac{V_{SWING}}{4V_T}\right)\right]$$

that, being a CML gate, can be also written in the form as in (5.1) for design purposes.

To further increase the estimation accuracy of the oscillation frequency, we have to evaluate the loading effect in each stage due to the subsequent gate more precisely. Indeed, the assumption that one has only the load capacitance given by (4.23) overestimates the loading effect of the subsequent CML input impedance, as already observed in Section 4.8. To this aim, let us consider the equivalent half circuit of a CML gate in Fig. 8.5 and, for simplicity, neglect capacitances C_{cs} and C_L.

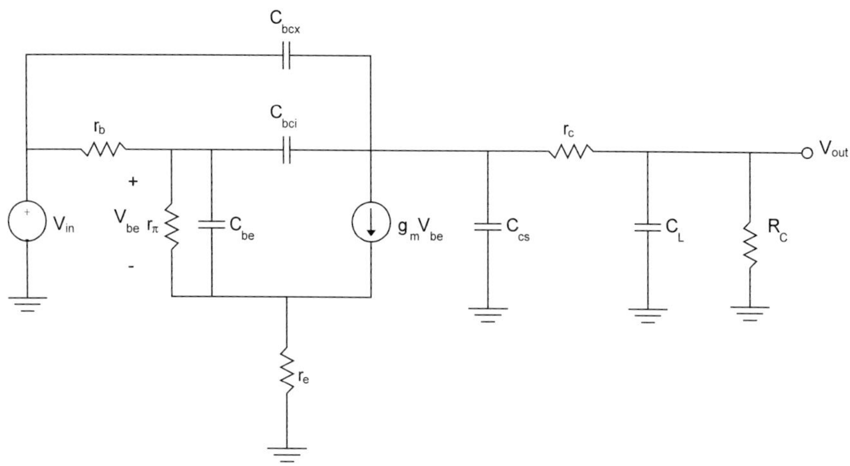

Fig. 8.5. Equivalent linear circuit of the CML inverter.

From Fig. 8.5, the impedance seen from the input of the gate (i.e., the base of the transistor) has a third-order transfer function that, after neglecting higher order terms, can be approximated to the first order impedance

$$Z_{in} \cong \frac{R_{in}}{1+sR_{in}C_{in}} \tag{8.14}$$

where

$$R_{in} \cong r_\pi (1+g_m r_e) \tag{8.15}$$

$$C_{in} \cong \frac{C_{be} + C_{bci} + C_{bcx}\left(1+\frac{V_{SWING}}{4V_T}\right)}{1+g_m r_e} \tag{8.16}$$

in which both r_e and r_b are assumed to be much lower than r_π, as usually occurs. It is apparent that (8.14) represents the impedance of resistance R_{in} in parallel with capacitance C_{in}. Therefore, we can include it in the time constant, and combining (8.13) and (8.14) we get

$$\tau_{PD} = 0.8\left[\frac{r_e + r_b}{1+g_m r_e} C_{be} + r_b C_{bci}\left(1+\frac{g_m\left(r_c + R_C \| R_{in}\right)}{1+g_m r_e}\right) + \right.$$
$$\left. + \left(r_c + R_C \| R_{in}\right)\left(C_{bci} + C_{bcx} + C_{cs}\right) + R_C \| R_{in} C_{in}\right] \tag{8.17}$$

that, along with relationship (8.6), models the oscillation frequency of a ring oscillator based on bipolar Current-Mode gates.

8.1.2 Validation of the oscillation frequency in a CML ring oscillator

The model of the oscillation frequency (8.6) with (8.17) was tested by means of SPICE simulations using the two bipolar technologies introduced in Chapter 4. As done before, the circuits were powered with 5 V and a 500-mV logic swing was set. Moreover, the bias current was varied from 100 μA to 1.4 mA and from 100 μA to 2.4 mA for the BiCMOS and HSB2 process, respectively, to avoid a degradation of transistor transition frequency.

Figures 8.6 and 8.7 show the simulated oscillation frequency of a CML ring oscillator with different number of stages and the theoretical one (given by (8.6) and (8.17)) versus the bias current I_{SS} for the BiCMOS and HSB2 technology, respectively. The oscillation frequency error for the BiCMOS and HSB2 process is summarized in Table 8.1 and 8.2, where the case with a simplified input impedance (4.23) and an accurate one (8.14) are considered. To evaluate the superiority of the model (8.14) compared to the simplified one (4.23), Table 8.3 summarizes the average errors for the analyzed cases.

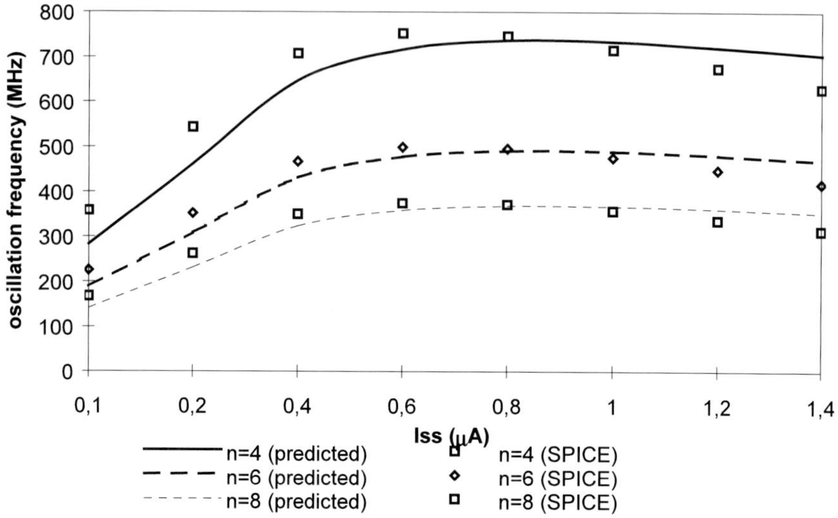

Fig. 8.6. Oscillation frequency with 4, 6 and 8 stages (BiCMOS process).

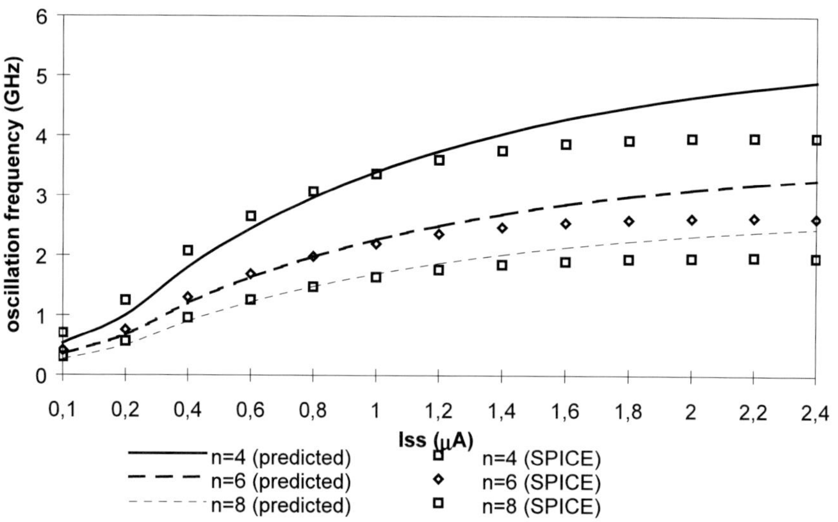

Fig. 8.7. Oscillation frequency with 4, 6 and 8 stages (HSB2 process).

TABLE 8.1

	BiCMOS technology			HSB2 technology		
I_{SS} (mA)	n=4	n=6	n=8	N=4	n=6	n=8
0.1	29 %	22 %	20 %	63 %	46 %	42 %
0.2	12 %	9.3 %	8.7 %	50 %	54 %	55 %
0.4	-2.7 %	-3.6 %	-3.8 %	29 %	22.2 %	20 %
0.6	-9.4 %	-9.9 %	-9.9 %	16 %	11.1 %	10.5 %
0.8	-13.5 %	-13.8 %	-13.8 %	5.9 %	2.6 %	2.1 %
1	-16.8 %	-17 %	-17 %	-1.9 %	-3.9 %	-4.2 %
1.2	-19.9 %	-20 %	-20 %	-7.8 %	-9.3 %	-9.6 %
1.4	-23.1 %	-23.3 %	-23.3 %	-13.2 %	-14.2 %	-14.5 %
1.6	/	/	/	-17.3 %	-18.3 %	-18.6 %
1.8	/	/	/	-21 %	-21.8 %	-21.8 %
2	/	/	/	-24.4 %	-25.1 %	-25.1 %
2.2	/	/	/	-27.2 %	-27.6 %	-27.6 %
2.4	/	/	/	-29.6 %	-30.3 %	-30.3 %

TABLE 8.2

	BiCMOS technology			HSB2 technology		
I_{SS} (mA)	n=4	n=6	n=8	n=4	n=6	n=8
0.1	26 %	19 %	18 %	32 %	18.1 %	15.1 %
0.2	17 %	14 %	13 %	24 %	14.3 %	12.1 %

0.4	9 %	8 %	8 %	15 %	8.6 %	7 %
0.6	5 %	4.4 %	4.4 %	8.5 %	3.9 %	3.2 %
0.8	1.2 %	0.9 %	0.8 %	3.4 %	0.2 %	0.2 %
1	-2.5 %	-2.8 %	-2.8 %	-0.8 %	-2.9 %	-3.1 %
1.2	-6.5 %	-6.7 %	-6.7 %	-3.7 %	-5.4 %	-5.6 %
1.4	-10.7 %	-10.9 %	-10.9 %	-6.9 %	-8 %	-8.3 %
1.6	/	/	/	-9.6 %	-10.7 %	-10.9 %
1.8	/	/	/	-12.3 %	-13.1 %	-13 %
2	/	/	/	-14.6 %	-15.4 %	-15.4 %
2.2	/	/	/	-16.9 %	-17.4 %	-17.4 %
2.4	/	/	/	-18.8 %	-19.5 %	-19.5 %

TABLE 8.3

	BiCMOS technology			HSB2 technology		
Load model	n=4	n=6	n=8	N=4	n=6	n=8
Simplified (4.23)	15.8 %	14.8 %	14.5 %	23.5 %	22.0 %	21.6 %
Accurate (8.14)	9.7 %	8.3 %	8.1 %	12.8 %	10.6 %	10.1 %

Inspection of Tables 8.1-8-3 shows that, considering the simplified load (4.23), the error is always lower than 30% for the BiCMOS technology but greatly increases in some cases to values higher than 50% for the HSB2 technology. This high error is heavily reduced by resorting to the accurate model in (8.14). Indeed, even for the HSB2 technology the accurate model has an error slightly higher than 30% in only one case, and is much lower for values of bias current that lead to an efficient power-delay trade-off. The improved accuracy achieved by the accurate model is also well confirmed in Table 8.3, where the accurate model's average error is shown to be reduced by around 6% for the BiCMOS technology with the average reduction being greater than 10% for the higher-speed technology. For both cases, the average error was reduced by more than 40%.

8.1.3 Remarks on the oscillation amplitude in a CML ring oscillator

In general, the evaluation of the oscillation amplitude in a ring oscillator is not an easy task, due to the non-linear behavior of its stages. Even for CML ring oscillators, no closed-form expression can be given to estimate the amplitude of the output voltage, that is equal for all stages by symmetry.

In practical cases, it is often of interest to maximize amplitude to improve the signal-to-noise ratio of the oscillator. Therefore, it is sufficient to understand when amplitude is close to its maximum achievable value, that is equal to half the logic swing of all stages, i.e. $R_C I_{SS}$. Intuitively, this is achieved when a high number of stages n is used, since in this case each gate has enough time to settle to the nominal high or low output voltages. This is because in a ring oscillator the input of each stage is constant for half a period $T/2= n\tau_{PD}$. When a smaller value of n is considered, the input of each gate is forced to change before the gate settles to the nominal output voltage, thus reducing amplitude to values lower than $R_C I_{SS}$.

To simplify analysis and evaluate the minimum number of stages that ensures amplitude to be close to $R_C I_{SS}$, it is reasonable to assume that this occurs when the output voltage of each gate is able to switch from 10% to 90% of the entire logic swing in a half period $T/2$. This means that the latter has to be greater than the output rise time, and we will analytically express this condition in the following. From (8.12) the half period is equal to $T/2=n\tau_{PD}=0.8*n\tau$, while the output rise time T_{RISE} can be approximated as that under a step input, which for a CML stage represented by a first-order system with time constant τ is equal to 2.2τ [MG87]. As a result, the condition that ensures an almost full-swing oscillation becomes

$$\frac{T}{2}=0.8\cdot n\cdot\tau>T_{RISE}\approx 2.2\cdot\tau\Rightarrow n>3 \qquad (8.18)$$

which means that a number of stages greater than 3 must be used to maximize amplitude. This result is confirmed by simulations, that for a 500-mV logic swing (i.e., $R_C I_{SS}$=250 mV) show an amplitude in the order of 100 mV for n=3, and around 245 mV for ring oscillators with a number of stages equal to 4 or greater.

8.1.4 CMOS SCL ring oscillators

The oscillation frequency model of ring oscillators developed in Section 8.1.1 can be analogously applied to SCL ring oscillators. In particular, the delay τ_{PD} of each stage, included in the frequency expression (8.6), is that of an SCL inverter (6.6) with a load capacitance equal to the input capacitance (6.13) of the subsequent stage (i.e. a purely capacitive input impedance, thus no change on its evaluation is needed). By remembering that delay of SCL gates is quite insensitive to the input rise time, it results in

$$\tau_{PD} = 0.8R\left[\left(C_{gd1,2} + C_{db1,2}\right) + \left(C_{gd3,4} + C_{db3,4}\right) + C_{in}\right] \tag{8.19}$$

that substituted into relationship (8.6) expresses the oscillation frequency of an SCL ring oscillator. Simulations confirm the validity of model (8.6) using (8.19). Indeed, under conditions used in the previous chapter, the maximum error found with respect to simulation results is lower than 25%, and in typical cases where an advantageous power-delay trade-off is achieved (i.e., delay is not very close to its asymptotic value, as explained in Chapter 7) it is much lower (in the order of 10%).

Obviously, relationship (8.19) could be expressed in the form (7.2), that is more suitable for design. Moreover, considerations in Section 8.1.3 regarding the minimum number of stages that ensures a full-swing oscillation hold for SCL ring oscillators.

8.2 CML FREQUENCY DIVIDERS

The static frequency divider is a fundamental block in various applications, such as mobile or satellite communication systems and multiple-Gb/s optic fiber systems. For such applications, high performance is essential, and low power consumption is highly desirable to extend battery lifetime in portable equipment and to make heat removal easier [K91], [I951], [F96]. As a consequence, design criteria to keep power dissipation as low as possible for an assigned speed requirement is of the utmost importance.

From an architecture point of view, a $1/2^n$ static frequency divider consists of n cascaded divide-by-two stages, as depicted in Fig. 8.8, where the operating frequency at the input of each stage is indicated as a fraction of the input signal frequency f_{IN}.

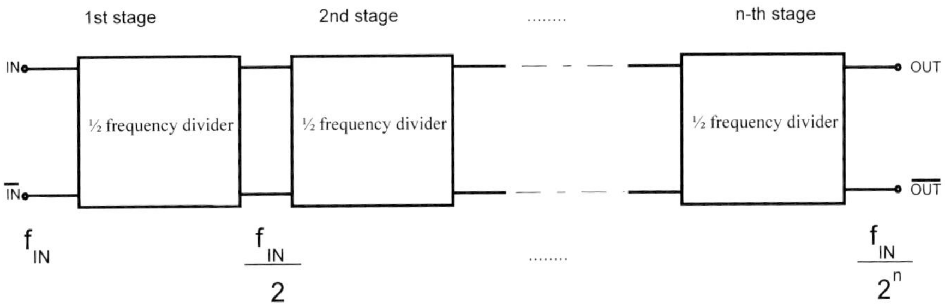

Fig. 8.8. Architecture of a $1/2^n$ frequency divider.

The topology of each 1/2 frequency divider consists of two cross-coupled CML D latches and a level shifter circuit, as shown in Fig. 8.9. The two cross-coupled D latches, realized in bipolar Current Mode logic as in Fig. 4.18, implement a Master-Slave T Flip-Flop since the output is fed back to the input after inversion. The level shifter is realized with a common-collector stage as usual, and is inserted to avoid that transistors Q1-Q2 in D latches in Fig. 4.18 work in the saturation region, as discussed in Chapter 2.

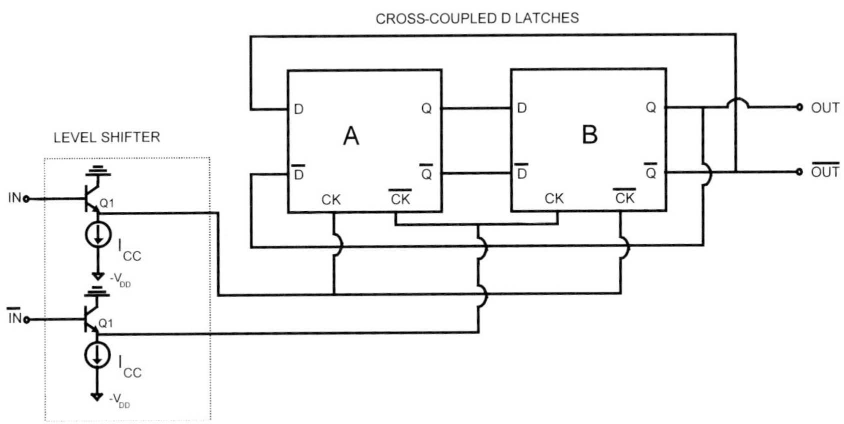

Fig. 8.9. Topology of a 1/2 frequency divider.

In the simple case of a single 1/2 frequency divider, the operation frequency is limited by the delay of the level shifter and D latch. More specifically, from Fig. 8.9, the input signal *IN* must propagate through the level shifter and the latch A within the positive half period, in order to set a correct input value in the latch B at the beginning of the successive half period. Since the same consideration holds in the negative half period for latch B, both latches must have the same speed, since otherwise it would be limited by the slower one.

From an analytical point of view, to let the output of latch A cross the logic threshold so that latch B has a correct input voltage in the following half period, the following condition on the input half period $T_{IN}/2$ must be satisfied

$$\frac{T_{IN}}{2} \geq \tau_{PD,level\_shifter} + \tau_{PD,latch} \tag{8.20}$$

where $\tau_{PD,level\_shifter}$ is the propagation delay of the level shifter and $\tau_{PD,latch}$ is the CK-Q latch delay, respectively. Therefore, the maximum operating frequency f_{max} of a single 1/2 frequency divider is equal to

$$f_{\max} = \frac{1}{2\left(\tau_{PD,level\_shifter} + \tau_{PD,latch}\right)} \tag{8.21}$$

To achieve a high-speed feature, the two delay contributions must be reduced as much as possible by properly setting the bias current of the latch I_{latch} and that of the level shifter $I_{level\_shifter}$.

Since the operating frequency of each stage is halved compared to that of the previous one, the delay of the generic stage can be twice as high as that of the previous one without degrading the speed of the divider. As a consequence, according to the model and design strategies developed in Chapter 5, it is possible to progressively reduce bias currents of each stage with respect to the previous one, allowing for a power saving [ADP02].

In the following, design equations are derived for bias currents in a frequency divider to achieve a high operating frequency by maximizing the speed performance of the first stage (Section 8.2.1), while minimizing overall power consumption in the successive stages that are unnecessarily fast (Section 8.2.2).

8.2.1 Design of the first stage

The first stage has to work at the highest frequency f_{max}, and accordingly its bias currents I_{latch} and $I_{level\_shifter}$ have to be properly sized. Delay of D latches $\tau_{PD,latch}$ pertains to the case where the switching input is applied to lower transistors Q1-Q2 in Fig. 4.18, as discussed in Section 5.5.3. Therefore, this delay contribution is modeled by relationship (5.1) with coefficients given by (5.29a), (5.42) and (5.29c). As demonstrated in Section 5.1, the latch delay can be minimized by setting bias current I_{latch} equal to

$$I_{latch,op} = \sqrt{\frac{b}{a}} \tag{8.22}$$

that from relationship (5.3) leads to the minimum delay achievable equal to $\tau_{PD,latch,op} = 2\sqrt{ab}$. A better trade-off between delay and power can be achieved by resorting to design criteria in Section 5.1 based on Fig. 5.1. In particular, an almost maximum speed is achieved by assigning I_{latch} equal to $0.6 \cdot I_{latch,op}$, that allows for a 40% power saving at the cost of only 10% increase of delay with respect to the optimum design.

As far as level shifters are concerned, their delay is given by relationship (5.17) with $I_{CC}=I_{level\_shifter}$, thus

$$I_{level\_shifter} = \frac{C_{je,buf}}{\left[\left(\frac{\tau_{PD,level\_shifter}}{\tau_{PD,level\_shifter,\min}}\right)^2 - 1\right]\tau_F} V_T = 4.76\frac{C_{je,buf}}{\tau_F}V_T \tag{8.23}$$

where ratio $\tau_{PD,level\_shifter}/\tau_{PD,level\_shifter,min}$ between the actual and the minimum delay achievable has been set to 1.1, as clarified in Section 5.3, thereby leading to a delay 10% greater than the minimum.

8.2.2 Design of successive stages

Once the first stage is designed, the bias current of subsequent ones can be downscaled to reduce the power consumption, since each stage has an input frequency which is half that of the previous one. Therefore, the delay of each stage can be set twice as high as the previous one without decreasing the maximum frequency. This choice allows the stage bias current to be set at the minimum value compatible with speed of the first stage. In particular, it is necessary to find the scaling law of $I_{latch}(i)$ and $I_{level\_shifter}(i)$ of i-th stage to double its delay $\tau_{PD}(i)$ with respect to the previous one.

Regarding D latches of the i-th stage, let us consider the normalized delay expression (5.5) versus bias current I_N normalized to $I_{latch,op}$, that is equal for all stages since they have the same load. From relationship (5.5), delay of the i-th stage's D latch is twice as high as that of the preceding (i-1)-stage if

$$2\left[I_N(i-1)+\frac{1}{I_N(i-1)}\right]=I_N(i)+\frac{1}{I_N(i)} \tag{8.24}$$

that, solved for $I_N(i)/I_N(i\text{-}1)$ (i.e., $I_{latch}(i)/I_{latch}(i\text{-}1)$), results in

$$\frac{I_N(i)}{I_N(i-1)}=1-\frac{1}{[I_N(i-1)]^2}\left(1+\sqrt{1+[I_N(i-1)]^2+[I_N(i-1)]^4}\right) \tag{8.25}$$

Design equation (8.25) shows how to scale currents from the second to the last stage by iteration, and can be approximated to

$$\frac{I_N(i)}{I_N(i-1)}\approx 0.5-0.24[I_N(i-1)]^{1.6} \tag{8.26}$$

where the approximation leads to an error lower than 3% in the range of interest (i.e., $I_N(i)\leq 1$, as discussed in Section 5.1). For the sake of clarity, relationship (8.26) is plotted in Fig. 8.10, from which the bias current scaling factor $I_N(i)/I_N(i\text{-}1)$ of i-th stage with respect to (i-1)-th one is derived from the normalized bias current $I_N(i\text{-}1)$ of the latter.

As far as the level shifter is considered, sizing of its bias current can easily be chosen by recursively doubling in (8.23) the ratio $\tau_{PD,level\_shifter}/\tau_{PD,level\_shifter,min}$ of the i-th stage with respect to the (i-1)-th in the evaluation of $I_{level\_shifter}(i)$.

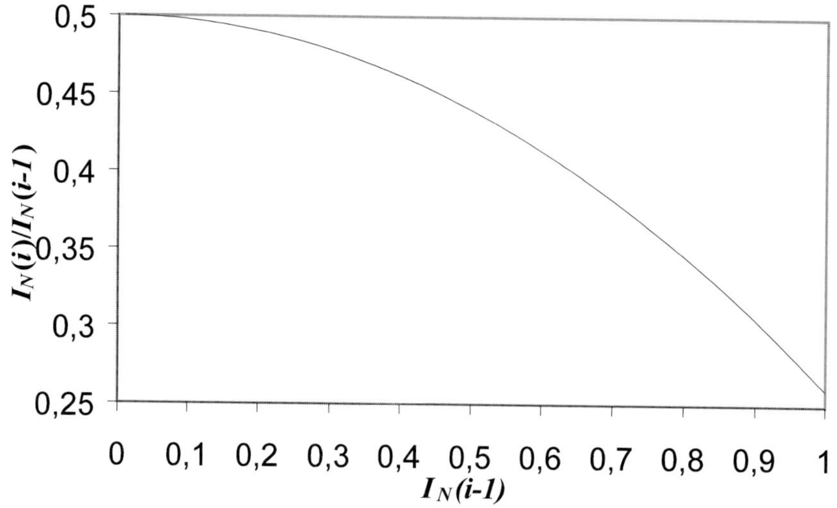

Fig. 8.10. Plot of relationship (8.26).

8.2.3 Design considerations and examples

It is worth noting that bias currents $I_{latch}(n)$ and $I_{level\_shifter}(n)$ of the last stage decrease as the number of stages n is increased, and for a high n they can become excessively low. Indeed, for an assigned logic swing $V_{SWING}=2R_C I_{latch}$, a very low I_{latch} (in the order of a few microamperes) leads to a high load resistance R_C (of the order of one hundred kΩ) that tends to occupy an excessive silicon area. Moreover, a very low $I_{level\_shifter}$ could be lower than the current level below which the transistor current gain β_F rapidly decreases (well below 1 μA, for the process considered).

From these considerations, when bias currents of the i-th stage reach values in the order of a few microamperes, one can avoid further downscaling bias currents of the successive stages, and this design option does not significantly affect the power saving offered by the design strategy. Indeed, bias currents of last stages whose current is not downscaled are negligible with respect to the power dissipation of first stages.

To illustrate the design strategy proposed and compare its results with simulations, let us consider the design of a 1/8 static frequency using the HSB2 bipolar process. By assuming V_{DD}=5 V and V_{SWING}=500 mV, coefficients a, b and c result to 2.2E-9, 6.69E-14 and 1.4E-12, respectively. From (8.22), the optimum latch current $I_{latch,op}$ is equal to 5.5 mA. For the first stage we use the power efficient design criteria $I_{latch}(1)=0.6\cdot I_{latch,op}=3.3$ mA. Moreover, the bias current of level shifters $I_{level\_shifter}(1)$ results 0.85 mA, from relationship (8.23) with delay ratio $\tau_{PD,level\_shifter}/\tau_{PD,level\_shifter,min}$ set to 1.1. For the second stage, from (8.26) we get $I_N(2)/I_N(1)=0.39$, that leads to $I_{latch}(2)=1.2$ mA, and doubling ratio $\tau_{PD,level\_shifter}/\tau_{PD,level\_shifter,min}$ with respect to the first stage (i.e. setting it to 2.2 in (8.26)), means that $I_{level\_shifter}(2)$ is equal to 50 μA. For the third stage we obtain $I_N(3)/I_N(2)=0.45$ and hence $I_{latch}(3)=0.5$ mA, while in the level shifter we set $\tau_{PD,level\_shifter}/\tau_{PD,level\_shifter,min}=4.4$, obtaining $I_{level\_shifter}(3)=10$ μA.

By using the evaluated bias currents, the maximum frequency obtained by SPICE simulations is 7.69 GHz, while that predicted by (8.21) and delay model used is equal to 7.35 GHz, that differs from the former by 4.4%. Input and output waveforms at maximum operating frequency are plotted in Fig. 8.11.

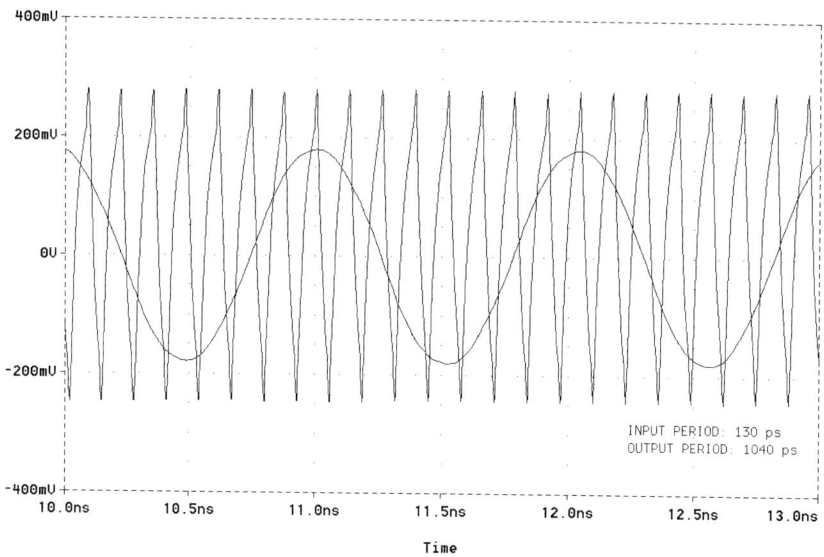

Fig. 8.11. Input and output waveforms at maximum operating frequency for the 1/8 frequency divider designed.

8.3 LOW-VOLTAGE BIPOLAR CURRENT-MODE TOPOLOGIES

Until now, the traditional series-gate approach was followed to implement CML gates. However, alternative approaches can be exploited to build CML gates with a reduced supply voltage [ROS94], [KKI97], [SMT98], [SPM00]. These alternative topologies are based on the consideration that the power consumption of a CML gate is equal to the product of bias current I_{SS} and supply voltage V_{DD}. Reduction of I_{SS} inevitably compromises the speed performance and in typical applications is not a viable solution, while reduction of V_{DD} does not significantly affect speed, provided that all transistors work out of the saturation region.

In the traditional series-gate approach, V_{DD} must be sufficiently high to ensure that all transistors work out of the saturation region, according to number of stacked levels, as discussed in Chapter 2. In [ROS94], [KKI97], [SMT98], some techniques that allow for a V_{DD} reduction with respect to traditional CML gates are proposed. Of these approaches, the one reported in [ROS94] appears to be the most promising since it operates under the lowest supply voltage. Therefore, in the following only that in [ROS94] will be considered and accordingly it will be referred to as the low-voltage topology.

8.3.1 Low-voltage CML by means of the triple-tail cell

The supply voltage V_{DD} of traditional CML gates can be lowered by reducing the number of series-gating levels. For example, according to considerations reported in Chapter 2, the minimum value of V_{DD} allowed by a two-level CML gate is about 2 V, while that of a one-level gate is 1.1 V. It is worth noting that, since the base-emitter voltage of a bipolar transistor is not affected by the technology scaling, these minimum values of V_{DD} do not depend on the process used.

The technique suggested in [ROS94] allows to reduce the number of series-gating levels by introducing the triple-tail cells concept. As shown in Fig. 8.12, a generic emitter coupled pair (Q1-Q2) in a series gate is activated (i.e., its input affects the output) when it is biased by the current I_{SS} steered by the pair lying at the lower level (Q3-Q4) (i.e., when the base voltage of Q3 is high). This requires two levels of series gating.

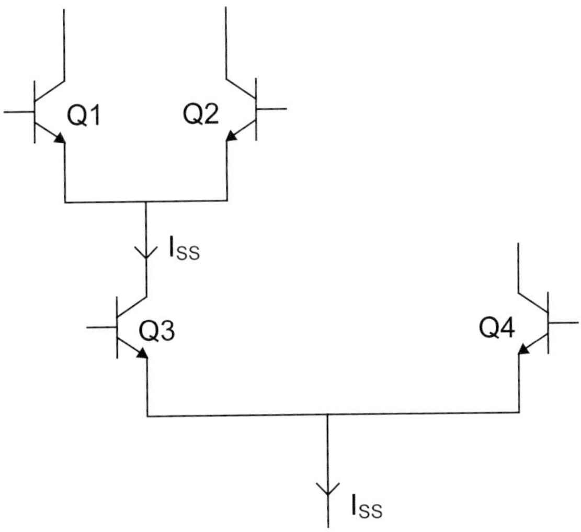

Fig. 8.12. Pair Q1-Q2 activated by transistor Q3 in a series gate.

A different strategy can be used to activate Q1-Q2 with transistors Q3-Q4 lying at the same level. To be more specific, transistor pair Q1-Q2 can be deactivated by connecting a third transistor Q3 (having its emitter in common with that of Q1-Q2) with a high base voltage, thus steering bias current to ground and turning off transistors Q1-Q2. On the contrary, when Q3 has a low base voltage, it does not affect operation of Q1-Q2, thereby activating it. The three emitter-coupled transistors that implement the pair

Q1-Q2 and the (de)activating transistor Q3 is called the *triple-tail cell* [ROS94] and is depicted in Fig. 8.13.

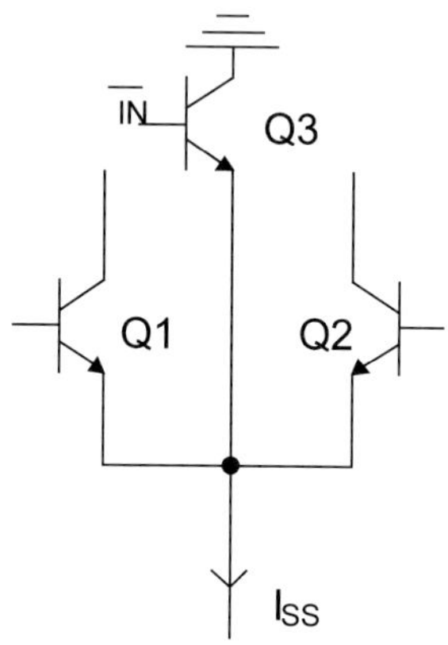

Fig. 8.13. The triple-tail cell.

When possible, by applying transformation in Fig. 8.12 that substitutes stacked emitter coupled pairs in a series gate with a triple cell (Fig. 8.13) a CML gate with a reduced number of series-gating levels is obtained. For example, low-voltage topologies of the MUX, XOR and D latch gates in Figs. 8.14-8.16 are obtained from the traditional CML topologies in Figs. 4.13, 4.14 and 4.18, respectively.

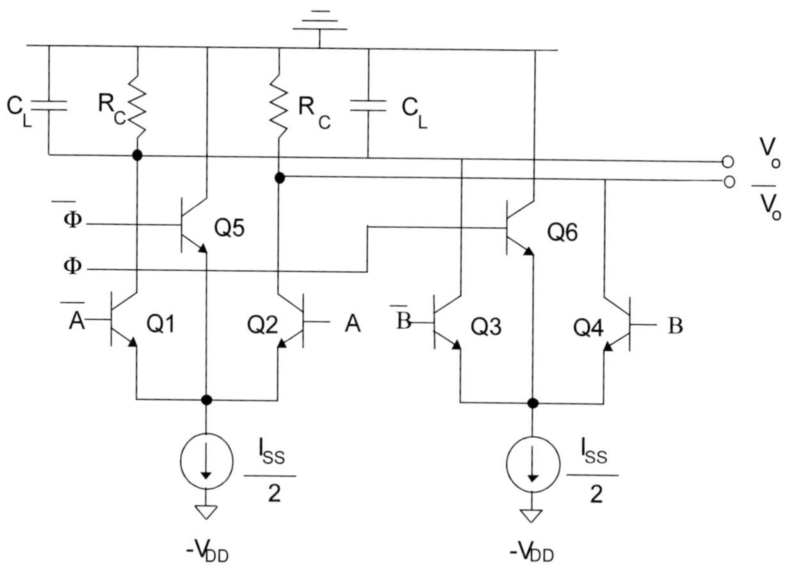

Fig. 8.14. Low-voltage MUX gate topology.

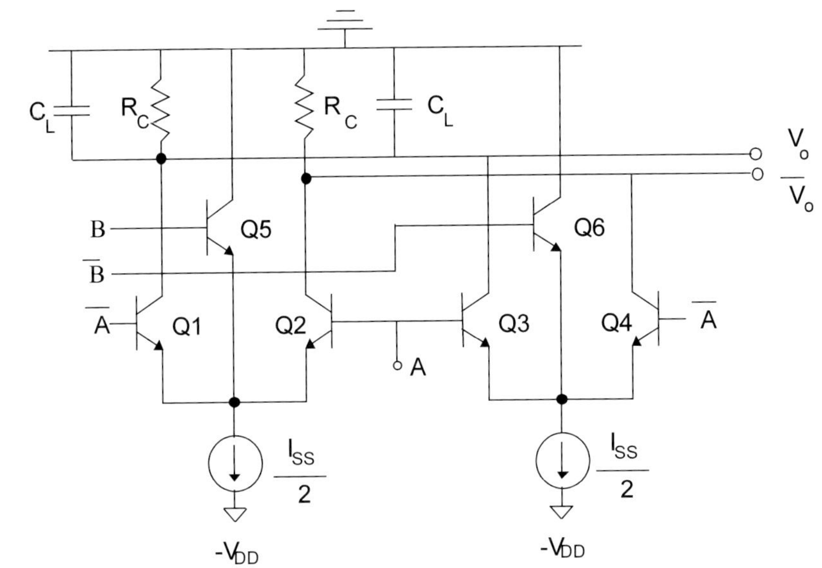

Fig. 8.15. Low-voltage XOR gate topology.

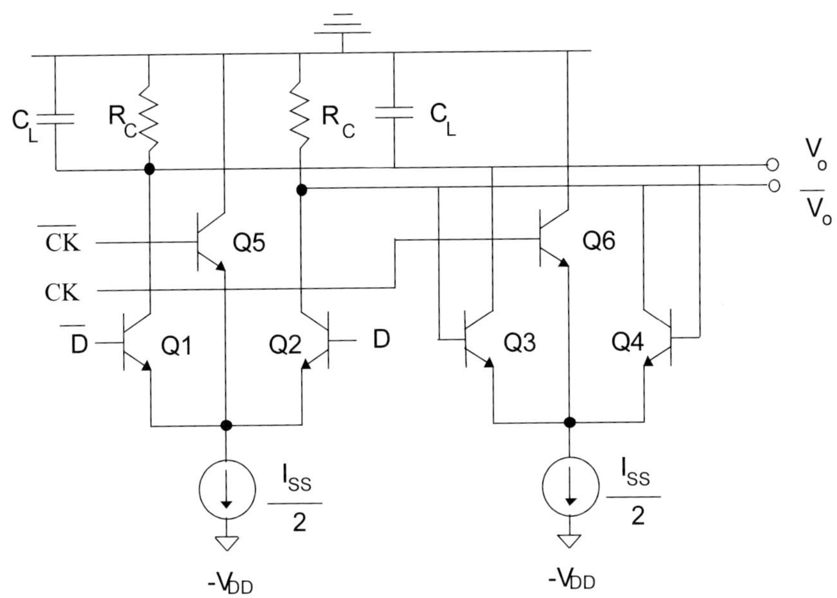

Fig. 8.16. Low-voltage D latch gate topology.

It is worth noting that, in a triple-tail cell, the deactivating transistor Q3 does not completely switch off transistors Q1-Q2. To better understand this point, assume without loss of generality that input of Q1-Q2 is such that Q1 is conducting and Q2 is not (i.e., base voltage of Q1 and Q2 are equal to zero and $-R_C I_{SS}$, respectively), and define N as the ratio between the emitter area of Q3 and Q1-Q2 (i.e., $A_{E3}=N\cdot A_{E1,2}$). When Q3 deactivates Q1-Q2 (i.e., its base is high), Q1 and Q3 have the same base-emitter voltage, therefore current of Q3 is greater than that of Q1 by the ratio N of their emitter area. Thus, the current flowing in Q1 assuming $\alpha_F=1$ results to

$$i_{C1}=\frac{I_{SS}}{2}\frac{1}{1+N} \tag{8.27}$$

and can be minimized by increasing factor N, or equivalently the area of transistor Q3.

In summary, the low-voltage CML gates in Figs. 8.14-8.16 can be implemented with a single-level logic like the simple inverter, thereby lowering the minimum V_{DD} by around a factor of 2 (i.e., from 2 V to 1.1 V). The effect of this supply voltage reduction on performance and power-delay trade-off will be analyzed in the next subsections for the particular case of a D latch gate, s it can be easily extended to the other low-voltage gates.

8.3.2 Analysis of the low-voltage CML D latch static operation

The low-voltage D latch topology, shown in Fig. 8.16, is made up of the two emitter-coupled pairs Q1-Q2 and Q3-Q4, biased with two current sources $I_{SS}/2$. The transistor pair Q1-Q2 is driven by the differential input D, while the cross-coupled pair Q3-Q4 implements the memory element.

Depending on the level of the differential clock signal CK, one of the two transistor pairs is alternatively deactivated by transistors Q5 and Q6, whose emitter area $A_{E5,6}$ is assumed to be greater than area $A_{E1,2,3,4}$ of the other transistors by a factor N (i.e., $N=A_{E5,6}/A_{E1,2,3,4}$). When CK is low, transistor Q6 is OFF and the cross-coupled pair Q3-Q4 holds the previous logic value thanks to the positive feedback, and the latch is in the hold state. When CK is high, transistor Q5 is OFF and Q6 is ON, thus deactivating the cross-coupled pair Q3-Q4. Hence, the output is set equal to input D by the transistor pair Q1-Q2, and the latch is in the transparent state.

In the following, the static behavior of the gate will be analyzed in terms of logic swing. As it will be demonstrated, the logic swing of the low-voltage D latch is given by

$$V_{SWING} = R_C I_{SS} \frac{N}{1+N} \tag{8.28}$$

that is lower than that of the traditional D latch by a factor greater than two for an assigned value of $R_C I_{SS}$.

To demonstrate relationship (8.28), consider a low-voltage D latch working in the hold state, i.e. with Q5 in the ON state and Q6 in the OFF state, where Q3-Q4 store the previous output. Ideally, in this condition the output should not be affected by input *D*, i.e., transistors Q1-Q2 should both be in the cut-off region. In reality, transistors Q1-Q2 are only partially deactivated by transistor Q5, and the conducting one sinks a current given by relationship (8.27), that influences the output voltage in a manner that depends on whether input *D* is equal or opposite to the stored value. For a better understanding of this aspect, assume input *D* to be low without loss of generality.

When the stored value is equal to *D* (i.e., at the low level, under the assumption made above) Q3 is ON and Q4 is OFF, hence the collector currents of Q3 and Q4 are $i_{C3}=I_{SS}/2$ and $i_{C4}=0$, respectively. Moreover, transistor Q2 is OFF and Q1 conducts the current (8.27), thus the differential output voltage $V_o - \overline{V_o}$ is equal to the low level given by

$$\begin{aligned} V_{OL,hold}\Big|_{D=OUT} &= -R_C\left(i_{C1}+i_{C3}\right)+R_C\left(i_{C1}+i_{C3}\right) \\ &= -\frac{R_C I_{SS}}{2}\left(1+\frac{N}{1+N}\right) \end{aligned} \tag{8.29}$$

By reiterating the same procedures and assuming input *D* to be high, it is demonstrated that the high output voltage $V_{OH,hold}\Big|_{D=OUT}$ when the stored value is equal to *D* is given again by (8.29) with an opposite sign. Therefore, the logic swing for *D=OUT* in the hold mode results in

$$V_{SWING,hold}\Big|_{D=OUT} = R_C I_{SS}\left(1+\frac{N}{1+N}\right) \tag{8.30}$$

In cases where the stored value is opposite to input *D* (i.e., $D=\overline{OUT}$), by repeating the same reasoning, the logic swing becomes

$$V_{SWING,hold}\Big|_{D=\overline{OUT}} = R_C I_{SS} \frac{N}{1+N} \tag{8.31}$$

Until now, analysis presented is based on the assumption that the latch operates in the hold mode. When the latch is in the transparent mode (i.e., when *CK* is high, hence transistor Q5 is OFF and Q6 is ON, thus deactivating pair Q3-Q4 and activating pair Q1-Q2, which sets the output equal to *D*), the logic swing can be shown to be equal to (8.30). By comparing (8.30) and (8.31), the worst-case logic swing (and noise margin) is achieved in the hold mode when *D* is opposite to the stored value *OUT*, and is equal to (8.31), which demonstrates (8.28).

From (8.31) the low-voltage topology has a lower logic swing, and hence a lower noise margin, than a traditional circuit for a given value of $R_C I_{SS}$. As a consequence, from (8.28) parameter *N* has to be chosen as high as possible to avoid an excessive noise margin degradation [SMP00]. However, a high value of *N* leads to a high emitter area of transistors Q5-Q6, thus increasing the input capacitance seen from input *CK*. As suggested in [ROS94], a good compromise between these opposing requirements is to set N=2. As the logic swing and parameter *N* are assigned before the design, from (8.28) the product $R_C I_{SS}$ also becomes a constant in the design.

8.3.3 Delay of the low-voltage CML D latch

Now let us evaluate the CK-Q and D-Q delay of a low-voltage D latch by applying the modeling strategy introduced in Chapter 4. It is useful to observe that, in practical cases such as in frequency dividers (see Section 8.2) or dual-modulus prescalers, the speed performance is usually limited by the CK-Q delay. As a consequence, in the following we shall focus on the CK-Q delay to describe the D latch speed performance, while the D-Q delay will only be briefly discussed.

To evaluate the CK-Q delay of a D latch, let us consider the output transition that occurs when the D latch goes from the hold mode to the transparent mode due to a low-to-high transition of the input *CK* (obviously *D* is assumed to be opposite to the previously stored value, otherwise no transition would take place). From a circuit point of view, the clock signal goes high and abruptly activates the transistor pair Q1-Q2, that can be thought of as a simple CML inverter. At the same time, transistors Q3-Q4 are deactivated, so that they affect the transient response only through their parasitic capacitances. As a result, the circuit can be schematized as the CML inverter Q1-Q2 loaded by an equivalent capacitance C_{eq} at the output nodes, that is equal to the capacitive contributions associated with transistors Q3-Q4 in parallel to the load capacitance C_L

$$C_{eq} = \left[C_{bcx3,4}\left(1+g_m R_C\right)+C_{cs3,4}\right]+C_L \tag{8.32}$$

where it was observed that Q3-Q4 load the output nodes through their collector-substrate capacitance and that seen from their base in (4.23). Therefore, the CK-Q delay is given by relationship (4.7) with a load capacitance C_{eq}

$$\tau_{PD} = 0.69\{(r_e + r_b)C_{be} + r_b C_{bci}[1 + g_m(r_c + R_C)]$$

$$+ (r_c + R_C)(C_{bci} + C_{bcx} + C_{cs}) \quad (8.33)$$

$$+ R_C[C_{bcx3,4}(1 + g_m R_C) + C_{cs3,4} + C_L]\}$$

By assuming $R_C I_{SS}$=500 mV and a minimum supply voltage V_{DD}=1.1 V, the capacitances obtained assuming the HSB2 technology are summarized in Table 8.4.

TABLE 8.4

C_{je}	44.9 fF
C_{cs}	17.4 fF
C_{bcx}	22.1 fF
C_{bci}	6.6 fF

By expressing the dependence on I_{SS} as usual, relationship (8.33) can be rewritten in the form (5.1) with coefficients given by

$$a = 0.69\left(2\frac{r_e + r_b}{V_{SWING}}\tau_F + \frac{r_b C_{bci1,2} r_c}{4V_T}\right) \quad (8.34a)$$

$$b = 0.69 V_{SWING}\left[C_{cs1,2} + C_{bc1,2} + C_{cs3,4} + \left(1 + \frac{V_{SWING}}{4V_T}\right)C_{bc3,4} + C_L\right] \quad (8.34b)$$

$$c = 0.69\left[(r_e + r_b)C_{je1,2} + r_b C_{bci1,2}\left(1 + \frac{V_{SWING}}{4V_T}\right) + r_c(C_{bc1,2} + C_{cs1,2})\right] \quad (8.34c)$$

By comparing results of delay model (5.1) using (8.34) to SPICE simulations under the conditions discussed above, error for the D latch implemented in the HSB2 process is always lower than 22%. Figure 8.17 plots the error versus the bias current assuming C_L=0 fF, 100 fF and 1 pF.

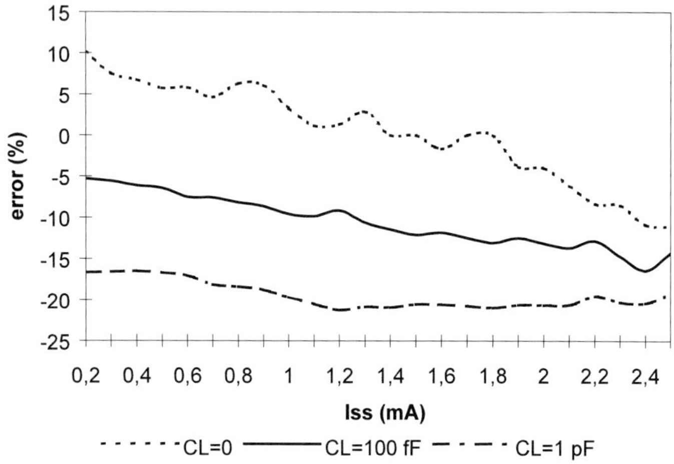

Fig. 8.17. Error of (5.1) using (8.34) vs. I_{SS} for the HSB2 process.

For the sake of completeness, let us evaluate the D-Q delay, when the D latch is in the transparent state. When *CK* is high, transistor Q5 is OFF and Q6 is in the linear region, thereby deactivating the emitter-coupled pair Q3-Q4. Therefore, when *D* switches, the circuit can be simplified into an inverter made up of transistors Q1-Q2, where the loading effect of Q3 and Q4 can be accounted for with a linear capacitance. The resulting delay expression is in the form (5.1), where coefficients are given by

$$a = 0.69\left(2\frac{r_e + r_b}{V_{SWING}}\tau_F + \frac{r_b C_{bci3,4} r_c}{4V_T} + 2\frac{r_c}{V_{SWING}}\frac{\tau_F}{1+N}\right) \tag{8.35a}$$

$$b = 1.38V_{SWING}\left[C_L + C_{cs3,4} + C_{cs1,2} + C_{bc1,2} + \left(1 + \frac{V_{SWING}}{2V_T}\right)C_{bc34} + C_{je}\right] \tag{8.35b}$$

$$c = 0.69\left[(r_e + r_b)C_{je} + r_b C_{bci}\left(1 + \frac{V_{SWING}}{2V_T}\right) + \right. \tag{8.35c}$$
$$\left. + r_c(C_{bci} + C_{bcx} + C_{cs}) + \frac{2}{1+N}\tau_F\right]$$

By comparing the delay model (5.1) and (8.35) with SPICE simulations under the previous conditions, error is always within 20%.

8.3.4 Comparison of the low-voltage and traditional CML D latch designed for high speed

Now let us compare performance achieved by the low-voltage and traditional D latch designed for a high speed, i.e. with an optimum bias current (5.2) that minimizes delay. To carry out a fair comparison, the supply voltage is set to the minimum value of 1.1 V and 2 V for the two topologies, and similar logic swing values are used. Therefore, from (2.9) using (8.28) and approximating their ratio to two, factor $R_C I_{SS}$ of the low-voltage latch will be assumed to be twice that of the traditional circuit.

The CK-Q delay of the low-voltage and traditional D latch is given by substituting coefficients (8.34) (low voltage topology) or (5.29) and (5.42) (traditional topology) into relationship (5.1). Assuming a logic swing of 500 mV and C_L=100 fF (i.e., about a unity fan-out), the numerical value obtained with the HSB2 process for both topologies is reported in Table 8.5. In the following, to avoid confusing the coefficients of the two topologies, subscript LV and TR will be used respectively to indicate the low-voltage and traditional D latch.

TABLE 8.5

	Low-voltage	Traditional
a	1.11 E-9	2.2 E-9
b	1.14 E-13	6.69 E-14
c	4.8 E-12	1.4 E-12

To understand and compare the speed performance and power-delay trade-off of the two topologies we can compare coefficients *a* and *b*. Comparing relationship (8.34) and (5.29a), ratio a_{LV}/a_{TR} is approximately equal to ½ (from Table 8.5, it results 0.51). Indeed, *a* is mainly determined by the base-emitter diffusion capacitance, that in the low-voltage D latch is halved since each transistor pair is biased by a current $I_{SS}/2$. As far as coefficient b is concerned, ratio b_{LV}/b_{TR} is equal to

$$\frac{b_{LV}}{b_{TR}} = 2\frac{C_L + C_{cs3,4} + C_{cs1,2} + C_{bc1,2} + \left(1 + \frac{V_{SWING}}{4V_T}\right)C_{bc3,4}}{C_L + C_{cs3,4} + C_{cs5,6} + 4C_{bc3,4,5,6} + C_{je3,4} + \frac{V_{SWING}}{4V_T}C_{bcx5,6}} \tag{8.36}$$

that turns out to be lower than 2 and approaches 2 for high values of C_L, since the denominator has essentially the same capacitive contributions as the numerator, but a slightly higher number of small addends. To be more specific, the denominator has the additional term $C_{je3,4}$, as well as a greater number of base-collector capacitances C_{bc}. As an example, for the HSB2 technology, ratio (8.36) ranges from 1.6 to 1.9 for C_L ranging from zero to 1 pF.

From eq. (5.3), the resulting ratio of minimum delay achievable $\tau_{PDop,LV}/\tau_{PDop,TR}$ in the low-voltage and traditional circuit is

$$\frac{\tau_{PDop,LV}}{\tau_{PDop,TR}} = \sqrt{\frac{a_{LV}b_{LV}}{a_{TR}b_{TR}}} \tag{8.37}$$

that, since $a_{LV}/a_{TR} \approx 1/2$ and b_{LV}/b_{TR} is slightly lower than 2, is slightly lower than unity and tends to 1 for very high load capacitances (since $b_{LV}/b_{TR} \rightarrow 2$). This means that the two topologies essentially exhibit the same maximum speed achievable, especially for high load capacitances.

Regarding the power-delay trade-off in the high-speed design case, let us consider the power-delay product expression in (5.4) for the optimum bias current (5.2), that results to $2V_{DD}b$. Therefore, setting the supply voltage to its minimum value for each latch topology, the power-delay product ratio of the low-voltage circuit and the traditional circuit results in

$$\frac{PDP_{LV}}{PDP_{TR}} \approx \frac{V_{DD,\min,LV}}{V_{DD,\min,TR}}\frac{b_{LV}}{b_{TR}} = 0.55\frac{b_{LV}}{b_{TR}} \tag{8.38}$$

From the considerations on ratio b_{LV}/b_{TR}, the two topologies have comparable power-delay product. To be more specific, for low values of C_L, the power-delay product ratio (8.38) tends to be lower than unity, which means that the low-voltage latch has a slightly better power efficiency. For high values of

C_L, relationship (8.38) tends to 1.1, which means that for high load capacitances the traditional D latch tends to be more efficient.

In actual design cases, as was demonstrated in Section 5.1, a more efficient design choicc is achieved by lowering the optimum bias current by a factor $I_N = I_{SS}/I_{SSop}$ according to the power-delay curve in Fig. 5.1. In particular, to carry out a consistent comparison, let us consider an equal factor I_N for the two latch topologies, that determines an equal delay increase compared to the optimum case. Therefore, delay and power-delay ratios are still equal to (8.37) and (8.38), respectively, and the above considerations still remain valid.

Considerations reported until now are valid for the CK-Q latch delay. By using the same approach, they can be extended to the D-Q delay, as briefly explained in the following. The D-Q delay of low-voltage and traditional D latch is given by (5.1) with coefficients (8.35) (low-voltage topology) or (5.32) and (5.46) (traditional topology), whose numerical values under previous conditions are summarized in Table 8.6.

TABLE 8.6

	Low-voltage	Traditional
a	1.61 E-9	2.2 E-9
b	1.24 E-13	5.32 E-14
c	8.78 E-12	4.53 E-12

When D-Q delay is considered, ratio a_{LV}/a_{TR} turns out to be higher, compared to that evaluated for the CK-Q delay. Indeed, from comparison of relationships (5.32) with (5.29), coefficient a_{TR} of the traditional gate is the same as the CK-Q delay, while from comparison of (8.34a) with (8.35a) coefficient a_{LV} for the D-Q delay has one addend more than the CK-Q delay. For the HSB2 process and under the previous conditions, ratio a_{LV}/a_{TR} results equal to 0.73. Moreover, ratio b_{LV}/b_{TR} is slightly greater than 2, due to the additional base-emitter capacitance C_{je} in b_{LV} expressed by (8.35b), and tends to 2 for high values of the load capacitance (ratio b_{LV}/b_{TR} ranges from 2.09 to 2.47 when C_L is varied from 0 F to 1 pF).

From these considerations, unlike the results obtained for the CK-Q delay, the minimum D-Q delay ratio is always greater than unity (this is also true for bias currents scaled with respect to optimum values). This means that, for a high-performance design, the traditional D latch is worse than the low-voltage one, regardless of the load capacitance.

8.3.5 Comparison of the low-voltage and traditional CML D latch designed for low power consumption

For gates that do not lie in the critical path of the circuit being designed, the speed performance is not a concern. They can therefore be designed with a much lower bias current than the optimum value I_{SSop}. In this case, the CK-Q delay expression (5.1) can be approximated as

$$\tau_{PD} \approx \frac{b}{I_B} \tag{8.39}$$

which shows that at low bias currents delay is inversely proportional to bias current.

To carry out a consistent comparison, let us consider the low-voltage and traditional D latch with the same bias current I_{SS}. As a consequence, the delay ratio between the former and the latter circuit becomes

$$\frac{\tau_{PD,LV}}{\tau_{PD,TR}} \approx \frac{b_{LV}}{b_{TR}} \tag{8.40}$$

which, as discussed in the previous subsection, is slightly lower than two. This means that in low-power designs, the traditional latch outperforms the low-voltage by a factor slightly lower than two.

To evaluate the power efficiency, we again consider the power-delay product ratio for the two topologies, that from (5.4) results in

$$\frac{PDP_{LV}}{PDP_{TR}} \approx \frac{V_{DD,\min,LV}}{V_{DD,\min,TR}} \frac{\tau_{PD,LV}}{\tau_{PD,TR}} \approx 0.55 \frac{b_{LV}}{b_{TR}} \tag{8.41}$$

which equals that obtained in the high-performance design, as can be found by comparison with relationship (8.38). Again, from relationship (8.41), the low-voltage circuit *PDP* is 10% worse than the traditional one for very high load capacitances. In contrast, for low values of C_L the power-delay product ratio (8.41) tends to be lower than unity, thus the low-voltage latch has a small advantage in terms of power efficiency.

To give an example, for the HSB2 process, a load capacitance of 100 fF and a bias current of 100 μA, the delay ratio predicted by relationship (8.40) and the simulated value are 1.7 and 1.56, respectively, while the power-delay product ratio (8.41) and the simulated ratio are 0.94 and 0.86, respectively.

Even though results discussed above refer to the CK-Q delay of the D latch, similar observations can be made for the D-Q delay. Indeed, relationship (8.40) states that the delay ratio is greater than two under a bias current which is much lower than the optimum value, meaning that the traditional latch outperforms the low-voltage one by a factor greater than two in low-power design.

8.3.6 Summary of results and remarks

In the previous sections, the low-voltage CML D latch topology has been analyzed and compared with the traditional implementation. The former allows a supply voltage reduction by a factor of about 0.55, which could be possibly exploited to achieve a power saving. However, the bias current must also be considered when making a significant comparison with the traditional topology in terms of the speed performance achievable and the power-delay trade-off.

By using the methodology developed in Chapter 5, the low-voltage D latch was designed and compared to the traditional topology in terms of the delay and power-delay trade-off, for a high-performance or low-power consumption design target. The results showed that the low-voltage D latch topology is advantageous in typical cases where a low CK-Q latch delay is required and a low fan-out is expected, since this circuit exhibits a moderate speed improvement (in the order of 20%) with respect to the traditional implementation. However, the power increase must be paid for this speed improvement, since the low-voltage and traditional topologies have roughly the same power-delay product. This is because the latch delay is that of inverter Q1-Q2 (or Q3-Q4) properly loaded whose biased current is only half of the total gate current I_{SS}, while the other half bias current is steered to ground by the deactivating transistor.

In the low-power design case, the traditional topology has a significant speed advantage over the low-voltage one (roughly by a factor of two), while the same considerations on the power-delay product as in the high-performance case hold.

As a result, the low-voltage topology never possesses a strong advantage in terms of delay or of power efficiency. In practical cases, the only significant advantage of the low-voltage circuit is that no output buffer (and thus its additive bias current) is required. In addition, when comparing the two D latch topologies, it should also be considered that the low supply voltage allowed by the low-voltage circuit imposes serious limits on the logic gates that can be implemented, since traditional series gates cannot correctly operate. Therefore, the low-voltage approach is a viable solution in low-fan-out high-speed applications requiring only gates that can be efficiently implemented with triple cells, such as MUX, XOR and D latches.

These results can be easily generalized to other low-voltage CML gates, such as the MUX/XOR gate.

8.4 OPTIMIZED DESIGN STRATEGIES FOR CASCADED BIPOLAR CURRENT-MODE GATES

Until now, criteria have been introduced to optimally design a single Current-Mode gate and consciously manage the power-delay trade-off. When cascaded gates are considered, an analogous trade-off exists between the overall delay of a path and the overall power consumption, both of which are equal to the sum of the contributions associated with each gate. As in the case of single gates, overall delay and power consumption strongly depend on the overall bias current I_{TOT} of gates belonging to the path considered.

Depending on the application, parameter I_{TOT} can be either a design variable or a constant assigned from considerations on power consumption at the system level. In the first case, where the maximum speed allowed by the technology is required, the overall delay τ_{PD} has to be minimized by properly setting all bias currents in cascaded gates to make each gate delay contribution minimum. Therefore, the problem of minimizing the overall delay is greatly simplified in the delay minimization of each single gate, whose bias current has to be optimized independently of the others[1]. This is achieved by resorting to the design strategies for single gates already discussed in Chapter 5. The resulting overall current I_{TOT} results equal to the sum of optimum currents, that for CML gates are given by relationship (5.2) and for output buffers by (5.18).

When the overall bias current is preliminarily set to meet a power consumption value assigned at the system level design, a significant effort is typically spent in sizing each bias current in cascaded gates to minimize overall delay. From an analytical point of view, this translates in the problem of minimizing the delay expression with the constraint that the sum of bias currents is equal to the desired value I_{TOT}. This can be done in a computationally efficient way by resorting to the delay model (5.1) and (5.17) of CML gates and output buffers as a function of bias currents. Indeed, these models have simple expressions and constant coefficients, therefore they are suitable for numerical optimization. From relationships

[1] Indeed, as for design strategies introduced in Chapter 5, delay of each gate is to be minimized for a fixed load capacitance, since it does not depend on other bias currents. This is because load of each gate consists of the input capacitances of subsequent gates, that do not depend on their bias currents.

(5.1) and (5.17), the resulting delay expression to be minimized is in the form

$$\tau_{PD} = \sum_{i=1}^{n}\left[a_i I_{SS,i} + \frac{b_i}{I_{SS,i}} + c_i + 1.6\sqrt{\left(\frac{V_{SWING}}{2I_{SS,i}} + r_{b,i}\right)\left(\tau_F + \frac{V_T}{I_{CC,i}} C_{je,i}\right) C_{L,i}} \right] \quad (8.42)$$

where n cascaded ECL gates were considered and subscript i refers to the i-th gate, whereas for CML gates the output buffer delay (and thus term under square root) is zero. Relationship (8.42) must be minimized under the constraint

$$\sum_{i=1}^{n}\left(I_{SS,i} + 2I_{CC,i}\right) = I_{TOT} \quad (8.43)$$

This approach is much more advantageous than the traditional one based on iterative simulations with a trial-and-error procedure, since the former is much less computationally expensive and time-consuming.

In some cases, the minimization of overall delay under a current constraint can be carried out in a pencil-and-paper manner. In particular, this is possible when only CML gates belong to the path considered, i.e. it does not include any output buffer. In the following sections, main results on analytical delay minimization under a bias current constraint for CML paths are illustrated.

8.4.1 Design of CML non-critical paths with a constraint on the overall bias current

Let us consider a specific path to be optimized that consists of n cascaded CML gates. As an example, this is the case of cascaded low-voltage circuits, since they do not include any output buffer, or a chain of traditional CML gates whose output drive the upper-level inputs of the following ones. From (5.1), the delay of each gate is

$$\tau_{PD,i} = a_i I_{SS,i} + \frac{b_i}{I_{SS,i}} + c_i \approx a_i I_{SS,i} + \frac{b_i}{I_{SS,i}} \qquad \text{for } i = 1 \ldots n \quad (8.44)$$

where subscript i refers to the i-th gate of the path. The overall delay of the path is given by

$$\tau_{PD} = \sum_{i=1}^{n} \left(a_i I_{SS,i} + \frac{b_i}{I_{SS,i}} \right) \tag{8.45}$$

which is minimized by properly setting bias current values $I_{SS,i}$, that must satisfy the following condition

$$\sum_{i=1}^{n} I_{SS,i} = I_{TOT} \tag{8.46}$$

When the limited power (i.e., bias current) budget represents a stronger constraint than speed, such as in a non-critical path where speed is not of concern, the bias current of each gate is made sufficiently lower than its optimum value to allow neglecting terms $b_i/I_{SS,i}$ with respect to terms $a_i I_{SS,I}$ in relationships (8.44) and (8.45).

Hence, the overall path delay can be simplified to

$$\tau_{PD} = \sum_{i=1}^{n} \frac{b_i}{I_{SS,i}} \tag{8.47}$$

Relationship (8.47) with constraint (8.46) is minimized if all terms $b_i/I_{SS,i}$ are equal

$$\frac{b_1}{I_{SS,1}^2} = \frac{b_2}{I_{SS,2}^2} = \ldots = \frac{b_n}{I_{SS,n}^2} \tag{8.48}$$

Indeed, by using (8.46), we rewrite (8.47) as

$$\tau_{PD} = \frac{b_1}{I_{TOT} - \sum_{i=2}^{n} I_{SS,i}} + \sum_{i=2}^{n} \frac{b_i}{I_{SS,i}} \tag{8.49}$$

and, setting to zero the derivatives of (8.49) with respect to terms $I_{SS,i}$ for i=2…n,

$$\frac{\partial \tau_{PD}}{\partial I_{SS,i}} = \frac{b_1}{\left(I_{TOT} - \sum_{i=2}^{n} I_{SS,i} \right)^2} - \frac{b_i}{I_{SS,i}^2} = \frac{b_1}{I_{SS,1}^2} - \frac{b_i}{I_{SS,i}^2} = 0 \qquad \text{for } i = 2 \ldots n \tag{8.50}$$

relationship (8.48) is demonstrated.

The specific value of the current in each stage, $I_{SS,i}$, can be evaluated after a few algebraic manipulations. In particular, let us consider the optimum value of terms $I_{SS,i}^2/b_i$ and define as X^* its square root

$$X^* = \frac{I_{SS,i}}{\sqrt{b_i}} \qquad i = 1 \ldots n \tag{8.51}$$

Evaluating $I_{SS,i}$ from (8.51) and substituting it into the current constraint (8.46), X^* results

$$X^* = \frac{I_{TOT}}{\sum_{i=1}^{n} \sqrt{b_i}} \tag{8.52}$$

that, once substituted into (8.51) and solving for $I_{SS,i}$, leads to the following expression of the optimum bias currents that minimize delay under the power consumption constraint

$$I_{SS,i} = \frac{\sqrt{b_i}}{\sum_{j=1}^{n} \sqrt{b_j}} I_{TOT} \tag{8.53}$$

From inspection of (8.53) we found that the overall bias current must be distributed in all gates proportionally to the weight of its term $\sqrt{b_i}$ with respect to the sum of all terms $\sum_{j=1}^{n} \sqrt{b_j}$. This result is confirmed by intuition, since, as observed in Section 5.7, coefficients b_i model the equivalent capacitance at the output node. This means that, to minimize the delay, gates having a greater parasitic capacitance must be provided with a greater bias current.

The resulting delay after optimization is obtained by substituting relationship (8.53) into (8.47)

$$\tau_{PD} = \frac{\left(\sum_{j=1}^{n} \sqrt{b_j} \right)^2}{I_{TOT}} = \frac{B}{I_{TOT}} \tag{8.54}$$

It is interesting to observe that the power-delay interdependence in cascaded gates designed for low power consumption is the same as in single CML gates designed according to the same criterion (see Section 5.1). Thus, the same considerations made in Section 5.1 for single gates are immediately extended to non-critical paths. In particular, the power-delay product results equal to $B \cdot V_{DD}$ (i.e., it has the same expression of single gates $b \cdot V_{DD}$) and does not depend on the overall bias current.

8.4.2 Design of CML critical paths with a constraint on the overall bias current

Let us consider again a path to be optimized that consists of n cascaded CML gates having a defined overall bias current, but unlike the previous case treated in Section 8.4.1, we assume that the bias current to be set in each stage can be close to the optimum value as speed, instead of power dissipation, is the main concern.

Following the same procedure discussed in Section 8.4.1 to reach (8.48), the delay is minimized if the following condition is satisfied

$$\frac{b_1}{I_{SS,1}^2} - a_1 = \frac{b_2}{I_{SS,2}^2} - a_2 = \ldots = \frac{b_n}{I_{SS,n}^2} - a_n \tag{8.55}$$

under constraint (8.46). From (8.55), the gates lying in the path must be biased with current $I_{SS,i}$ such that their associate factors ($b_i/I_{SS,i}^2$-a_i) are all equal. In the following, this common value is referred to as X

$$X = \frac{b_i}{I_{SS,i}^2} - a_i \qquad \text{for } i = 1 \ldots n \tag{8.56}$$

hence, by inverting relationship (8.73), the expression of bias currents is

$$I_{SS,i} = \sqrt{\frac{b_i}{a_i + X}} \qquad \text{for } i = 1 \ldots n \tag{8.57}$$

Therefore, factor X is evaluated from the constraint (8.46), which can be written as

$$\sum_{i=1}^{n} \sqrt{\frac{b_i}{a_i + X}} = I_{TOT} \tag{8.58}$$

and can be solved for X by using standard numerical methods for one-variable equations. Once factor X is evaluated, bias currents are easily calculated by using relationship (8.57).

Evaluation of factor X can also be carried out analytically in a high-speed design, where each bias current is assumed to be a significant fraction of the optimum bias current (5.2). This can be shown by rewriting (8.58) as

$$\sum_{i=1}^{n}\left[I_{SSop,i}\,f\left(\frac{X}{a_i}\right)\right]=I_{TOT} \tag{8.59}$$

where it was observed that $\sqrt{\frac{b_i}{a_i}}=I_{SSop,i}$, and function $f(x)$ was defined as

$$f(x)=\frac{1}{\sqrt{1+x}} \tag{8.60}$$

Inversion of (8.76) for X becomes much easier if $f(X/a_i)$ is not much lower than unity (in the order of 0.2 or greater) as in the considered case of high-speed design, where bias currents are comparable to optimum values (i.e., $f(x)$ is not much lower than unity). Under this assumption, function (8.77) can be approximated as

$$f(x)\approx 1.65-x^{0.12} \tag{8.61}$$

that, compared to expression (8.60), has an error lower than 10% for x ranging from 0.01 to 25, or equivalently for $f(x)\in[0.2, 0.99]$ (i.e., from (8.59)-(8.60) for a bias current ranging from 20% to 99% of the optimum value (5.2)), as shown in Fig. 8.18.

By substituting approximation (8.61) in (8.59) and performing some simple calculations, factor X results to

$$X=\left[\frac{1.65\sum_{i=1}^{n}I_{SSop,i}-I_{TOT}}{\sum_{i=1}^{n}\frac{I_{SSop,i}}{a_i^{0.12}}}\right]^{\frac{1}{0.12}}=\left[\frac{1.65\sum_{i=1}^{n}\sqrt{\frac{b_i}{a_i}}-I_{TOT}}{\sum_{i=1}^{n}\frac{b_i^{0.5}}{a_i^{0.62}}}\right]^{8.3} \tag{8.62}$$

Once X is evaluated, the bias current of each gate is found by resorting to relationship (8.57).

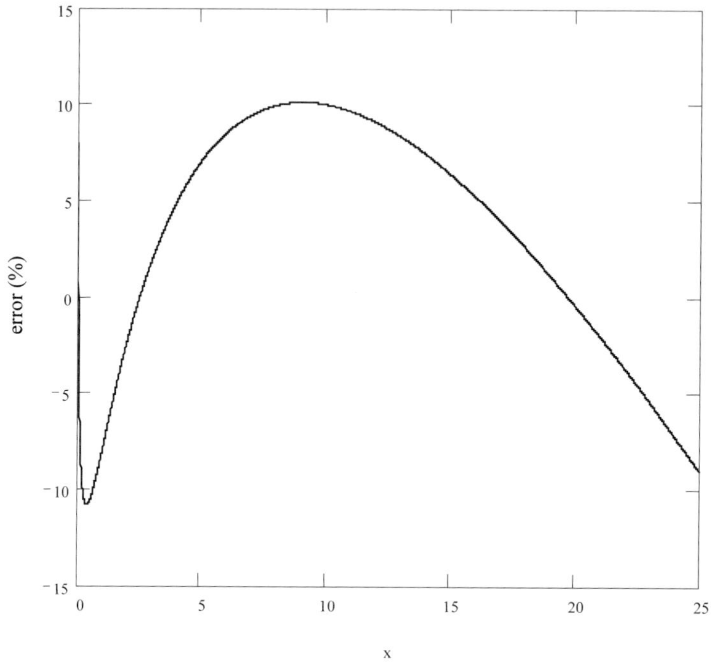

Fig. 8.18. Error of (8.61) with respect to (8.60) versus x.

Until now, general design criteria (i.e. for arbitrary coefficients a_i and b_i) to optimize bias currents under a power consumption constraint in CML critical paths have been discussed. However, such results can be further simplified in the frequent case where all transistors belonging to the critical path have an equal emitter area, as will be shown in the following section.

8.4.3 Design of CML critical paths with a constraint on the overall bias current and equal transistors' emitter area

Let us consider n cascaded CML gates having their transistors lying in the critical path with equal emitter area. This implies that coefficients a_i of all gates are equal

$$a_i = a \qquad \text{for } i = 1 \ldots n \tag{8.63}$$

because coefficients a_i can always be expressed in the form (5.32a) regardless of the gate considered, and only depend on the transistor emitter area, for a given logic swing. In practical cases, equal emitter area values (or, equivalently, coefficients a_i) are often used, since, as observed in Section 5.7, an increase in the emitter area determines a proportional increase in the input capacitance, and is thus rarely beneficial.

Under assumption in (8.63), delay (8.45) can be simplified into

$$\tau_{PD} = a\sum_{i=1}^{n} I_{SS,i} + \sum_{i=1}^{n} \frac{b_i}{I_{SS,i}} = aI_{TOT} + \sum_{i=1}^{n} \frac{b_i}{I_{SS,i}} \tag{8.64}$$

where the first term is a constant. As a consequence, delay (8.81) is minimum under the same condition found in Section 8.4.1, which occurs when (8.48) is satisfied.

Rewriting relationship (8.48) in a more expressive way by multiplying all ratios by coefficient a

$$\frac{I_{SS,1}^2}{\frac{b_1}{a}} = \frac{I_{SS,2}^2}{\frac{b_2}{a}} = \ldots = \frac{I_{SS,n}^2}{\frac{b_n}{a}} \tag{8.65}$$

and substituting (5.2) of optimum current

$$\frac{I_{SS,1}}{I_{SSop,1}} = \frac{I_{SS,2}}{I_{SSop,2}} = \ldots = \frac{I_{SS,n}}{I_{SSop,n}} = i_N \tag{8.66}$$

an interesting interpretation can be given. Indeed, from (8.66), the delay is minimized when all bias currents $I_{SS,i}$ are set to an equal fraction i_N of their optimum value $I_{SSop,i}$

$$I_{SS,i} = i_N I_{SSop,i} \tag{8.67}$$

where factor i_N is evaluated from the constraint on the overall bias current (8.46), and from (8.83) results to

$$i_N = \frac{I_{TOT}}{\sum_{i=1}^{n} I_{SSop,i}} \tag{8.68}$$

By substituting (8.67) into (8.64) and using relationship (5.3), the resulting overall delay is

$$\tau_{PD} = \sum_{i=1}^{n} \left(a i_N I_{SSop,i} + \frac{b_i}{i_N I_{SSop,i}} \right) =$$

$$= \sum_{i=1}^{n} \left(i_N \frac{\tau_{PD,op,i}}{2} + \frac{\frac{\tau_{PD,op,i}}{2}}{i_N} \right) = \tag{8.69}$$

$$= \frac{1}{2}\left(i_N + \frac{1}{i_N} \right) \tau_{PD,op}$$

where $\tau_{PD,op}$ is the minimum achievable path delay, i.e. the sum of minimum achievable gate delay

$$\tau_{PD,op} = \sum_{i=1}^{n} \tau_{PD,op,i} \tag{8.70}$$

After defining T_{PD} as the path delay normalized to its minimum value in (8.70), relationship (8.69) results to

$$T_{PD} = \frac{1}{2}\left(I_N + \frac{1}{I_N} \right) \tag{8.71}$$

and analytically describes the power-delay trade-off in the path. It is worth noting that (8.71) is formally equal to (5.5) found for single CML gates and depicted in Fig. 5.1. This means that a path made up of CML gates with equal transistor emitter areas (or, equivalently, equal coefficients a_i) has a power-delay interdependence equal to that of a single gate, and the desing techniques developed in Chapter 5 for single gates still apply to cascaded gates.

REFERENCES

[ACK93] D. Allstot, S. Chee, S. Kiaei, M. Shristawa, "Folded source-coupled logic vs. CMOS static logic for low-noise mixed-signal ICs," *IEEE Trans. on CAS – part I*, vol. 40, no. 9, pp. 553-563, Sept. 1993.

[ADP02] M. Alioto, G. Di Cataldo, G. Palumbo, "Design of Low-Power High-Speed Bipolar Frequency Dividers", *Electronics Letters*, Vol. 38, No. 4, pp. 158-160, February 2002.

[ADR00] Auvergne, D., Daga, J.M., Rezzoug M. "Signal Transition Time Effect on CMOS Delay Evaluation,", *IEEE Trans. on Circuits and Systems – part I*, vol. 47, no. 9, pp. 1362-1369, Sept. 2000.

[AM88] P. Antognetti, G. Massobrio, *Semiconductor Device Modeling with Spice*, McGraw-Hill, 1988.

[AMP03] M. Alioto, R. Mita, G. Palumbo, "Performance Evaluation of the Low-Voltage CML D-Latch Topology", *Integration – The VLSI Journal*, Vol. 36, No. 4, pp. 191-209, November 2003.

[AP99] M. Alioto, G. Palumbo, "Highly Accurate and Simple Models for CML and ECL Gates", *IEEE Trans. on CAD*, Vol. 18, No. 9, pp. 1369-1375, September 1999.

[AP991] M. Alioto, G. Palumbo, "CML and ECL: Optimized Design and Comparison", *IEEE Trans. on CAS part I*, Vol. 46, No. 11, pp. 1330-1341, November 1999.

[AP00] M. Alioto, G. Palumbo, "Modeling and Optimized Design of Current Mode MUX/XOR and D Flip-Flop", *IEEE Trans. on CAS part II*, Vol. 47, No. 5, pp. 452-461, May 2000.

[AP01] M. Alioto, G. Palumbo, "Oscillation Frequency in CML and ESCL Ring Oscillators", *IEEE Trans. on CAS part I*, Vol. 48, No. 2, pp. 210-214, February 2001.

[AP03] M. Alioto, G. Palumbo, "Design Strategies for Source Coupled Logic Gates," *IEEE Trans. on CAS – part I*, Vol. 30, No. 4, pp.640-654, May 2003.

[AP031] M. Alioto, G. Palumbo, "Design of MUX, XOR and D-Latch SCL Gates", *IEEE ISCAS 2003*, Bangkok, pp. V261-V264, May 2003.

[APP02] M. Alioto - G. Palumbo - S. Pennisi, "Modeling of Source Coupled Logic Gates," *International Journal of Circuit Theory and Applications*, Vol. 30, No. 4, pp.459-477, July, 2002.

[ARL95] L. Andersson, B. Rudberg, P. Lewin, M. Reed, S. Planer, S. Sundaram "Silicon Bipolar Chipset for SONET/SDH 10 Gb/s Fiber-Optic Communication Links," *IEEE Journal of Solid-State Circuits*, vol. 30, no. 3, pp. 210-217, March 1995.

[BNK98] L. Bisdounis, S. Nikolaidis, O. Koufopavlou, "Analytical transient response of propagation delay evaluation of the CMOS inverter for short channel devices," *IEEE Jour. of Solid-State Circuits*, vol. 33, no. 2, pp. 302-306, Feb. 1998.

[C92] C. Chuang, "Advanced Bipolar Circuits," *IEEE Circuits & Devices*, pp. 32-36, November 1992.

[CBA88] E.-Chor, A. Brunnschweiler, P. Ashburn, "A Propagation-Delay Expression and its Application to the Optimization of Polysilicon Emitter ECL Processes," *IEEE Jour. of Solid-State Circ.*, Vol. 23, No. 1, pp. 251-259, February 1988.

[CB95a] A. Chandrakasan, R. Brodersen, "Minimizing Power Consumption in Digital CMOS Circuits," *Proc. of the IEEE*, vol. 83, no. 4, pp. 498-523, 1995.

[CB95b] A. Chandrakasan, R. Brodersen, *Low Power Digital CMOS Design*, Kluwer Academic Publisher, 1995.

[CBF01] A. Chandrakasan, W. Bowhill, F. Fox, (Eds.), *Design of High-Performance Microprocessor Circuits,* IEEE Press 2001.

[CCK95] C. Choy, C. Chan, M. Ku, "A feedback control circuit design technique to suppress power noise in high speed output driver," *Proc. of ISCAS'95*, pp. 307-310, April 1995.

[CCK97] C. Choy, C. Chan, M. Ku, J. Povazanec, "Design procedure of low-noise high-speed adaptive output drivers," *Proc. of ISCAS'97*, pp. 1796-1799, 1997.

[CG73] B. Cochrun, A. Grabel, "A method for the Determination of the Transfer Function of electronic Circuits," *IEEE Trans. on Circuit Theory*, Vol. CT-20, No. 1, pp. 16-20, Jan. 1973.

[CH99] Y. Cheng, C. Hu, *MOSFET modeling & BSIM3 user's guide*, Kluwer Academic Publishers, 1999.

[CHL92] C. Choy, C. Ho, G. Lunn, B. Lin, G. Fung "A BiCMOS Programmable Frequency Divider," *IEEE Trans. on Circuits and Systems part II*, Vol. 39, No. 3, pp. 147-154, March 1992.

[CJ89] C. Choy, P. Jones, "Minimisation Tachnique for Series-Gated Emitter-Coupled Logic," IEE Proc. Part G, vol. 136, no. 3, pp. 105-113, June 1989.

[CP86] K. Chu, D. Pulfrey, "Design Procedures for Differential Cascode Voltage Switch Circuits," *IEEE Jour. of Solid-state Circ.*, vol. SC-21, no. 6, pp. 1082-1087, December 1986.

[CW98] H. Chang, J. Wu, "A 723-MHz 17.2-mW CMOS Programmable Counter," *IEEE Journal of Solid-State Circuits*, vol. 33, no. 10, pp. 1572-1575, Oct. 1998.

[DKS90] B. Del Signore, D. Kerth, N. Sooch, E. Swanson, "A monolithic 20-b delta-sigma A/D converter," *IEEE Jour. Solid-State Circuits*, vol. 25, pp. 1311-1317, Dec. 1990.

[DS03] B. De Muer, M. Steyaert, *CMOS Fractional-N Synthesizers (Designfor High Spectral Purity and Monlithic Integration)*, Kluwer Academic Publishers, 2003.

[E48] W. Elmore, "The transient response of damped linear networks," *J. Appl. Phys.*, vol. 19, pp. 55-63, Jan. 1948.

[EBR94] S. Embabi, D. Brueske, K. Rachamreddy, "A BiCMOS low-power current mode gate," *IEEE Jour. of Solid-State Circuits*, vol. 29, no. 6, pp. 741-745, June 1994.

[F90] W. Fang, "Accurate Analytical Delay Expressions for ECL and CML Circuits and Their Applications to Optimizing High-Speed Bipolar Circuits," *IEEE Jour. of Solid-State Circ.*, Vol. 25, No. 2, pp. 572-583, April 1990.

[F96] A. Felder, et al., "46 Gb/s DEMUX, 50 Gb/s MUX, and 30 GHz Static Frequency Divider in Silicon Bipolar Technology," *IEEE Jour. Of Solid-State Circ.*, Vol. 31, No. 4, pp. 481-486, April 1996.

[F97] I. Fujimori *et al.*, “A 5-V single chip delta-sigma audio A/D converter with 111 dB dynamic range,” *IEEE Jour. of Solid-State Circuits*, vol. 32, pp. 329-336, Mar. 1997.

[FMP96] A. Felder, M. Moller, J. Popp, J. Bock, H. Rein "46 Gb/s DEMUX, 50 Gb/s MUX, and 30 GHz Static Frequency Divider in Silicon Bipolar Technology," *IEEE Jour. of Solid-State Circ.*, Vol. 31, No. 4, pp. 481-486, April 1996.

[FBA90] W. Fang, A. Brunnschweiler, P. Ashburn, “An Analytical Maximum Toggle Frequency Expression and its Application to Optimizing High-Speed ECL Frequency Dividers,” *IEEE Jour. of Solid-State Circ.*, Vol. 25, No. 4, pp. 920-931, August 1990.

[GM93] P. Gray, R. Meyer, *Analysis and Design of Analog Integrated Circuits (third edition)*, John Wiley & Sons, 1993.

[GMC91] H. Greub, J. McDonald, T. Creedon and T. Yamaguchi, “High-performance standard cell library and modeling technique for differential advanced bipolar current tree logic,” *IEEE Jour. of Solid-State Circuits*, vol. 26, no. 5, pp. 749-762, May 1991.

[GMO90] M. Ghannam, R. Mertens, R. Van Overstraeten, “"An Analytical for the Determination of the Transient Response of CML and ECL Gates,” *IEEE Trans. on Electron Devices*, Vol. 37, No. 1, pp. 191-201, January 1990.

[GT86] R. Gregorian, G. Temes, *Analog MOS Integrated Circuits for signal processing*, John Wiley & Sons, 1986.

[H95] Y. Harada, “Delay Components of a Current Mode Logic Circuit and Their Current Dependency,” *IEEE Jour. of Solid-State Circ.*, Vol. 30, No. 1, pp. 54-60, January 1995.

[H96] J. Hauenschild, et al., “A Plastic Packaged 10 Gb/s BiCMOS Clock and Data Recovering 1:4-Demultiplexer with External VCO,” *IEEE Jour. of Solid-State Circ.*, Vol. 31, No. 12, pp. 2056-2059, December 1996.

[H97] G. Hurkx, “The relevance of f_T and f_{MAX} for the Speed of a Bipolar CE Amplifier Stage,” *IEEE Trans. on Electron Devices*, Vol. 44, No. 5, pp. 775-781, May 1997.

[HFP01] C. Hung, B. Floyd, B. Park, K. O, “Fully Integrated 5.35-GHz CMOS VCOs and Prescalers,” *IEEE Trans. on Microwave Theory and Techniques*, vol. 49, no. 1, Jan. 2001.

[HLL99] A. Hajimiri, S. Limotyrakis, T. Lee, "Jitter and Phase Noise in Ring Oscillators," *IEEE Journal of Solid State Circuits*, Vol. 34, No. 6, pp. 790-804, June 1999.

[HR99] F. Herzel, B. Razavi, "A Study of Oscillator Jitter Due to Supply and Substrate Noise," *IEEE Trans. on Circuits and Systems part II*, Vol. 46, No. 1, pp. 56-62, January 1999.

[I95] N. Ishihara, et al., "3.5-Gb/s x 4-Ch Si Bipolar LSI's for Optical Interconnections," *IEEE Jour. of Solid-state Circ.*, Vol. 30, No. 12, pp. 1493-1500, December 1995.

[I951] K. Ishii, et al. "Very-High-Speed Si Bipolar Static Frequency Dividers with New T-Type Flip-Flops," *IEEE Jour. Of Solid-state Circ.*, Vol. 30, No. 1, pp. 19-24, January 1995.

[IIS89] H. Ichino, N. Ishihara, M. Suzuki, S. Konaka, "18-GHz 1/8 Dynamic Frequency Divider Using Si Bipolar Technologies," *IEEE Journal of Solid-State Circuits*, vol. 24, no. 6, pp. 1723-1728, Dec. 1989.

[IIT95] K. Ishii, H. Ichino, M. Togashi, Y. Kobayashi, C. Yamaguchi "Very-High-Speed Si Bipolar Static Frequency Dividers with New T-Type Flip-Flops," *IEEE Jour. of Solid-State Circ.*, Vol. 30, No. 1, pp. 19-24, January 1995.

[JM97] D. Johns, K. Martin, *Analog Integrated Circuit Design*, John Wiley & Sons, 1997.

[JMS97] S. Jantzi, K. Martin, A. Sedra, "Quadrature bandpass ΣΔ modulator for digital radio," *IEEE Jour. of Solid-State Circuits*, vol. 32, pp. 1935-1949, 1997.

[JR97] K. Jayabalan, S. Rezaul Hasan, "Current-mode BiCMOS folded source-coupled logic circuits," *Proc. of ISCAS'97*, pp. 1880-1883, June 1997.

[K91] M. Kurisu, et al., "A Si Bipolar 21-GHz/320-mW Static Frequency Divider," *IEEE Jour. of Solid-state Circ.*, Vol. 26, No. 11, pp. 1626-1630, November 1991.

[K98] K. Koike, et al. "High-Speed, Low-Power, Bipolar Standard Cell Design Methodology for Gbit/s Signal Processing," *IEEE Jour. of Solid-state Circ.*, Vol. 33, No. 10, pp. 1536-1544, October 1998.

[KA92] S. Kiaei, D. Allstot, "Low-noise logic for mixed-mode VLSI circuits," *Microelectronics Journal*, vol. 23, no. 2, pp. 103-114, Apr. 1992.

[KB92] G. K. Konstadinidis, H. H. Berger, "Optimization of Buffer Stages in Bipolar VLSI Systems," *IEEE Jour. of Solid-State Circ.*, Vol. 27, No. 7, pp. 1002-1013, July 1992.

[KBW01] H. Knapp, J. Bock, M. Wurzer, G. Ritzberger, K. Aufinger, L. Treitinger, "2-GHz/2-mW and 12-GHz/30-mW Dual-Modulus

Prescalers in Silicon Bipolar Technology," *IEEE Jour. of Solid-State Circ.*, Vol. 36, No. 9, pp. 1420-1423, September 2001.

[KCA90] S. Kiaei, S. Chee, D. Allstot, "CMOS source-coupled logic for mixed-mode VLSI," *Proc. Int. Symp. Circuits Systems*, pp. 1608-1611, 1990.

[KDN91] B. Kup, E. Dijkmans, P. Naus, J. Sneep, "A bit-stream digital-to-analog converter with 18-b resolution," *IEEE Jour. of Solid-State Circuits*, vol. 26, pp. 1757-1763, Dec. 1991.

[KH00] J. Kundan, S. Hasan, "Enhanced folded source-coupled logic technique for low-voltage mixed-signal integrated circuits," *IEEE Trans. on CAS – part II*, vol. 47, no. 8, pp. 810-817, Aug. 2000.

[KH97] J. Kundan, S. Hasan, "Current mode BiCMOS folded source-coupled logic circuits," *Proc. of ISCAS'97*, pp. 1880-1883, June 1997.

[KKI97] K. Kishine, Y. Kobayashi, H. Ichino, "A High-Speed, Low-Power Bipolar Digital Circuit for Gb/s LSI's: Current Mirror Control Logic," *IEEE Jour. of Solid-state Circ.*, Vol. 32, No. 2, pp. 215-221, February 1997.

[KOS91] M. Kurisu, M. Ohuchi, A. Sawairi, M. Sugiyama, H. Takemura, T. Tashiro "A Si Bipolar 21-GHz/320-mW Static Frequency Divider," *IEEE Jour. of Solid-State Circ.*, Vol. 26, No. 11, pp. 1626-1630, November 1991.

[KUO92] M. Kurisu, G. Uemura, M. Ohuchi, C. Ogawa, H. Takemura, T. Morikawa, T. Tashiro "A Si Bipolar 28-GHz Dynamic Frequency Divider," *IEEE Jour. of Solid-State Circ.*, Vol. 27, No. 12, pp. 1799-1804w, November 1992.

[L95] Z. H. Lao, et al., "A 12 Gb/s Si Bipolar 4:1-Multiplexer IC for SDH Systems," *IEEE Jour. of Solid-state Circ.*, Vol. 30, No. 2, pp. 129-132, February 1995.

[L96] Z. Lao, et al., "Si Bipolar 14 Gb/s 1:4-Demultiplexer IC for System Applications," *IEEE Jour. of Solid-State Circ.*, Vol. 31, No. 1, pp. 54-59, January 1996.

[LHG84] H. Lee, D. Hodges, P. Gray, "A self-calibrating 15-bit CMOS A/D converter," *IEEE Jour. of Solid-State Circuits*, vol. 19, pp. 813-819, Dec. 1984.

[LL96] Z. Lao, U. Langmann, "Design of a Low-Power 10 Gb/s Si Bipolar 1:16-Demultiplexer IC," *IEEE Jour. of Solid-State Circ.*, Vol. 31, No. 1, pp. 128-131, January 1996.

[LR00] C. Lam, B. Razavi, "A 2.6-GHz/5.2-GHz Frequency Synthesizer in 0.4-μm CMOS Technology," *IEEE Jour. of Solid-State Circ.*, Vol. 35, No. 5, pp. 788-794, May 2000.

[LS94] K. Laker, W. Sansen, *Design of Analog Integrated Circuits and Systems*, Mc Graw-Hill, 1994.

[LWO91]H. Leopold, G. Winkler, P. O'Leary, K. Ilzer, J. Jernej, "A monolithic CMOS 20-b analog-to-digital converter," *IEEE Jour. of Solid-State Circuits*, vol. 26, pp. 910-916, July 1991.

[M89] R. Middlebrook, "Null Double Injection and the Extra Element Theorem", *IEEE Trans. on Education*, vol. 32, no. 3, pp. 167-180, August 1989.

[M92] S. Masui, "Simulation of substrate coupling in mixed-signal MOS circuits," *Proc. Of Symp. VLSI Circuits*, pp. 42-43, 1992.

[M97] C. Maier, et al., "A 533-MHz BiCMOS Superscalar RISC Micorprocessor," *IEEE Jour. of Solid-state Circ.*, Vol. 32, No. 11, pp. 1625-1633, November 1997.

[M971] J. McNeill, "Jitter in Ring Oscillators," *IEEE Journal of Solid State Circuits*, Vol. 32, No. 6, pp. 870-879, June 1997.

[MG87] J. Millman - A. Grabel, *Microelectronics (Second Edition)*, McGraw-Hill, 1987.

[MK86] R. Muller, T. Kamins, *Device Electronics for Integrated Circuits*, John Wiley & Sons, 1986.

[MKA92]S. Maskai, S. Kiaei, D. Allstot, "Synthesis techniques for CMOS folded source-coupled logic circuits," *IEEE Jour. of Solid-State Circuits*, vol. 27, no. 8, pp. 1157-1167, Aug. 1992.

[MSO92]M. Mizuno, H. Suzuki, M. Ogawa, K. Sato, H. Ichikawa, "A 3-mW 1.0-GHz Silicon-ECL Dual-Modulus Prescaler IC," *IEEE Jour. of Solid-State Circuits*, vol. 27, no. 12, pp. 1794-1798, December 1992.

[NA97] H. Ng, D. Allstot, "CMOS current steering logic for low-voltage mixed-signal integrated circuits," *IEEE Trans. on VLSI Systems*, vol. 5, no. 3, pp. 301-308, Sept. 1997.

[NIE03] H. Nosaka, K. Isshii, T. Enoki, T. Shibata, "A 10-Gb/s Data-Pattern Independent Clock and Data Recovery with a Two-Mode Phase Comparator," *IEEE Jour. of Solid-State Circuits*, vol. 38, no. 2, pp. 192-197, February 2003.

[O99] V. Oklobdzija, *High-Performance System Design (Circuits and Logic)*, IEEE Press 1999.

[P01] H. Partovi, "Clocked Storage Elements", in A. Chandrakasan, W. Bowhill, F. Fox, (Eds.), *Design of High-Performance Microprocessor Circuits,* IEEE Press 2001.

[PBS97] P. Poras, T. Balsara, D. Steiss, "Performance of CMOS differential circuits," *IEEE Jour. of Solid-State Circuits*, vol. 31, pp. 841-846, June 1997.

[PM91] O. Pederson, K. Mayaram, *Analog Integrated Circuits for Communication (principles, Simulation and Design)*, Kluwer Academic Publishers, 1991.

[R98] B. Razavi, *RF Microelectronics*, Prentice Hall, 1998.

[R96] J. Rabaey, *Digital Integrated Circuits (A Design Perspective)*, Prentice Hall, 1996.

[R961] B. Razavi (Ed.), *Monolithic Phase-Locked Loops and Clock Recovery Circuits (Theory and Design)*, IEEE Press 1996.

[R02] B. Razavi, "Prospect of CMOS Technology for High-Speed Optical Communication Circuits," *IEEE Jour. of Solid-State Circ.*, Vol. 37, No. 9, pp. 1135-1145, September 2002.

[RDR01] M. Reinhold, C. Dorschky, R. Pullela, P. Mayer, F. Kunz, Y. Beeyens, T. Link, J. Mattia, "A Fully Integrated 40-Gb/s Clock and Data Recovery IC With 1:4 DEMUX in SiGe Technology," *IEEE Journal of Solid-State Circuits*, vol. 36, no. 12, pp. 1937-1945, Dec. 2001.

[RM96] H. Rein, M. Moller, "Design Considerations for Very-High-Speed Si-Bipolar IC's Operating up to 50 Gb/s," *IEEE Jour. of Solid-State Circ.*, Vol. 31, No. 8, pp. 1076-1090, August 1996.

[ROS94] B. Razavi, Y. Ota, R. Swartz, "Design techniques for low-voltage high speed digital bipolar circuits," *IEEE Jour. of Solid-State Circ.*, Vol. 29, No. 2, pp. 332-339, March 1994.

[RP96] J. Rabaey, M. Pedram (Eds.), *Low power design methodologies*, Kluwer Academic Publishers, 1996.

[S81] S. Sze, *Physics of Semiconductor Devices*, John Wiley & Sons, 1981.

[S02] R. Singh (Ed.), *Signal Integrity Effects in Custom IC and ASIC Design*, IEEE Press 2002.

[SD93] C. Stout, J. Doernberg, "10-Gb/s Silicon Bipolar 8:1 Multiplexer and 1:8 Demultiplexer," *IEEE Journal of Solid-State Circuits*, vol. 28, no. 3, pp. 339-343, March 1993.

[SE94] K. M. Sharaf, M. Elmasry, "An Accurate Analytical Propagation Delay Model for High-Speed CML Bipolar Circuits," *IEEE Jour. of Solid-State Circ.*, Vol. 29, No. 1, pp. 31-45, January 1994.

[SE96] K. M. Sharaf, M. I. Elmasry, "Analysis and Optimization of Series-Gates CML and ECL High-Speed Bipolar Circuits," *IEEE Jour. of Solid-State Circ.*, Vol.31, No. 2, pp. 202-211, February 1996.

[SK97] R. Saez, M. Kayal, M. Declercq, M. Schneider, "Design guidelines for CMOS current steering logic," *Proc. of ISCAS'97*, pp. 1872-1875, 1997.

[SKD96] R. Sàez, M. Kayal, M. Declercq, M. Schneider, "Digital circuit techniques for mixed analog/digital circuits applications," *Proc. of ICECS'96*, Rodos, Greece 1996.

[SL98] Shing-Tag Yan, H. Luong, "A 3-V 1.3-to-1.8-GHz CMOS Voltage-Controlled Oscillator with 0.3-ps Jitter," *IEEE Trans. on Circuits and Systems part II*, Vol. 45, No. 7, pp. 876-880, July 1998.

[SLM93] D. Su, M. Loinaz, S. Masui, B. Wooley, "Experimental results and modeling techniques for substrate noise in mixed-signal integrated circuits," *IEEE Jour. of Solid-State Circuits*, vol. 28, pp. 420-430, Apr. 1993.

[SP93] R. Senthinatan, J. Prince, "Application specific CMOS output driver circuit design techniques to reduce simultaneous switching noise," *IEEE Jour. of Solid-State Circuits*, vol. 28, no. 12, pp. 1383-1388, Dec. 1993.

[S96] F. Sato, et al., "A 2.4 Gb/s Receiver and a 1:16 Demultiplexer in One Chip Using a Super Self-Aligned Selectively Grown SiGe Base (SSSB) Bipolar Transistor," *IEEE Jour. of Solid-State Circ.*, Vol. 31, No. 10, pp. 1451-1456, October 1996.

[SMS94] T. Seneff, L. McKay, K. Sakamoto, N. Tracht, "A Sub-1 mA 1.5-GHz Silicon Bipolar Dual Modulus Prescaler," *IEEE Jour. Of Solid-State Circ.*, Vol. 29, No. 10, pp. 1206-1211, October 1994.

[SMT98] G. Schuppener, M. Mokhtari, H. Tehnhunen, "A Divide-by-4 Circuit Implemented in Low Voltage, High Speed Topology," *Proc. ISCAS'98*, Monterey, pp. 215-221, June 1998.

[SPM00] G. Schuppener, C. Pala, M. Mokhtari, "Investigation on Low-Voltage Low-Power Silicon Bipolar Design Topology for High-Speed Digital Circuits," *IEEE Jour. of Solid-State Circ.*, Vol. 35, No. 7, pp. 1051-1054, July 2000.

[SPW91] N. Sheng, R. Pierson, K. Wang, R. Nubling, P. Asbeck, M. Chang, W. Edwards, D. Phillips, "A High-Speed Multimodulus HBT Prescaler for Frequency Synthesizer Applications," *IEEE Jour. of Solid-State Circ.*, Vol. 26, No. 10, pp. 1362-1367, October 1991

[SR01] J. Savoj, B. Razavi, *High-Speed CMOS Circuits for Optical Receivers*, Kluwer academic Publisher, 2001.

[SS91] A. Sedra, K. Smith, *Microelectronic Circuits (third edition)*, Saunders College Publishing, 1991.

[SVR94] B. Stanistic, N. Verghese, R. Rutenbar, L. Carley, D. Allstot, "Addressing substrate coupling in mixed-mode IC's: simulation and power distribution synthesis," *IEEE Jour. of Solid-State Circuits*, vol. 29, pp. 226-238, Mar. 1994.

[T98] Y. Tsividis, *Operation and modeling of MOS transistors*, McGraw-Hill, 1998.

[T89] R. Treadway, "DC Analysis of Current Mode Logic," *IEEE Circuits and Device Magazine*, pp. 21-35, March 1989

[TS79] D. D. Tang, P. M. Solomon, "Bipolar Transistor Design for Optimized Power-Delay Logic Circuits," *IEEE Jour. of Solid-State Circ.*, Vol. SC-14, No. 4, pp. 679-684, August 1979.

[TUF01] A. Tanabe, M. Umetani, I. Fujiwara, T. Ogura, K. Kataoka, M. Okiara, H. Sakuraba, T. Endoh, F. Masuoka, "0.18-µm CMOS 10-Gb/s Multiplexer/Demultiplexer ICs Using Current Mode Logic with Tolerance to Threshold Voltage Fluctuation," *IEEE Jour. of Solid-State Circuits*, vol. 36, no. 6, June 2001.

[VK95] C. Vaucher, D. Kasperkovitz, "A Wide-Band Tuning System for Fully Integrated Satellite Receivers," *IEEE Jour. of Solid-State Circ.*, Vol. 33, No. 7, pp. 987-997, July 1995.

[W90] G. Wilson, "Advances in Bipolar VLSI," *Proceeding of IEEE*, Vol. 78 No. 11, pp. 1707-1719, November 1990.

[WKG94] T. Weigandt, B. Kim, P. Gray, "Analysis of Timing Jitter in CMOS Ring Oscillators," *IEEE Proc. ISCAS'94*, London, pp. 27-30, June 1994.

[YTC92] A. T. Yang, Y. Chang, "Physical Timing Modeling for Bipolar VLSI," *IEEE Jour. of Solid-State Circ.*, Vol. 27, No. 9, pp. 1245-1254, September 1992.

[YEC83] P. Yang, B. Epler, P. Chatterjee, "An Investigation of the Charge Conservation Problem for MOSFET Circuit Simulation", *IEEE Jour. of Solid-State Circuits*, vo. 18, no. 1, pp. 128-138, February 1983.

ABOUT THE AUTHORS

Massimo Alioto was born in Brescia, Italy, in 1972. He received the laurea degree in electronics engineering and the Ph.D. degree from the University of Catania, Italy, in 1997 and 2001, respectively.

Since 2001 he has been teaching undergraduate and graduate courses on basic electronics, microelectronics and digital electronics.
In 2002, he joined the Dipartimento di Ingegneria dell'Informazione (DII) of the University of Siena as a research associate and in the same year as an assistant professor.

His primary research interests include the modeling and optimized design of bipolar and CMOS high-performance digital circuits in terms of high-speed or low-power dissipation, as well as arithmetic circuits. He has authored or co-authored more than 40 journals and conference papers.

Gaetano Palumbo was born in Catania, Italy, in 1964. He received the laurea degree in Electrical Engineering and a Ph.D. degree in 1988 and 1993, respectively, from the University of Catania.

Since 1993 he conducts courses on Electronic Devices, Digital Circuit and Systems and basic Electronics. In 1994 he joined the DEES (Dipartimento Elettrico Elettronico e Sistemistico), now DIEES (Dipartimento di Ingegneria Elettrica Elettronica e dei Sistemi), at the University of Catania as a researcher, subsequently becoming associate professor in 1998. Since 2000 he is a full professor in the same department.

His initial research interest has been devoted to analog circuits with particular emphasis on feedback circuits, compensation techniques, current-mode approach and low-voltage circuits. After several years, his research has also embraced digital circuits with emphasis on bipolar and MOS current-mode digital circuits, adiabatic circuits, high-performance building blocks focused on achieving optimum speed within the constraint of low power operation, and arithmetic circuits. In all these fields he is developing some research in collaboration with ST-microelectronics of Catania site.
He was the co-author of the books "CMOS Current Amplifiers" and "Feedback Amplifiers: theory and design", both by Kluwer Academic Publishers, in 1999 and 2002, respectively. He is a contributor to the *Wiley Encyclopedia of Electrical and Electronics Engineering*. In addition, he is

the author or co-author of more than 200 scientific papers on international journals (over 85) and conferences, and of several patents.

From June 1999 to the end of 2001 he served as an Associated Editor of the *IEEE Transactions on Circuits and Systems part I* for the topic "Analog Circuits and Filters". Since 2004 he is serving as an Associated Editor of the *IEEE Transactions on Circuits and Systems part I* for the topic "digital circuits and systems".

In 2003 he received the Darlington Award[1].
Prof. Palumbo is an *IEEE Senior Member.*

[1] best paper bridging the gap between theory and practice published in *IEEE Trans. on CAS I* or *II*, during the two calendar years preceding the award.